AF679197

# Design Science

## Introduction to the Needs, Scope and Organization of Engineering Design Knowledge

Springer
*London*
*Berlin*
*Heidelberg*
*New York*
*Barcelona*
*Budapest*
*Hong Kong*
*Milan*
*Paris*
*Santa Clara*
*Singapore*
*Tokyo*

Vladimir Hubka and W. Ernst Eder

# Design Science

## Introduction to the Needs, Scope and Organization of Engineering Design Knowledge

With 68 Figures

Springer

Vladimir Hubka, Dipl.-Ing., Dr., Dr.h.c.
*Retired from* Eidgenössische Technische Hochschule, Zürich, Switzerland

W. Ernst Eder, MSc, P.Eng.
Department of Mechanical Engineering, Royal Military College of Canada,
Kingston, Ontario, Canada K7K 5LO

Dedicated to Dr. Dagmar Hnik

ISBN 3-540-19997-7 Springer-Verlag Berlin Heidelberg New York

British Library Cataloguing in Publication Data
Hubka, Vladimir
Design Science: Introduction to the Needs, Scope and
Organization of Engineering Design Knowledge - 2Rev.ed
I. Title II. Eder, W. E.
620.0042
ISBN 3-540-19997-7

Library of Congress Cataloging-in-Publication Data
Hubka, Vladimir.
[Einführung in die Konstruktionswissenschaft. English]
Design science : introduction to the needs, scope and organization of
engineering design knowledge / Vladimir Hubka and W. Ernst Eder.
p. cm.
Includes index.
ISBN 3-540-19997-7 (pbk.)
1. Engineering design. I. Eder, W. E. (Wolfgang Ernst)
II. Title.
TA174.H79813 1995 95-37137
620'.0042'01 - - dc20 CIP

This book is a completely revised and augmented English edition of "Einführung in die Konstruktionswissenschaft" by Vladimir Hubka and W. Ernst Eder, Springer-Verlag Berlin, Heidelberg 1992

Apart from any fair dealing for the purposes of research or private study, or criticism or review, as permitted under the Copyright, Designs and Patents Act 1988, this publication may only be reproduced, stored or transmitted, in any form or by any means, with the prior permission in writing of the publishers, or in the case of reprographic reproduction in accordance with the terms of licences issued by the Copyright Licensing Agency. Enquiries concerning reproduction outside those terms should be sent to the publishers.

© Springer-Verlag London Limited 1996
Printed in Great Britain

The publisher makes no representation, express or implied, with regard to the accuracy of the information contained in this book and cannot accept any legal responsibility or liability for any errors or omissions that may be made.

Typeset by authors in TrueType Times New Roman with Publish-It! ver. 3.0 for Windows by Timeworks
Graphics with AutoSketch Release 2 for Windows by Autodesk
Word processing with WordStar 7.0C
Translation assisted by Power Translator Professional 3.0 by Globalink
Printed and bound at the Athenæum Press Ltd, Gateshead
69/3830-543210 Printed on acid-free paper

# Preface

*Nothing that has not yet been done, can be done, except by means that have not yet been tried*

***Francis Bacon**, Novum Organum, Aphorisms, Book 1, IV, **Advancement of Learning and Novum Organum**, New York: Collier, 1990*

In preparing this new edition of *Konstruktionswissenschaft* (Design Science) [229], two considerations prevailed. One is that the knowledge about engineering design has obviously improved in the three years between the editions. Many new works have been published in this field, but also our own understanding of design has developed.

The other consideration is that the situation with respect to engineering design and Design Science is distinctly different in continental Europe and the USA (typically, but similar conditions exist in most of the other English-language regions). These differences, and the states of affairs and knowledge in the regions, have been surveyed and outlined in several publications. Self-generated views for the region's own consumption are seen in [9,10,67]. Views from English-speaking regions to Europe appeared in [123,124,133,134]. The opposite direction was explained in [56,127,129]. The differences range from historic, linguistic, cultural and economic outlooks, and regional power relationships, via employment patterns (e.g., for academics), to attitudes about scope and validity of research.

A brief sketch to highlight these differences can draw the criticism that the circumstances have been exaggerated. Nevertheless, an attempt should be undertaken. The following can be considered as a purely personal and biassed view.

Continental Europeans tend to being outward-looking and trans-national, but also more formal and systematic; English-speakers tend to be more insular and isolationist, with any "foreign" language as a perceived cultural barrier, but also more intuitive and casual, and less formal. As one example, consider technical handbooks for engineering: the main German handbooks (Dubbel, Hütte) are more general, usable, and better structured to find both theoretical and applications information; the main English-language handbooks (Machinery's, Kent's, Mark's, Perry's, etc.) are more focussed on the individual branches of industry, and their structure and ease of use leave something to be desired. In engineering design literature, each of the German works (E. Reuleaux [365], R. Koller [257], K. Roth [376], G. Pahl and W. Beitz [333], etc.) tend to include many facets, e.g., design philosophy, generalized methodology, higher levels of abstraction, applied examples from engineering practice, computer applications, taxonomies, systematizations, etc. The English-language works (J.C. Jones [239], N. Cross [98], J.R. Dixon [111], B. Archer [53], G. Dieter [109], T.T. Woodson [459], L. Harrisberger [188], C.L. Dym [119], W.H. Middendorf [298], G. Nadler [317], etc.) tend to be one-dimensional, focused on single themes, fragments masquerading as wholes, and designing (if treated at all) isolated from the object being designed and from engineering practice.

The *importance* of engineering design, the activity, lies mainly in its contribution to the world, country and local economy. If good products are designed and brought to market at an economic price, the economies will prosper. Part of the task connected with designing is to coordinate among at least three aspects of management:

- managing the forms and capabilities of the product mix for a company or enterprise, which includes product development;
- managing the process of designing (engineering) and its information sources and outputs; and
- managing the company and its organization.

## Objectives

Each knowledge area represents a mixture of objectifiable and tacit knowledge. It needs a certain order and arrangement of its recorded knowledge. The order should show the elements of that knowledge, and their relationships -- how they fit together. In this way we can create the necessary overview and to open ways to purposeful improvements -- the area needs a science base. Engineering design is such a knowledge area where a science base is needed. Designing can be improved if we know more about the processes of designing. Changes are needed to set engineering design into a broader context, to better define that context, and to reinstate the needed status and prestige of engineering design and designers. Neither side of the German/English divide is either fully right or wrong, both need to be brought together. Collecting the knowledge, making it available with good orientation, and providing a more comprehensive overview is among the main envisaged roles of Design Science, and therefore of this book.

Although Design Science is primarily intended as tool for engineering practice, it can hardly be expected revolutionize the present state of design work in industry. In engineering practice only a few designers are involved in sufficient philosophical depth, most of them have more urgent tasks. A few let themselves be inspired and satisfied by such an enterprise of a fundamental investigation of design knowledge.

The current state of knowledge about engineering design shows a distinct lack of order. Large amounts of knowledge have been gathered, many observations made, and much formal research undertaken, but the results remain dispersed. This may be a result of attempted self-protection, not wanting to acknowledge other contributions. Exactly for this reason, we have set the following objectives (goals, aims) for this book:

- to improve the state of knowledge about designing, to create a basis for order;
- to constitute and outline *General Design Science*;
- to explain a structure for General Design Science, to outline contents and relationships;
- to help in making progress towards rationalizing engineering design work in practice;
- to outline laws, patterns, and paradigms, and to establish a uniform terminology;
- to build a starting point for developing *Specialized Design Sciences*, especially in view of attempting to help the practice;
- to build a basis for engineering design education;
- to provide a framework for coordinating engineering design research knowledge.

The objective of this book is directed especially at the future; there must first be designers educated in the way of thinking implied by Design Science. Further or continuing education (after first graduation) is of relatively less use here. It can however create the conditions that some designers in practice “find” the existing attempts at creating Design

Science for themselves, from the literature, and introduce its application to rationalize the engineering design practice. Rationalizing this design process can mean:

(a) being more rational, with deeper and more logical thought;
(b) being more effective and economic, especially in use of time and resources;
(c) being systematic, going step by step through a process, and having a useful set of methodologies and methods; but these also imply
(d) having a theory about the objects and processes involved, as a basis for rationalizing.

The main object of this book is to outline and define Design Science. Many people are familiar with parts of the problem, and various issues related to it, so this is no simple task. This English translation and revised edition encounters additional problems. We therefore think it important to set this work into a more general and narrative context, before discussing the stricter issues. One characteristic of good engineers is that they tend to reserve judgment on any proposals until enough information is available -- which is advised for this book.

A broad discussion of many facets of engineering (and other) design in the context of society is presented. The discussions show a complex picture of elements and their interactions. The headings and contents in Chapter 1 are intended as an advance survey of the field, and present a possible organizational pattern for the contents of the book. Most of the factors reappear later in the book, with more extensive discussions and improved foundation, and in a different arrangement. Little of this is really new in its outline and detail. We think we have brought it together in a different and better way that can provide a more comprehensive understanding of the engineering design process, and the processes and objects to be designed. The treatment is as comprehensive as possible, but cannot be complete.

Our purpose is therefore to describe a *paradigm*, an example, pattern, abstraction, and model of what structures and typical contents we should expect in science about designing. As in most situations, the phenomenon of *engineering design* as a process is best described by a mixture of art and science [131], both of which should be considered.

On the other hand, we usually expect to find some new elements of knowledge, and we should not be surprised if we have to revise our ideas and models as a result of each new discovery.

The utility of taxonomies, hierarchies and other forms of arrangement needs to be discovered, especially if we wish to compare various design processes or models. On such a basis we can also compare different descriptions and claims for tools and methods.

## Background

The work presented here is the result of more than 35 years of effort. Both authors have spent several years in industry, working in and/or leading design teams. V. Hubka has 25 years of experience in five branches of Czech industry, mainly mechanical, but including some time in a company producing electro-heating appliances. W.E. Eder spent most of his four appointments over 10 years with power transformers and their associated switchgear, in companies in Austria and England. Both authors were then involved in education for engineering design, abstracting from their own experience and formalizing their thoughts, V. Hubka in Denmark and Switzerland, W.E. Eder in Great Britain and Canada.

This work could only be realized by close cooperation of the authors, and with assistance from many of their colleagues. Many already existing directions and research projects were explored and essential thoughts incorporated into the schemes to be developed in this book, to obtain as complete a survey as possible. The events run by **WDK-Workshop Design-Konstruktion**, an international society for investigating the science of engineering design, have contributed substantially.

WDK is active in building a unified and comprehensive science about design. This involves research on the design process, on the nature of technical systems, and on education for engineering design. Various special groups have been formed to consider particular topics within this range, and to investigate in what ways the broader concepts are applicable to these topics. Further groups are planned, and suggestions for organizers, co-ordinators and participants are welcome.

Members of WDK are scientists, engineers and educators who take an active part in the investigations, meetings and publications of the society. The main goal of WDK is to rationalize engineering design work through application and development of the science about engineering design, and to bring together the world community of engineering designers, both theoretical and practical.

The events run under WDK comprise the congresses "International Conference on Engineering Design" -- ICED 81 to ICED 95 and continuing [29,125,207,208,209,213,215, 221,226,227,230,372] --, as well as over 85 seminars, workshops and courses led by the authors and their colleagues in places as far afield as Zürich and Rigi Kaltbad (Switzerland), Pilsen and Prague (Czech Republic), Copenhagen (Denmark), Rome and Milan (Italy), Delft (The Netherlands), Egolfstein (Germany), New Orleans and San Luis Obispo (USA). We are very grateful to all partners for the resulting inspirations. Especially however it was the friendly atmosphere of the events, which has contributed much to the "positive catalysis." Nevertheless, the presented results respect the opinions and works of many other researchers and authors world-wide.

WDK has also published books under the imprint of Heurista, Zürich, to make the results of design-scientific investigations more widely available. These include several volumes of reprints to various related subjects [72,128,205,206,213,216,371,377], and some foundation books [204,217 (reprint),228]. Related books have been issued by other publishers [47,48,49,199,202,203,214,217 (original),219,220,229].

## Remarks to Presentation

A further objective for this book was to obtain the best possible intelligibility and clarity. This has involved a complete revision and updating from the German original. The following remarks to the text, to the figures and the content structure should explain our conception.

***To the text***: general experience teaches that it is difficult for a reader to understand a longer theoretical essay. Therefore we have arranged the text into smaller sections, which describe the content of the section not only through the headings, but also through questions to the subject. The questions at the beginning of a section should help to foster understanding and to make the text accessible for study in smaller portions. Nevertheless, the relationships among these items of knowledge are essential for a holistic understanding of engineering design, the processes of designing, the designed products, and the scope of their sciences. USA spelling is used, except in quotations.

The book contains some historic sidetracks which should supply a general understanding of the present situation. However, it does not deal with a proper historic study which could be judged by the scales of the science of history.

In this connection the *figures* have also been developed from a new viewpoint. Each figure, which often contains extensive text, should treat a closed range of problems as completely as possible. Thus each figure becomes a unit and unity of knowledge. However, the formal order should not lead to schematizing the knowledge contents. Thus cross-references to other figures are often indicated and above all to literature sources and references. Careful study of these figures is recommended, not only for each individual figure, but also for their mutual relationships. The contents of the figures are usually not repeated in the text of the chapters.

***To the structure of the contents***: The book is arranged into three major parts, each sub-divided into several chapters.

Part I consists of Chapters 1 to 3, and describes design processes in general, the general requirements for science, their range of problems, the situation in some geographical regions and the development tendencies and trends towards improvement. An orientation figure to Part I should illustrate todays defective "order" in the knowledge for and about designing, but also its complexity. This should help the reader to better understand the need for a model of General Design Science, it is derived from the defects of the present situation.

Part II includes the Chapters 4 to 7. An orientation figure in Chapter 4 should illustrate the path from a loose accumulation of knowledge towards order, clarity and overview -- therefore the way to Design Science. The discussion about contents, boundaries, structure, sub-areas and sources for Design Science presented in Chapters 4, 5 and 6 should allow recognition of the possibilities for the construction of such a knowledge system. Chapter 7 presents the partial areas of this knowledge region, along with their positioning questions.

Part III, with the Chapters 8 to 11, deals with the specialized knowledge systems derived from General Design Science, from the aspect of the complexity of technical systems, or other aspects such as their functions. Chapters 9 shows the forms of Design Science for different recipients. The final chapters deal with the judgment of quality of the existing knowledge, and with the ways to improving the standard of knowledge.

## Terminology

Some linguistic factors need to be discussed now. Words used in their conventional and ambiguous senses can easily lead to confusion. Consistent and well-defined meanings and terms are needed, but they should not be so rigid and exclusive that conventional usage of the word is violated.

Firstly, we must distinguish clearly between two interpretations of the word "design." Only by recognizing this distinction can we rationally discuss engineering design, abstract the discussions into a science about designing, and lay foundations for design teaching and learning.

One of the interpretations is "design" as a *noun*, meaning the outward appearance and pattern of (artificial) objects, artifacts, systems, *products*. This involves studying how and why things look and behave the way they are, including the possibility of mathematics or computer-based analysis. The usual content of books (typically) on "Design of Aircraft" or similar systems (refer for instance to figure 7--1), and on historical developments of technologies, convey the meaning of "design" as a noun.

The other interpretation, and probably more important in the context of engineering, is "design" as a *verb*. This means a *process* of establishing which of several alternative ways (and with what tools) things could be done, which of these is most promising, and how to implement that choice, with continual reviews, additions and corrections to the work --

*designing*. Of course, in some ways the two (object and activity) must be coordinated.

We will also distinguish two interpretations of the word "descriptive." This word is used in a casual interpretation (see for instance [154,155]), and implies a quality of any informal description or narrative. Used as a formal term, as in the following chapters of this book, *descriptive* means the structure and theory that are the subject of discussion.

The word "technology" also has two interpretations for us. Technology in the general sense is intended to cover not only "the science of practical or industrial arts" [2], but also the experimental (craft-based) "know-how" that in part defies description by scientific means. In this sense the word "technology" is written here without an article. General statements about technology are therefore intended to convey all the functions of any device, and all the functions of the means for manufacturing that device (working means), including the functions of commerce, transportation, management, etc. needed to realize and distribute it. This interpretation is closer to the German word "*Technik*," the general manifestation of technological means and their significance for humans and society, compared to the quoted dictionary definitions of technology. In a narrower sense, the term "a technology" written with a definite or indefinite article can be interpreted as the means and knowledge needed to perform particular manufacturing and other operations [228].

Optimizing has been an aim in various fields, including designing. When designing, we can never be certain that we have reached an optimum (an absolutely best, which may even try to imply a permanent best). A different principle, concept or embodiment, which may remain to be discovered, may be better. Even though the goal of the optimum can never be reached, it is a good aim to approach -- an *optimal* solution is an achievable aim.

Because of this uncertainty in designing, we will distinguish between "determining" and "establishing." A value can be *determined* by measuring from an existing phenomenon. In designing, we usually have several alternative solutions available, and the choice among them is not clear. We will therefore try to *establish* a solution, reflecting the uncertainty. "Determine" is too deterministic for our purpose.

A recent distinction that may prove to be valuable is between "verify" and "validate." *Validation* is used when the subject is a model, to be checked for reasonable fidelity to the behavior of a modeled phenomenon. *Verification* is used to describe checking the information and data used as input to the model, and that produced by using the model.

Normally, when we speak of *cause and effect*, we indicate that a certain set of causes will, under repeatable circumstances, always have the same effects. Most phenomena are in this sense predictable. A chain of causes and effects, a causal chain, can frequently be observed. The question of whether there is or can be an ultimate cause from which all other causes and effects derive is far beyond the scope of this book.

A different concern is that a tendency exists in English usage to circumscribe a concept by deliberately using different and unusual terms, even freshly coined words. Some recently introduced techniques can illustrate this point:

- Solid components can now be made under computer control, by hardening a suitable plastic material layer by layer. This may be termed "rapid prototyping," but other names have been used, such as "solid freeform manufacturing," "shape deposition manufacturing" and "stereolithography." Industrial companies find a need to create wordings such as these to distinguish their product (in this case manufacturing machines and processes) from others, and protect a company's market. Yet in principle the task of all three processes is the same.
- "Concurrent engineering" means designing a large-quantity product, and at the same time designing a manufacturing process, including tooling, processing and transportation equipment. The term "simultaneous engineering" is used in the same sense. Yet some people want to recognize a difference between the two expressions.

Alternatively, some distinctions are useful in their context. *Knowledge* and *information* are two such words. Knowledge can be interpreted as only that part that is held in a person's memory maps. This will include actual knowledge of facts, interpretations, opinions etc., but also knowledge about where to find such items. Information, in this interpretation, is everything recorded external to a human mind. This distinction will be used in this book, but only where it shows advantages. Otherwise, the word "knowledge" will be assumed to cover both aspects.

We see a general need to define a set of words as a terminology, and use them in a consistent way. Thus we may hope that our interpretations can be accepted. The definitions used in this book have been selected with this in mind, but also to be compatible with terminology used in many German-language publications. Nevertheless, we want to avoid unnecessary strictness of these definition. We do not want to fall into the trap of pseudo-scientific exaggeration, that way lies folly [144].

## Closure

Science is a living formation, for that reason it can never be complete. The cited literature can only present a cross-section and at best a severely limited selection of the existing works. A fuller bibliography can be supplied by the authors on request. The science and literature gets revised and developed again and again through new impulses, faster or slower according to circumstances. Therefore each suggestion for the expansion or improvement of our view, as presented here, is welcome.

We are obligated to all our colleagues and the numerous participants of the conferences, seminars, workshops and courses, who have contributed with references, commentaries and impulses to educate our opinions. Support from the Academic Research Programme, Department of National Defence, Canada, should be acknowledged. Dr. L. Badoux undertook a linguistic revision of the German edition, and we have received further valuable inspirations to improvements from the publishing house. We owe thanks Dr. Dagmar Hnik, who processed the text of the German edition with particular engagement. Therefore we dedicate our book to her. Nevertheless, any errors are our responsibility.

We thank the Springer-Verlag publishing house for their traditionally good cooperation and the exemplary completion of the book.

June 1995

V. Hubka, Greifensee, Switzerland
W.E. Eder, Kingston, Ontario, Canada

# Contents

## Part I. General Survey

## Part III. Derivations from Design Science in Other Disciplines -- Specialized Design Science

*Part I.*

# General Survey

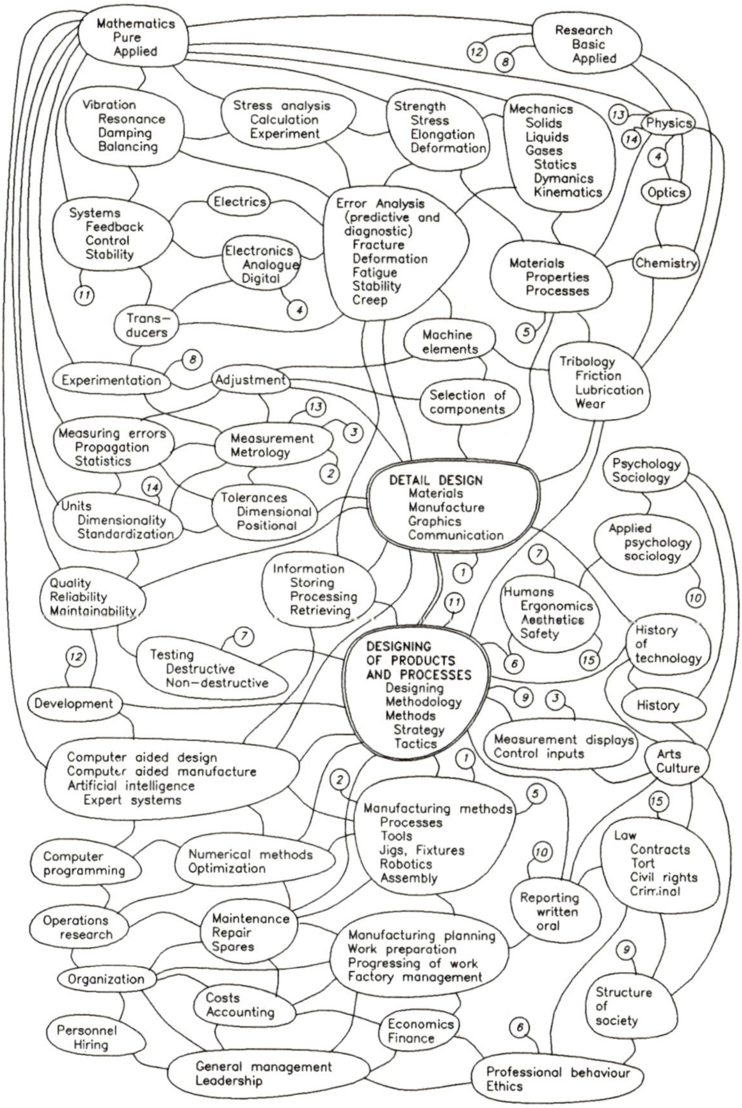

# 1 INTRODUCTION -- DESIGNING

*It is through knowledge that I gain understanding -- and understanding lets me do by choice what others do by constraint of fear.*
***Aristotle***

*A little learning is a dang'rous thing;*
*Drink deep, or taste not the Pierian spring;*
*There shallow draughts intoxicate the brain;*
*And drinking largely sobers us again.*
***Alexander Pope***

## 1.1 What Do We Understand by Designing?

The object of this book is a new *branch of science*, namely Design Science. Why should this science be called "new," when we have known for a long time that active persons have been designing, as we understand it today? Somebody always had to have a first idea for building a house, a cathedral, or a mill, and afterwards had to implement the idea and realize the object. Whether they were artists or craftsmen -- even in those days they solved the problems and successfully created outstanding works. In the manufacturing industries, designing has been formally acknowledged as a separate activity for at least 150 years. This happened during the industrial revolution, as the concept of division of labor was implemented in industrial organizations.

We stated that an idea about an object (product or process) which should be created must have been preconceived and developed, before it can be built or realized. Somebody must think ahead. According to our view, the "somebody with an idea" is the *designer* of a house, a cathedral, a mill, their interior fittings, etc.

What have designers done? Using the words of the technological sciences (compare Section 2.4), designers have found and described a certain structure that is suitable to perform a certain function. From this description and representation a house could be built which should protect humans -- it will possess many other properties, some that are desired, others that are not.

Cooks must also "design" their processes, namely derive a structure from individual operations, before they begin to cook, particularly if a new dish is to be prepared. The process then is described in a recipe, so that others can reproduce the dish from the plan.

Why do we use relatively trivial examples here, when we already design and build large moon rockets. It is just such simple and understandable examples which contribute to

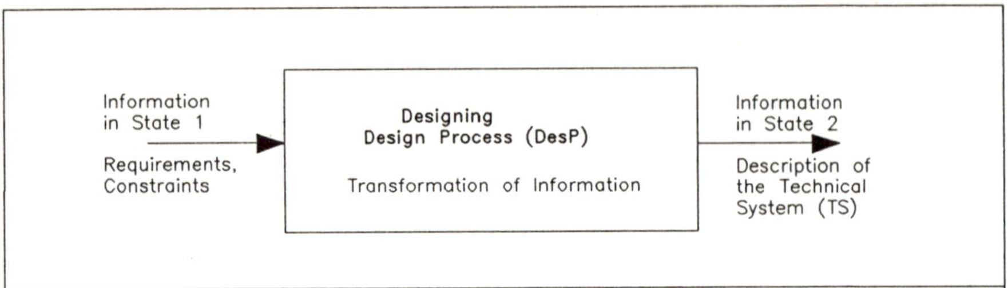


Figure 1--1 "Black Box" Block Diagram of the Design Process

understanding. More complicated cases are burdened with so much additional information which disguise the substance.

Let us summarize: The task of the designing consists of thinking ahead and describing a structure, which appears as (potential) carrier of the desired characteristics (properties, particularly the functions). One can express this statement also in process terms: designing is defined as the transformation of information from the condition of needs, demands, requirements and constraints (including the demanded functions) into the description of a structure which is capable of fulfilling these demands. The demands must include the wishes of the customers, but also all stages and requirements of the life cycle and all intermediate states that the product must pass through.

The black-box representation of this definition is shown in figure 1--1. The design process is composed of several activities.

## 1.2 To the Term "Designing"

### 1.2.1 *Terminus Technicus* "Designing"

With the term "designing" we designate the entirety of all design activities. But "designing" as a collective term can be controversial. Already in our example of the cooking we meet an unusual statement: Cooks "design" a cooking process, if they plan a new dish and prepare it to experimental completion. We certainly meet similar heterogeneous applications of the term "designing" in other areas, where conventional terms like projecting, engineering, planning or organizing are covered by this collective term.

This term "designing" will be accepted with difficulty in an enterprise, in which the "design" section is responsible only for the last stages of designing and its work leads into the work of the "development" section. The collective term also creates difficulty in "industrial design," where "designers" are mainly concerned about the appearance and ergonomics of a solution, but almost not about its inner workings (functioning). This, however, remains a matter of habits and a question of the time.

This problem emerges again and again, when science begins to formulate its own language and *termini technici*, and must give a conventional term a different, usually more precisely defined content than is the usage in colloquial language. Of course equal or still larger difficulties would appear if a different collective term than "designing" were to be chosen for this complex of activities.

The advantage of the word "designing" lies in its general intelligibility, even if it sounds unusual, for example as a metaphor. In addition, this word is widely used internationally, it is understood in the defined context in Germanic, Romance and Slavic languages.

The word "design" can be used as noun or a verb, compare [131] and the preface. We will explore and describe only this latter complex, seeing that these activities (processes) do not proceed uniformly in practice. From here on in this book we will use "designing" as a generic designation, and avoid "design," with or without an article.

### 1.2.2 Designing in the Statements of Other Authors

Many authors have tried to define the *terminus technicus* "designing." Some examples follow (several of them in translation by the authors from German originals):

Taylor [421] (1959): Engineering design is the process of applying various techniques and scientific principles for the purpose of defining a device, a process, or a system in sufficient detail to permit its physical realization.

Asimow [64] (1962): Engineering design is a purposeful activity directed towards the goal of fulfilling human needs, particularly those which can be met by the technology factors of our culture. *And: (ibid.)* Decision making, in the face of uncertainty, with high penalty for error.

Feilden [149] (1963): Mechanical engineering design is the use of scientific principles, technical information and imagination in the definition of a mechanical structure, machine or system to perform pre-specified functions with the maximum economy and efficiency. The designer's responsibility covers the whole process from conception to the issue of detailed instructions for production and his interest continues throughout the designed life of the product in service.

Alexander [39] (1963): Finding the right physical components of a physical structure.

Kesselring [246] (1964): Designing means to find a technically perfect, economically favorable and esthetically satisfactory solution for a given task.

Booker [79] (1964): Simulating what we want to make (or do) before we make (or do) it as many times as may be necessary to feel confident in the final result.

Archer [51] (1964): A goal-directed problem-solving activity.

Reswick [363] (1965): A creative activity -- it involves bringing into being something new and useful that has not existed previously.

Hansen [186] (1966): Developing a technical construct is determined through prior visual thinking out.

Jones [238] (1966): The performing of a very complicated act of faith.

Page [332] (1966): The imaginative jump from present facts to future possibilities.

Farr [147] (1966): The conditioning factor for those parts of the product which come into contact with people.

Gregory [175] (1966): Relating product with situation to give satisfaction.

Matchett [287] (1966): The optimum solution to the sum of the true needs of a particular set of circumstances.

Nadler [315] (1967): Planning and design (P & D) is a process to create or restructure a situation-specific solution. The result may be a house, legislation, an information system, a corporate plan, an appropriate technology transfer, a regional housing plan, a product design, a course outline, a factory layout -- almost anything.

VDI 2223 [25] (1973): Designing is a predominantly creative activity, founded on knowledge and experience and striving for optimal solutions by thinking ahead about technical products, in order to determine the functional and structural construction and create documents that are ready for manufacture. As part of developing, designing includes intellectual and representational form-giving, the choice of raw materials and production processes, and makes possible a technically and economically justifiable material realization. Designing is executed in two well characterized, but not strictly separated stages of laying out and detailing. (Note that despite the reference to two stages, VDI 2222 [23] also refers to "establishing and clarifying the problem," and "conceptualizing" as two other important design stages).

Alexander [40] (1979): Design is the process of inventing physical things which display new physical order, organization, form, in response to function.

Jones [239] (1980): ... the chain of events that begins with the sponsor's wish and moves through the actions of designers, manufacturers, distributors and consumers to the ultimate effects of a newly designed thing upon the world. All one can say with certainty is that society or the world is not the same as it was before the new design appeared.

Katz [243] (1984): ... we normally think of *design* as the activity involved with actually constructing the system; i.e., given a specification of the system, we map that specification into its physical realization (e.g., an integrated circuit chip, a computer program, a physical plant or airplane). The design task, however, extends throughout a *system life cycle*, from the initial commitment to build a new system to its final full scale production.

Suh [415] (1989): ... the creation of a synthesized solution in the form of products, processes or systems that satisfy perceived needs through mapping between the functional requirements (FRs) in the functional domain and the design parameters (DPs) of the physical domain, through proper selection of the DPs that satisfy the FRs.

The nature of designing is reflected in many other statements which at best capture only parts of the truth. Typical for such statements (each made with the personal conviction of its author to have fully explained designing with one simplistic phrase) are, that "design is ...":

(A discussion of the use of words, e.g., "is" and "science" in this context appeared in [131].)

- an art, not a science,
- problems solving,
- decision making,
- applying science,
- creativity and imagination,
- heuristic search,
- learning,
- evolution,
- selecting suitable patterns and adjusting,
- dealing with people,
- team-building,
- collecting and processing data,
- negotiating to achieve a satisfactory solution,
- accepting adequate solutions, satisficing,
- optimizing,
- transferring and transforming knowledge,
- drawing and calculating,
- directing, leading, organizing,

- considering the "bottom line" of costs and profit,
- satisfying needs, satisfying the customer,
- ethical and professional conduct,
- foresight towards production, assembly, testing and other processes,
- etc.

The above definitions show some of the necessary elements of design processes, but none of these is sufficient without the others.

Many of the quoted statements have been taken out of context, so that an interpretation of these statements possibly distorts the opinion of the author. The quotations should however imply more the manner in which the literature deals with the term "designing," they do not present a semantic or systematic study. In addition the examples show how difficult it is for a reader to gain understanding and clarity from this multitude of statements. On the other hand we can see by deeper study that many of these definitions also cover the opinions which we show in figure 1--1.

### 1.2.3 The "Breadth" of Designing

The social dimension of designing also occurs in the above definitions, in particular the fulfillment of human needs. We should therefore question, with respect to the designers of a baking oven, whether the problem to be solved was hunger. Without doubt this is not the case, for the clear task of these designers consisted of designing a baking oven with certain characteristics. The fundamental problem actually happened long ago, and the first baking oven was built in those days as an immediate means for the satisfaction of a human need of stilling hunger. We will meet other such cases where designers are set a starting situation that is unclear and must solve from the beginning, namely to decide about the transformation process that should be implemented.

For clarity we wish to distinguish between:

- the "real" (narrower, immediate) design process, in which we start from the demands (needs, requirements, constraints) on the technical system (e.g., as given by customers or their company-internal representatives), and end with the description (ready for manufacture and/or implementation) of the technical system, and
- the design process in the wider sense, in which additional solution operations (e.g., product planning) are undertaken before (and possibly also after) the real design process, according to the given situation.

The graphic model in figure 1--2 (compared with figure 1--1) offers a good orientation about this breadth of designing. In this way, engineering design interacts with (integrated) product development as a management function, as discussed in [9,47,92,194,195,373].

In the following chapters of this book our explanations refer to the immediate (narrower, real) design process, without special emphasis. Particular remarks may be directed to the design process in the wider sense.

The design process is modified in its breadth according to the object to be designed and/or the stages in the design process. To designate such cases, particular terms have become traditional. Some examples are:

- *planning* -- emphasizes the temporal dimension of the problems (mostly of longer duration) and the scale of the solution suggestions (mostly larger) which result in a large

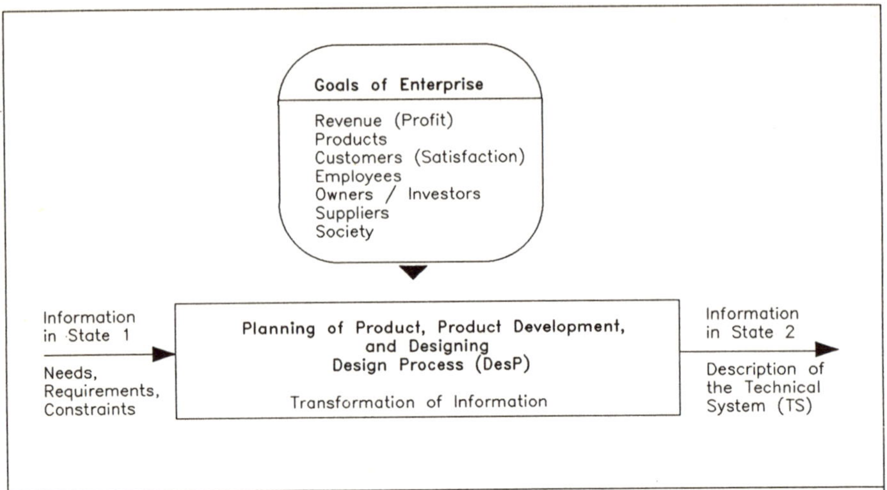


Figure 1--2 Designing -- Extended and Narrow Interpretation

number of smaller problems for other more specialized areas; this expression is used especially in connection with the planning of regions, cities, etc. and the overall construction of large plant. "Designing" is often described as "part the planning process," but the relationship is reciprocal;

- *project engineering* -- the object to be constructed is usually a large to medium industrial plant, where many of the elements are selected as independent machines from suitable manufacturers' catalogs or are produced by manufacturers to special order ("one-of-a-kind" or small series);
- *developing* -- the object should be a new, unknown system (at that time) -- caution is needed in this connection, the same word is also used for the different process of testing and modifying an existing, newly built (realized) system (e.g., a prototype) to achieve satisfactory operational parameters;
- *organizing* -- the object consists mostly of humans (and technical means to support human activities), the task is composed of generating goals, and giving the instructions (guidelines) which should be suitable for reaching these goals.

Other activity words refer to partial areas of designing, but they can be considered under some circumstances as equivalent to designing:

- *conceptualizing* -- conceiving one or several possible structures which could fulfill the given needs, demands, requirements and constraints. A *concept* can be a sketched interpretation of a proposed solution, but also an intellectual abstraction with relationships for a class of objects or phenomena;
- *laying out* -- representing and deciding about the possible arrangements of physical elements and their functionally conditioned sizes and forms, normally by producing drawings (similarly to an assembly drawing) in which only the main dimensions (and connecting dimensions between the elements) are indicated and which deliver the information for producing detail drawings (or their analogs on data carriers -- e.g., computer media);

- *detailing* -- producing the individual drawings (or their computer analogs) for the components which should be manufactured;
- *form-giving* -- establishing the possible and necessary forms (and sizes) for the complete assembly and the components.

Let us emphasize again that, in our interpretation, these terms are subordinate to the collective term *designing*.

### 1.2.4 Needs for Design and Designed Products

Any job (with its tasks, duties, *purposes, needs, requirements*) that someone wants done (if it can be done, *constraints*) can be done in various ways (*processes*) using various tools (*means, products, systems*). It is characteristic that we can usually find different products which can be used to perform the same duty. The usual situation is that someone wants to achieve a result from a given state of affairs, and tries to find a tool (process systems and object system) to help in getting that task done and produce a desired effect. If such a tool does not exist, it needs to be designed and realized. In designing, we frequently look for *candidate causes* that are likely to be able to achieve the *desired effects*. The causal chain (from cause to effect) familiar from analysis and science needs to be reversed. This raises the question of whether we buy (or sell) a product, or whether we buy (or sell) the service (a process) that the product can perform.

The main question of designing is to find what structures are needed and adequate (or even best) so that the product and/or process is suitable for the intended purpose.

The process (or service) to achieve a result must make changes (transformations) in some material, energy, information and/or living being (an operand), from an available starting state to a (preferably more desirable) finishing state. Some disturbing quantities enter the process, and others are emitted by the process. The process, if it is to produce the desired results, needs to be operated by a combination of humans, object systems (tools), information, and management (goal setting, resourcing, directing, etc.), under the influence of its environment. The desired (future) process or service drives the search for an appropriate object system. All these factors should be considered when designing a process, and/or its object system.

Almost all processes and products must be designed before they can be implemented or manufactured. This is mainly because in modern times two (groups of) people with different abilities, skills, knowledge, and tools will be involved. One designs, the other implements or manufactures. Processes and products are of different kinds, with various properties, characteristics, qualities.

Processes may be *natural*, therefore evolved rather than designed, and difficult and/or unpredictable to affect by our attempts to control them. Other processes, the ones of interest to us, are *artificial*, intended to fulfill human needs, and mostly under our control. Nevertheless, the norm tends to be a hybrid process.

Products must be able to perform the desired task (drive the transformation, achieve the results of the process, they must function). They must do this with the demanded performance, operability, sufficiently long life, safety, reliability, adjustability, maintainability, and so on. The above groups of *properties* are sometimes collected under the term of *functionality*. In addition, products must be manufacturable to the appropriate quality, packaged, distributed, commissioned, etc. The time scale of planning, designing, manufacturing, and delivering must be suitable. Recycling and disposal have recently become more important. Products must be suitable for humans to operate them (ergonomics) and live with them (esthetics). They must conform to the laws, regulations, standards, codes, moral and ethical considerations applicable for the place of origin and of use. Cash flow and financing must be considered. They must also be economical, offered at the right price, with appropriate services and support, and acquired at a suitable cost (e.g., to the manufacturer).

All the *properties* of systems as listed in the last paragraph should be considered when designing, not only in their positive sense of trying to achieve them, but also in their negative connotation of anticipating, detecting, avoiding, and maybe correcting possible failures. Only in this way can we be sure to satisfy our customers, and achieve the appropriate quality of processes and products. That quality (the mixture and values of the properties) is established in the first place by designing, and is then realized by suitable manufacturing. Therefore all intermediate stages that the product is anticipated to pass through must also be considered. The conditions of usage, including maintenance and repair, can affect the actually achieved quality of the product in use. In this context, Kuesel [265] writes with respect to bridges: "we should not now be debating how much further we can reduce safety margins (for loads and strength), but rather how much we should increase safety margins (for wear and corrosion)."

#### 1.2.4.1 Types of Product

At one end of the scale of products are *artistic works*. The artist is usually both the designer and manufacturer. Appearance is the primary property, esteem value (beauty in the eyes of the beholder) tends to be very high compared to costs. Usage of these product tends to be relatively trivial. The designed (and made) work frequently consists of only one or a few parts (components), its structure tends to be simple, but its form can be complicated. Designing can and does take place during manufacture. Especially evaluation for suitability of the product is continuous, and depends almost purely on the judgment of the artist. There is almost no scientific knowledge base to consider, except that resident in the materials, tools and knowledge of manufacture. Design methods hardly come into consideration. Nevertheless, research activity into design takes place, e.g [97]. We can regard as typical products such items as: coins and banknotes, decorations, conversation pieces, and jewelry.

*Consumer products* are frequently consumable items and materials. Designing involves to some extent the product, but of probably more importance is the packaging and advertising. Again the knowledge and science base is minimal, especially regarding the packaging, unless the product itself is something like a chemical compound which requires design of the composition and the manufacturing process/equipment. Methods for designing tend to be marginal. Products typical of this group are: packaged butter, motor oil, purified water delivered by pipe into a home, and newspapers.

*Consumer durables* must have appropriate appearance and operability. They must project the "right" image, of the product, but also of the manufacturing (or selling) company. This is independent of whether the product is to be used typically in the household, on the road, or elsewhere. These products must also perform useful tasks, they must function with suitable performance parameters, and be made available at a (usually pre-defined) suitable cost. The designed product tends to have many components, where those providing the main means of functioning are frequently not visible or accessible to the casual observer. Two separate (groups of) designers are usually employed. The primary and most advertised design task is that of the "industrial designer" and/or "stylist," to establish the outward appearance and operability conditions -- esthetics and ergonomics. Within that "envelope" and usually of at least equal importance for the product, the technical designers must establish the means for the product to function. Conflict can arise between the two, where it becomes difficult to create the means because of the constraints demanded by appearance.

Cooperation is called for. There is usually a substantial knowledge base involved, and for the technical aspects much of the information can be scientific. Design methods have been discussed, but substantially more for the technical aspects than for the appearance "styling." Typically such products include: lighting fixtures for the home, domestic appliances, motor cars, personal computers, and furniture.

*Bulk or continuous engineering products* are generally raw materials for other manufacture, e.g., lubricating oils, fuels, metal rolled sections or rod and bar stock, plastics in pellet form for moldings, sheet and foil woven or roving, laminates, etc. The processes and tools for manufacturing need to be designed. Sizes, shapes, tolerances and material properties of the product are mostly laid down in standards. Imprints or embossings may be added to alter the surface texture and appearance, and these may need to be designed.

*Industry products* are generally items or assemblies that are bought by a manufacturing company for assembling into their own products. They include machine elements, purchased OEM goods (products intended for "original equipment manufacturers" to build into their own devices), and other supplies. Appearance matters only in some cases. If two items of equivalent functionality, performance and price are offered, appearance makes a difference. These products include: ball and roller bearings, electric motors and controllers, circuit boards (e.g., computer motherboards), crane hooks, electronic controllers for passenger elevators in high-rise building, high-voltage insulator hangers for power lines, gas turbine engines for aircraft propulsion, etc.

*Industrial equipment products* are self-contained devices (e.g., machines) which can perform a complex function, and are intended for use within industry. Appearance is distinctly secondary, functioning and performance are primary. Among these products are: personal computers and industrial workstations, machine tools, goods vehicles, hydro-electric turbine-generator sets, chemical reactor columns, earth moving machines, and passenger aircraft.

*Special purpose equipment*, including jigs, tooling, fixtures, and specialized manufacturing and assembly machinery, special-purpose robotics, handling and packaging machinery, but also ocean-going ships [71]. These are usually produced to special order, as single items ("one-of-a-kind") or a small series. Designing and developing takes place specifically for one customer, and in most cases these products are newly designed. There is no opportunity for redesign, they must be directly (and usually quickly) integrated into the customer's usage and/or production facilities. Mechanical, pneumatic, hydraulic, electrical, electronic and software components are used in equal importance. Delivered (purchased from suppliers, OEM) components are intensively used to control the risks, configuration tasks are more frequent than custom designing of items. Modularity of the structures can be an advantage. Sensor technology, and user or developer software are particularly important. The requirements include extreme demands for cycle times, reliability and availability, ease and flexibility of re-tooling for a different production process or product, and anticipated possibilities for retro-fitting to improve performance. The phases of generating preliminary design proposals for cost estimating and tendering (to obtain a customer contract) are extensive, and need substantial time and financial expenditure. Tenders have to be offered for complete machinery, including delivery time and price, before all details of the devices have been worked out. Many of these tenders are unsuccessful, they do not lead to a supply contract. A particularly careful analysis of the task requirements is needed, to adjust the design specification to the customer's specific needs. Even so, frequent changes are required during designing and developing the machinery because of progressive increase in the amount of knowledge about the task, and additional wishes of the customer. Software is often developed and adapted during commissioning, before the customer's final acceptance of the devices. A major task consists of coordinating deliveries from suppliers, including interfacing with other equipment and components, performance data, drawing and delivery standards, etc. The existing know-how of the customer must be used, by providing and obtaining extensive information, and using the customer's personnel as consultants in the development. Training and support for the operating personnel must usually be arranged before and during commissioning. This must especially concern motivation, acceptance, automation psychology and qualification of personnel. Any future developments and trends that can be foreseen must be thought out in advance and considered. Engineering designers employed in this kind of product need high flexibility, the types of tasks change rapidly from one contract to the next. They need good communication capabilities and negotiating styles. Close cooperation between design and production for these machines is essential. Companies involved in this class of product are usually small to middle-sized enterprises (or subsidiaries).

*Industrial plant* usually consists of collections of industrial equipment products, and devices to provide control and/or connections among them. The plant (and some of the connecting devices) is designed to special order, most of the items are bought from other suppliers. This is probably the extreme case of a product incorporating other products, and the (recursive) task of designing them and their components is mainly transferred to the suppliers. Typically: water purification plant for a city, electric power station, oil refinery, telephone network, and so on.

These classifications refer both to usage and to the complexity of the product. They are neither unique nor complete, the boundaries are fluid and overlap. It is of little use to try to make these boundaries more definite, common usage of words cannot easily be altered or forced. Individual products may appear in more that one classification, or cross the boundaries. Obviously, a product from one company may well be input material for another.

#### 1.2.4.2 Some Implications and Contexts

Products as described can imply choices among possible strategies for designing. Among these strategic ways are those known as *top down* (progressing from a holistic view to the detail) or *bottom up* (from detail to overall definition), and *inside out* (defining the inner working means first, thus constraining the outside) or *outside in* (defining the envelope, thus constraining the inner workings).

As the above examples may show, the traditional divisions among engineering disciplines (mechanical, electrical, civil, chemical, but also aeronautical, mineral, electronic, computer, etc.) are not particularly useful, if they ever were. Is a speed-controlled electric motor (e.g., for an audio tape drive) a product of electrical engineering (the electrical and magnetic sections), or of mechanical engineering (the shafts, bearings and housings), or of electronics (the controller circuits)? The word "or" is obviously misplaced. Many products are made under the direction of one (group of) discipline(s), and used under another. Examples are machines for mining, road construction, or robotic assembly of printed-circuit boards. The "high-tech" products usually contain a hybridization of components related to mechanical, computer, and other disciplines. Trends such as miniaturization and integral hybridization have appeared -- mechatronics is the result of one such trend.

It should be clear now that professional engineers are not necessarily involved in all design tasks. A company making relatively "low-tech" items is unlikely to employ or even consult a professional engineer. There is hardly a need for professional engineering expertise to design a security lock to be fitted to a sliding door in a house, or for a lighting fixture. Mostly, engineering designers tend not to be university graduates.

The currently best incorporation of properties in a product is usually called the *state of the art*. This state usually exhibits different levels, according to whether we try to observe it in research, speculative design and development, production design and development, advanced production, or everyday use. The role of *invention* is to improve the state of the art in research and speculative design and development, mainly by suggesting new embodiments of operational principles into a candidate product. Invention may or may not lead to a commercially viable product or process. If a new process or product appears (e.g., on the market) and is implemented, this is termed an *innovation*.

A recently introduced trend in industry and other organizations is to try to achieve *continuous improvement*, relating to a product or process, and to the organization. Even though the improvement tendency is continuous, a product or process can and should only be *upgraded* at distinct time intervals, to preserve some continuity.

Each product goes through a *life cycle*, which can briefly be characterized by stages such as: product planning, designing and developing, preparing for manufacture or implementation (organization and resources), implementation or manufacture and assembly, distribution (packaging, transporting, etc.), commissioning (getting ready for use), exploitation (using the product to perform the intended task, but including maintaining, repairing, renewing, etc. the product), and disposal (recycling, reusing, destroying, storing waste, etc.). Within these broad stages, any product passes through many intermediate stages.

The useful life of a product is ended after exploitation, and in some cases even before. Criteria for when *life ended* occurs range among:

- *wear out* (time-dependent, environment-dependent, etc., but life can be extended by repair and partial replacement);
- *catastrophic failure* (time-dependent, environment-dependent, etc.);
- declared *obsolescent* (overtaken by development in the state of the art, e.g., by a newly possible change of working principle, but still usable or upgradable) or *obsolete* (no longer usable);
- intended as a *one-shot device* (only a single use is foreseen); or
- experience *replacement* before completed life expectancy (e.g., aircraft gas turbine blading).

Even after normal "life ended," some products find re-use. Examples are: a different type of organization may extend or revive the life (e.g., steam railways run as tourist attractions by volunteers); a partially gutted item may be used as a display item or monument. Otherwise, these products can be (partly) recycled into raw material, or sent to scrap or waste disposal.

### 1.2.5 Characterizing Designing

Designing as a process is more or less *creative*. This usually includes: intuitive, iterative, recursive, opportunistic, innovative, ingenious, unpredictable, refined, striking, novel, reflective [385], searching for elegance, beauty, etc. Iteration is always necessary, no design problem is simple enough to fit within the mental limitations of short-term memory in humans [301,302,303,304].

For a particular problem, any two designers will usually produce solutions that show substantial differences, designing is to some extent *idiosyncratic*, depending on the experience background of that designer. But the solutions will also be substantially similar.

Designing usually takes place in answer to a perceived *need*. Especially in engineering, designing is *goal-directed*. Goals include attempting to resolve an issue, usually to achieve satisfaction for the customers. Trying to find a set of reasonably logical steps and progressions that can suggest ways to rationalize designing would thus make sense.

The aim may be a *novel* product, with little or nothing taken over from a previous product. Risks tend to be relatively large, especially if untried principles are used. The process or product may be more or less of an *adaptation* to new conditions, one form of redesign. It can be a *renewal* of a previous product or process, using modified principles, performance values, etc. Designing can result in an *alteration* or *updating* of products. Many design projects relate to *variants*, designing another member of a discrete range of size or performance, by scaling up or down with appropriate adaptation of size relationships.

Designing is an activity of humans (individuals, groups and teams) together with their assisting tools (e.g., computers), performed using information, under direction and with goals delivered from management, in a working environment. All these factors affect the quality of the designed product. Responsibility rests with the designers, but information is needed by and from many others. Designing as a process accepts input information about the task and the requirements for its output. Designing delivers as its output a set of instructions for implementing or manufacturing the designed system. The next direct customer for this output information is typically a manufacturing planning department, and its manufacturing facilities.

Designing for a *novel system* or process accounts for 0.1 to 10% of design projects. This form of designing has been broken down into smaller sub-tasks. The degree of novelty and complexity of the project account for the highest demands made on a designer (or design team). A typical arrangement from a management view consists of (iteratively) clarifying the task, conceptualizing, embodying, and detailing. *Clarifying the task* leads to a design specification. *Conceptualizing* uses abstract models of the system to be designed, to generate a framework and outline of the product and its ways of functioning. *Embodying* places material where needed around the outline, as a preliminary and/or dimensional layout -- a *configuration* and its *parametrization*. The final stage of *detailing* produces the complete manufacturing information, typically detail drawings, parts lists, and instructions for assembly, testing, adjustment, maintenance, etc.

Redesign may involve some conceptualizing, but is more likely to consist of relatively routine tasks within embodying. Redesign can be performed by abstracting from a given prior system, searching for possible and preferable alterations, and concretizing back into the embodiment and details.

During designing, a system may need to be broken down (*decomposed*) into sub-systems. Each sub-system can be regarded as a different design problem. In analogy to computer programs, we speak of design projects being *nested* -- a subroutine within a subroutine, e.g., problem solving called many times within designing, or evaluation and decision making called many times within either designing or problem solving. The process of designing is normally *recursive* -- a subroutine calling itself, e.g., problem solving calling problem solving for a sub-problem, or a design process calling a second design process for a less complex sub-system, going downwards in a hierarchy of complexity. Designing must also be *iterative* -- exploring forward into more advanced (usually concrete) design stages, to repeat (backwards) for review, expansion, completion and correction. *Opportunistic* actions can and should be taken at any time when a likely avenue opens, but designers should not just go off at any tangent.

*Nested* in the design process is a group of frequently occurring operations which we collectively call *problem solving*, described e.g., in [228,318,397,398,408,437,458]. This group consists of defining a problem, searching for solutions, evaluating and deciding among candidate solutions, and communicating -- either to the next more detailed level, or to a sub-problem, or to an implementation process. Searching for solutions can involve literature and other existing knowledge and information, from own or outside sources, advisory systems, information banks, knowledge-based systems, and other artificial intelligence (AI) applications, but it can also imply thinking of new or transferred applications. Evaluating can use analysis and its tools, including computer programs such as finite elements (FEM), boundary elements, computational fluid dynamics, circuit analysis, system simulations and so on. Reasons for evaluating include checking to make sure that the proposed system is unlikely to fail, comparing proposed systems to find the best for the situation, and/or comparing a candidate system to an ideal to establish its potential quality. Evaluation needs criteria, statements about acceptable performance related to the properties -- tolerance limits, maxima or minima. Some *objective* evaluation criteria will contain numbers or mathematicizable relationships, but others will be (more or less) *subjective*, depending on human judgment. Evaluation criteria should appear in the design specification, but only a limited selection of these criteria is applicable at any design stage. Decision processes may point out possibilities for optimization and improvement of a candidate system, but their main purpose is selection. Decision theory [74,101,300,407,409,415, 422,446] has been developed for that purpose, to make evaluative decisions more rational, providing that the criteria and goal functions can be simplified and formulated in mathematical and/or symbolic terms.

The process of designing can be performed in systematic, methodical, structured ways, but also intuitively with no obviously followed structure. These apparently different ways are not in conflict. Structuring can be and is used to support the intuitive and creative leaps, design procedures, knowledge and information, it is the basis for rationalizing.

Many methods exist to help the designer, most of them are applicable in the concrete stages of embodiment, layout and detailing. They include computer applications, including simulations, artificial intelligence (AI), finite element methods, and computer-aided engineering (also CAD, CAM, and CIM). Methods can be divided into *strategies* to be used as guidelines and principles, and *tactics* giving details of operations. A coordinated grouping of methods is usually called a *methodology*. Any method should be approached in a flexible manner, and must be adapted by the user to the problem in hand.

The need to bring products to the market as quickly as possible arises from trying to shorten the time to economic break-even and profit. There is consequently a need to rationalize, and preferably systematize the processes of designing. The design process must be managed (e.g., [147,182,272], made accountable, and made more reliable and predictable in timing and quality of output. Further consequences are:

- a need to search for and evaluate alternative solution candidates at several levels of abstraction, and to make informed decisions among these alternatives;
- a need for best possible knowledge about objects that have been designed, available principles of operation, and derivations from science, but also of failures that have occurred, and their causes;
- a need for best possible knowledge about design process -- preferably embodied, learned and internalized by the designers, but if possible also brought onto more permanent storage media (e.g., paper) and algorithmized;
- a need to evolve the proposed solution, the designed product, before making it, so that as many of the potential faults can be avoided;
- a need to audit the work of the designers, to validate the models, verify the data (input and output) and check the implementation and manufacturing information -- with respect to all properties and the accepted design specification.

Rationalizing the design process also implies that the product and its manufacturing process (tooling, transfer arrangements, equipment layouts, etc.) should be designed concurrently (simultaneously) where possible [269,414,418]. For a novel product, concurrency usually needs to wait until at least the conceptual stages of product design are substantially completed. During embodiment and detailing of the product, concurrent engineering of the manufacturing process can proceed. But there is a need to "freeze" the development and design of the product at a suitable stage, before designing on the manufacturing process can be completed -- otherwise an expensive running redesign situation can occur. *Running redesign* is an iterative chain that occurs when a change in the product or its manufacturing process necessitates a change in the other, and too few restrictions are placed on continuing the chains without end.

From the viewpoint of a company, the mixture and marketable properties of products needs to be managed. The satisfaction of external customers can lead to continuing operation and economic viability of the enterprise. Therefore the product family, its scope, variety, innovation, advertising, packaging, and all other features need to be planned, kept under review, directed and implemented. This requires some integration -- helped by management organization, philosophies and techniques such as quality assurance and its international recognition in the standards of the ISO 9000 series [11], Total Quality Management (TQM) [94,105,232,233,240,297], and its continuation as Total Quality Development (TQD) [91], Benchmarking [445], Quality Function Deployment (QFD) [82, 134,189,192,278,279], VA/VE [28,190,299], cost-benefit analysis [162,271], Taguchi experimentation [279,347,375,416,417], emphasizing team-building and cooperation [396] and many others. The expression *integrated product development* [9,47,92,194,195,373] covers these aspects. Designing is one of the necessary constituent functions of this process. Some of these management techniques can also be useful and directly applied within designing, to assist the process.

Even if a process or product is well designed, manufactured, and marketed, it can fail to reach its sales potential under circumstances. A revised strategy can sometimes repair the situation and result in a successful launch [166].

This narrative description is equally valid for generating a piece of art (literature, painting, sculpture, book, etc.), a tactical move in sports, a tool, an organization, a management procedure, or any other useful or desirable (to someone) item. The products, processes and services related to *technology* are only special because they are usually associated with industry. In this book, we will restrict considerations to design within technical contexts, *engineering design*, used to create such products, processes and services.

### 1.2.6 Roles for Failure and Development

Designing has the main task of developing solutions to a problem. Two aspects of *development* are usually acknowledged. One involves testing an existing product, prototype, sub-system, or simulation, in order to modify the product and improve its performance up to the required level. If designing has not been performed well enough to anticipate the performance requirements and failure modes, a dangerous and costly (in money and time) situation of "running redesign" can set in. The requirements are usually defined in a requirements or design specification.

The other aspect of *development* describes advances in the state of the art for a class of process system or object system over a period of time.

*Failure* is defined as a lack of adequate functioning, which may be a result of slow degradation or of a catastrophic event [448]. Any perceived failure may need application of diagnostics, detecting and establishing the presumed causes for failure. This can lead to additional research to find new knowledge. But a failure usually needs correction, by adjustment, repair, renewal or replacement of the process system or object system, and/or by designing a successor system to overcome the faults [453,454]. One of the tasks of designing is to try to anticipate, predict and avoid any possible failures (e.g. [75]).

A time sequence for developments in the state of the art has been proposed (e.g., [343,344,345,346]). For the studied systems, there have been consistent attempts to make the process and/or product cheaper and better. One typical way has been by progressive reduction in safety margins and usually also increase in performance, functionality and complexity. As successive systems have been implemented or built, confidence in the principles and configurations has increased.

Eventually, a system has developed to a state that reaches the limits of a previously undiscovered failure mode, or it has reached the limitations of a simplifying convention that overlooked effects of a different failure mode. This has often resulted in a catastrophic failure of a system. If this failure was sufficiently noteworthy (and newsworthy), a period of renewed conservatism has set in, with retrenching (e.g., using wider safety margins), and subsequent attempts to evade the failure causes, possibly by using a different principle or concept. Therefore improvements in the state of the art have taken place in a succession of failure and renewal cycles, including re-learning of some lessons from the past which were forgotten in the period of confidence. The cycle apparently takes about 30 to 40 years.

John Hersey in a novel (*A Single Pebble*, New York: Knopf 1957), wrote about a young engineer. In his travels up the Yang Tse Kiang, a big river in China, the engineer remarked that there could not be any new ways of traversing the dangerous rapids, especially considering that the Chinese junk (a square-rigged sailing boat) had not changes much over the centuries. The old boat owner informed the engineer that: "There is a new way every forty years. It is an old way but it is called new. River pilots wait only for their grandfathers, who know better, to die before they claim they have found a new way." It could well be that the improvement cycle mentioned in the previous paragraph lasts about a generation, during which the old knowledgeable people tend to be replaced by younger and less experienced people. Capturing the old knowledge in a recorded and retrievable form (so that it does not get lost with the person) is an important task for rationalizing designing.

### 1.2.7 Societal Conditions

Designing and manufacturing, buying and selling, take place within a culture, and its related moral, ethical, economic and political climate, both within a region or nation, and trans-national. Some of the factors in play are competition, the nature and influences of money markets, stability, and general organization of society.

There has always over the centuries been a trend to produce better tools and products. This raises the question of "better for whom?" Engineering has been defined as "using the resources of nature to improve the living conditions of mankind." Such a sweeping statement is of little use in practice, because what benefits a part of mankind may damage another. Many questions of moral, ethical, environmental, and sustainable development need to be considered.

In a narrower sense, the persons who should be satisfied by a product are mainly the company-external *customers*, the buyers, the users, and the directly affected people. They are all interested in a suitable range and mixture of the *properties* of the product -- and the various people involved each have a somewhat different expectation and judgment.

Management also needs to give a different answer to the question "who is the customer?" Within a company, anyone downstream in the *life cycle* of a product is a customer, e.g., manufacturing is the obvious customer for engineering information, delivered by designing (e.g., in the form of engineering drawings) as the manufacturing specification. But also anyone upstream in the life cycle is a customer, e.g., designers as customers for information and advice about easy and economic manufacturing. The consequence is a need to exchange information, up and down a management hierarchy, but also more importantly at any one level across the hierarchical structure. This would seem to be facilitated by reducing the number of levels in a management hierarchy. But according to Kumar [268], referring to [191,286]: "Hobbes, not Marshall, would seem to be the better guide to understanding the modern corporation. At every level in the hierarchy there appears a competitive struggle for power and control, with consequential disregard for the ideal rational function of the organization as a whole."

This need for lateral communication and cooperation has always existed, but is now more urgent because markets are more internationalized, economic competition is world-wide, more rapid systems exist for exchanging information, and other factors. When some people report a "change from a capitalist to the information society" [118], they seem to forget that information has always been necessary for designing and using objects, and important as the basis for power. The current situation represents only a shift of emphasis, and a quantum jump in the speed of transmission and quantitative availability of information. But the current situation is maybe not as suitable for (internalized) knowledge as it is for (externalized) information.

Over the centuries, the trend to produce better tools and products, has led to novelty, improvement, and increase of capabilities and functionality of many product ranges. More recently, several products have become too complex for comfort, users are not willing or able to exploit these products to use even a fraction of their functionality. In fact, they can even be frightened of attempting to use the systems, because they are too complex. This is also a form of failure.

## 1.3 Technical Knowledge about Designing: Current Situation

Designing is a process. The technical knowledge about designing in its highest form, that is in the theory of designing, is *process knowledge* within the meaning of the technological sciences (compare Section 2.1.1.2).

Study of the literature lets us recognize very quickly how little of this knowledge is contained even in specialized publications about designing. How a certain design process was executed, and why just this result was obtained and not another, could hardly be report. Even successful designers could rarely say why they had executed a work just in this way and not in another. Almost in vain Reuleaux [364,365] critically wrote in 1856:

> *The knowledge of those principles borrowed from mechanics does not suffice in any way to generate a layout of a machine to be constructed ...*

Only some partial areas, for instance in mechanical engineering, have been developed more deeply in connection with design knowledge. For instance, strength calculations -- initially a very problematic branch -- were developed (starting with Bach, 1847-1931) into a reliable operation as part of the design process. Also the design and calculation of machine elements (e.g., bolts, brakes, linkages, etc.) and some specialties were developed, e.g., cranes, steam turbines. Representing (usually defined as making a visual/graphical representation of a real object) within the framework of technical drawing has also reached a high technical level, especially as wire-frame, surface, solid and features-based modeling on computer systems.

It is astonishing how much this situation has influenced the idea of designing in common opinions. Some technical people (especially in education) have considered designing simply a more or less sophisticated calculation activity. This can include analysis on the basis of knowledge from engineering science, but also more complex operations such as finite element methods, optimization, statistical processing and decision theory applications. For other, especially for outsiders, the task of technical drawing characterized designing.

The remainder -- and exactly the central core -- of design work, the prior thinking out of the technical objects and process, has been removed from consideration merely as "getting an idea" or an inspiration, be it good or bad. People believed (from these viewpoints) that these ideas, as well as other decisions, could be obtained just with a portion of "intuition" and "creativity" -- without defining these terms or their implications. In any case one had to have talent. Tacitly, but more often quite clearly, designing was held to be an "art form" rather than involving a scientific activity (refer [131]).

Only with the movement towards rationalizing and streamlining design in the 1950's and 60's was the activity of designers explained to be cognitive and scientific-technical (and this perhaps at first unconsciously). This was rediscovered in the late 80's in the USA (e.g., as reported by Dixon and others [112,113,154,155], with practically no reference to previous European work). Designing therefore needed technical knowledge, and a new task was to find, explore and process this knowledge to a theory. The discipline known under the terms "methodical or systematic designing," or sometimes "structured procedures," has gathered much material and provided considerable preparatory work during the last thirty years.

Was the design knowledge completed in this way? Can it now fulfill its momentous task as an element of Design Science? Unfortunately, the answer is largely negative. Only a model of Design Science (as presented in Chapter 7) delivers an impression of the more complete problem of this area and offers a basis for comparison.

The slight interest shown by engineering practice concerning the solution of this problem seems surprising in this connection. It appears almost as if designers were satisfied, and considered the quality of the designed solution as adequate. With the present craft, the impression was that they could eliminate or compensate for all errors and neglect from designing during the subsequent development and production of the product. Designers in practice mostly do not find the time and motivation to locate within their activity the research about designing, to understand it and to combine it into their own work. In part, after Müller [311], they conduct a self-defense against these newer results. A different explanation for the weak response of practice to the research results of methodical (systematic) designing can hardly be found, but some additional reasons are mentioned in Chapter 8.

In Chapter 7 we will deal with design knowledge and the research results, for that reason we will not enter into further details at this point. It is, however, useful to concern ourselves more closely with some terms which help in understanding designing.

These are the ideas of creativity, intuition, innovation, etc. Immediately we would like to emphasize that these ideas do not appear as competition to methodical (systematic, structured) designing. Only their wrong application could arouse such an impression.

## 1.4 Relationships Between Designing and some Other Terms

Surprisingly some words or terms remain closely connected with others for long periods. In the course of the time such conglomerates are imbued with fundamentally different contents than originally intended. Understanding is thereby hurt, as is well known.

We want to explore some of these combinations here and give appropriate explanation of their content. For people who wish to study this problem more deeply, some passages from the relevant literature are quoted and further literature references added.

### 1.4.1 Designing and Intuition (Inspiration)

For many designers and design researchers the word "intuition" is a key-word. For that reason we will deal comprehensively with this term.

#### 1.4.1.1 What is Intuition?

Intuition has been defined in various ways. Examples are:

- Spontaneous mental registering, insight resting on knowledge and experience. In the irrational knowledge theories: An insight not resting on experience knowledge, but produced emotionally (sub-consciously) by "internal" inspiration [13];
- Original viewpoint, meditation, later mental vision, of the type of a revelation, not obtained through experience or consideration, but in a mystical way through direct registering of the nature of a perceived reality, similar to a revelation; emotional discovering of final truths, which are unprovable and do not need proofs. In another sense intuition is also understood as experience thinking, of which the individual stages do not become fully conscious, as is the case for example in a diagnosis made by an experienced doctor [114].

The German poet and playwright Goethe knew the idea of *scientia intuitiva* originated by Spinoza. He called intuition an "exact fantasy." Kant describes it in *Kritik der Urteilskraft* (Criticism of Judgment Power) as "the spontaneous, free agreement of the imagination with the laws of understanding." According to this view it is desirable to have not only the right thoughts at ones disposal, but also clear impressions. Both the sensations of the reality and the right visualization of the reality are important information sources for the intuition. Those who feel clearly and think clearly have the chance that both abilities meet and that, "like with two electric poles, the spark ignites" [283].

#### 1.4.1.2 Conditions for Intuition -- Inspiration

Investigating the conditions of the state in which the knowledge emerges by means of an intuition (in other words in an inspiration situation), three typical features are discovered, according to Hubka [201]:

- The persons who received the ideas have dealt consciously or unconsciously with the problem (disputation, pursuing the thought, continual search); they have tried repeatedly to consciously solve the problem; they have extensive knowledge and experience in the problem area; they have great interest in solving the problem (they are strongly motivated); they are personally disposed towards intuitive work; they have sufficient conceptual knowledge and experience of the field in which they can be intuitive.

  It is also remarkable, in harmony with the nature of intuitive knowledge, that the persons who received the ideas can hardly report about the endured process. They usually say things like: "and after some time the idea struck me ... ," which perhaps only shows the incessant search and continual pursuance of the thought. Newton answered the question about how he was led to the law of gravitation: "In that I thought continually about it." Goethe hit the nail on the head with the words: "What is inventing? It is the conclusion of searching."

- The idea comes mostly in a period of relaxation, when the problem is not in conscious thought. Frequently the conditions of space (bath tub, bed, hearing music) and time (when falling asleep or awakening) are the same for different persons. Exploring this phenomenon it has been shown that only a third of the involved persons have indicated arbitrary and unimportant conditions.
- The intuitive registering of a thought or an idea is accompanied by strong feelings of overwhelming clarity, which gives an extremely effective impression contrasted to the previously felt mental tension and uncertainty.

### 1.4.1.3 Results of Intuition

An appraisal of the results of intuitive thinking shows that often brilliant ideas were achieved quickly. But on the other hand intuition shows itself to be extremely unreliable, both in timing and in usefulness of the idea. Should someone wait humbly for the fruitful moment? What is to be done in case intuition does not occur or proves unworkable?

In any case, an idea gained by intuition must be examined (logically, systematically, analytically) for the possibility of its realization. And if the idea is capable of functioning, it must be quantified. Frequently ideas must be rejected. No investigation is known which can demonstrate, for example, the numbers or ratios of useful to useless cases of intuitive ideas. Some analyses of definitive results of solutions which allegedly have emerged from intuition show that almost nothing has remained of the original ideas, since many improvements and corrections were needed. Experience shows generally that hard work has the largest share in the effort of solving a problem (see also the second quote from Edison in Section 1.4.5).

### 1.4.1.4 Kinds of Intuition

According to Lüscher [283] there are different kinds of intuition. The one of the chess master is unlike that of the composer, the manager's unlike that of the inventor. Each possesses a differently constituted *archive of conscious experiences*, and disposes over a way of thinking that is combined from different visualization-forms.

The terminologically thinking person can have *theoretical intuitions*. If the available visualization-forms combine intuitively into a holistic reference system, a theoretical insight emerges, as for example the law of gravitation by Newton.

Persons who think by visualizing whilst executing an activity can have *practical intuitions*. Many *technical inventions* -- from the hammer to the computer -- spring from the wish to simplify the execution of an activity, to make a happening simpler, or to increase effectiveness. Because such product innovations yield profit, methods are employed to advance intuitive ideas by "brainstorming" [330], "synectics" [173], "breakthrough thinking" [317], "contradiction-orientation," and many others, especially in the area of technology and marketing (and in some cases the purveyors of a method get immediate profit from consulting about the method).

Those who receive ideas (thoughts) from feelings can have *psychological intuitions*. Knowledge of human nature, judging and understanding others, presuppose a differentiated understanding of one's own personality.

People who are capable of imaginative thinking and can express a fantasy pictorially or musically, can have *artistic intuitions*. All artistic persons, whether painters, composers, architects or authors, create their works from a wealth of internalized experience-figures.

#### 1.4.1.5 How Can Intuition be Brought About and Perfected?

It is a fact that engineers think intuitively and are generally inclined to think so (Hubka [201]). How can one use the advantages of intuition and avoid the disadvantages, instead of merely either to damn intuition generally or to praise it as a miracle means? Three requirements must be fulfilled, namely:

- the first work experiences of the engineering designer should be conducted with a systematic mode of operation, to create a *stereotype* of the systematic operational mode;
- the intuitive and systematic modes of operation complement each other in all situations, for example when searching for a solution and during the following validation, verification and checking;
- the records and documents resulting from systematic work must be demanded by a higher (legal and/or management) authority.

The important factor is to obtain the right intuitions and their justifiable share in the work. Within the meaning of the explanation for the intuition mechanism, further conditions can help to achieve the successful result. A certain amount of regulation and control of mental activity is achievable. For advancing good results with intuition, the following advice can be recommended:

- perceptions and knowledge which can be of importance to the specialist and expert should always be connected with essentially categorizing ideas. For example, if we see a machine, then we can (mentally) connect it through its abstract function with other members of the machine family;
- systematic preparation work should be accomplished by a series of conscious attempts to find solutions;
- preparations should be accomplished well ahead of any deadline, so that a suitable time span (incubation period) can develop;
- thorough work should alternate with relaxation periods, not only to achieve good results but also for reasons of psychic hygiene;
- favorable work conditions must be available, particularly a favorable atmosphere.

Goldberg [171] (among others) reports on how one can further ones own intuition.

Psychology brings several insights about intuition, but these insights are not based on explored physical processes in the brain. The person is considered as a black box. Figure 1--3 shows one such result, in which the participation of thinking in the conscious, subconscious and unconscious modes is qualitatively shown for different forms of thinking.

Intuition assumes that the person shows *attention and open-mindedness* (Lüscher [284]). To be capable of intuition, we must be unencumbered and self-assured. However, how can a creative person, e.g., an art creator, a politician, a manager or a creative housewife and mother, be unencumbered and self-assured?

Despite all tasks which are to be fulfilled, people will feel unencumbered as long as they are filled by a feeling of internal freedom and satisfaction. Despite all uncertainty which rules in the environment, they will feel self-assured if they have acquired self-confidence and self-esteem.

Intuition is to some extent learnable, if one learns to give oneself fully to the creative task and to eliminate all self over-valuation (self admiration or arrogance) and all self devaluation (self doubt and feelings of inferiority). The four normal self-feelings

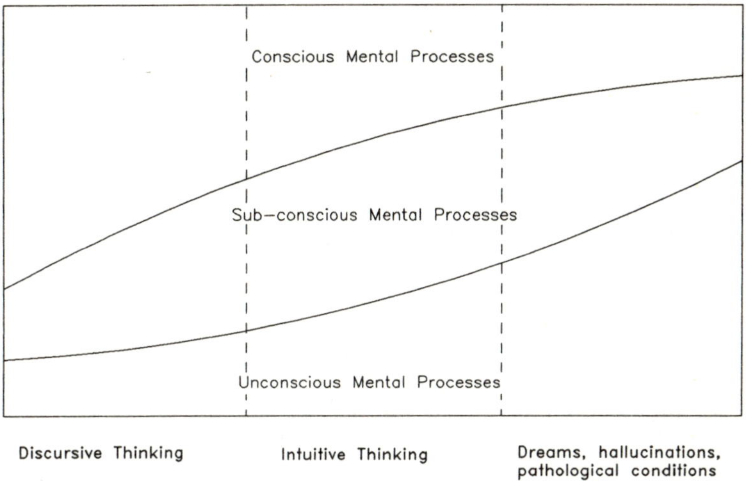


Literature:
Hubka, V., 'Intuition und Konstruktionsgefühl' (Intuition and Design Feel), Schw. Masch., vol. 75, No. 50, p. 42--44

| Figure 1--3 | Involvement of Conscious, Sub–conscious and Unconscious Mental Processes in Various Forms of Thought |
|---|---|

-- self-esteem;
-- self-confidence;
-- satisfaction;
-- internal freedom

are not lasting properties. They are unconscious experiences which result from adequate relationships with the environment. The normal self-feelings undergo a constant dynamic process. They change into negative attributions, if a "semi-skilled helplessness" and depression emerges instead of self-confidence, or if an acquired inferiority feeling becomes fixed instead of self-esteem. (About attributions, see e.g., Kelley [244], Seligman [391]).

Since intuition is blocked by wrong self-feelings (such as neurotic feelings), these must be avoided. The four normal self-feelings must be regarded as inalienable requirement for intuition and must be held in balance.

### 1.4.1.6 Examples of Intuition

#### *Management and Intuition*

Leadership personnel use intuition mainly in the following situations (Lüscher [283]):

- if the task involves judgment on persons;
- if a problem must be treated quickly;

- if they must progress on the basis of partial information to generate a general view; and
- if new solutions for organizational concepts, for marketing strategies or for technical innovations are to be found.

***Genius and Intuition***

Many persons who are recognized as geniuses -- Heinrich von Kleist and Vincent van Gogh are typical -- are driven to unusual performances because they must escape their intolerable self-feelings (e.g., self-doubt) and look for refuge in the world of intuition and harmony. During their creative work, the neurotic self-feelings remain largely switched off. For that reason everyone feels well -- not only geniuses -- if they are "wholly absorbed in their task."

*Rich experiences belong to the prerequisites for each intuition.* Lack of experience, and thus a small basis for comparison, can only lead to a risky guessing game, never however to an intuition with confident feelings. Geniuses like Mozart disposed of an unusual reserve of differentiated experience-figures. Albert Einstein possessed especially clear visualization-figures. In a private conversation he once called them "aesthetic pictures." According to Simon [398], a good chess player in class "A" disposes over about 2000 patterns (visualization-figures) of chess situations together with the following moves. A chess master, however, has about 50 000 chess patterns in recallable memory.

Experienced doctors give a diagnosis in intuitive way much more frequently than they personally admit. Medical test results, which must also be interpreted, are often used only as a rationally founded confirmation of the intuition. Psychotherapists often have a similar experience when they put questions to the patient which occur to them apparently without reason on the spur of the moment. Frequently such ideas target the kernel of the problem.

### 1.4.2 Designing and "Feel for Design"

If one observes experienced designers at work, a fact can soon be seen (Hubka [201]). An expert is able to decide an appropriate wall thickness, screw size or form of a channel, initially without calculating. The subsequent calculation, in orders of magnitude and with refined processes, mostly confirms the correctness of the estimate. This ability is called "feel for design," and it is often regarded as a talent. It is obvious in this case, if the following considerations are valid, that this "feel" is based on experience and conceptual knowledge (understanding), rather than the person being "born with" that something innate.

In contrast to a relatively abstract idea as result of an intuition, the products of "feel for design" are quantified statements about different dimensions of formed parts, or more exact statements about their form. Both kinds of these more concrete data are dependent variables, usually even depending on several quantities.

Incidentally, persons can exploit (utilize) only that knowledge that they have stored and processed (internalized) into their mental maps. Talent certainly influences the processes and productions in several aspects, especially in reference to duration and effectiveness. A further and more important element is the working method. Such a "feel" decision must contain three (not necessarily conscious) operations:

- determine the dependence;
- determine the influencing sizes; and
- determine the size $x$ from these influencing sizes and dependencies:
  $x = f$ ***(influential factors)***

The dependencies are not always quantitatively available, be it in the form of an exact or approximate derived formula, or of a quantified experience value (heuristic "rule of thumb"). Often only qualitative statements are possible: larger or smaller, sharper or blunter.

In the course of time proportional relationships between dimensions emerge for the experienced designer, as for example between the diameter of a screw and thickness of the parts to be fastened together, diameters and length of a motion guidance, thickness of a box wall and dimension of the connection positions. Designers then project these values into new layouts, independent of whether they are now right or wrong. The necessity for reviewing these decisions when unusual conditions are encountered is clear (compare [346]), because only the normal, average cases are relatively reliably covered.

Similarly a "feel" for the right form develops through experience, for example with regard to strength: gentle transitions, larger dimensions at more highly stressed places, larger radii for curved beams; with respect to favorable fluid flow: gradual changes of cross-sections, largest possible channel radius; or for "industrial design": appearance, color, balance, texture, symmetry, and so on. These and other rules, which are presented as formulae, heuristic advice or guidelines, or were gained intuitively through experience, begin subconsciously to influence the thinking processes after frequent, repeated use. In this way, but always supported by a certain size, the developed "design feel" delivers quick decisions. These decisions are of limited validity, the experience must remain within a "conventional" range, and can break down if this range is exceeded. Mostly such decisions are supported through checking in "orders of magnitude" calculations, and possibly confirmed through refined calculations. Changes are introduced if the "feel" is not confirmed by calculations. The danger exists that a designer is not aware of exceeding the "conventional range" limits, and neglects the checking calculations.

These phenomena can be explained with the help of figure or field theory. Experience, generalized as a summary of perceptions of correct models, forms the favorable relationships of the characteristics of a perception (compare Booker [76]). If a set of visually perceived elements (for example sketches of machine details) cause certain changes of some elements relative to the impressed model of the relationships, so the designers impressions of an esthetically unsatisfactory formations emerge. The field vectors cause an internal tension in the face of an unbalanced condition. Similar processes are evident in education for *artistic* activities, where students gain experience under guidance in studio classes.

### 1.4.3 Designing and Creativity

The combination of these two ideas can be formulated in a simple sentence: Designers should be creative. Design methodology offers the appropriate methods to support creativity.

The word "creative" means: capable of creation, inventive, imaginative, showing imagination in addition to habitual skill and knowledge. Being creative [431] is quite common among humans, but well developed in only a few. In attempts to measure this ability and to make it measurable, the word "creativity" was coined. The frequent abuse of the term "creativity" in the literature (particularly in the American) may give a somewhat frivolous impression.

Guilford [178], in his attempts to discover attributes of human intelligence, discovered that the mental abilities of a person can be presented on three axes, with some sub-groups:

- *intellectual operations* on information -- cognition (perception), memory (short-term and long-term), convergent production (thinking), divergent production (thinking), judgment;
- *intellectual contents* (types of information; presentation in verbal, graphical, or

mathematical form) -- concrete (figural), symbolic, semantic (meaning), behavioral (conduct);

- *products of thinking*, results of thought processes -- units, classes, relations (relationships, interactions), systems, changes (transformations), implications (consequences, conclusions).

According to this model, cognitive thinking is practiced by using intellectual operations on suitable intellectual content (therefore knowledge) to generate the products of thinking. This thinking can occur and be represented in different media: verbal (in writing or orally, using words), numerical-mathematical-symbolic, or graphical-spatial-visual. A thought process, and a human ability, are composed of an element of each of these three axes, altogether 120 different combinations. Each person can master only a limited fraction of these abilities fully or partially.

In the context of this section, the most important of these abilities is *divergent production*. Some, or many, other possibilities for solving relevant situations can be developed by this thought mode, starting from a given situation (thought process) and using the available (mainly internalized) knowledge. This would be the creative portion of thinking. This ability can be divided into four sub-groups, according to the characteristics:

- *fluency* of production, measured by the number of thoughts produced in a given time;
- *flexibility*, diversity of produced thoughts (thought groups);
- *originality*, novelty;
- *elaboration*, extending and detailing of individual thoughts.

Some determinants of creativity [130,135] need to be explored, with the emphasis on ingenuity and functionality, rather than on sheer novelty. They include extensive sections of knowledge of various types.

Creativity does not "just occur." Recent investigations in psychology have shown that creativity, generating novel ideas, occurs as a result of a natural tension between intellectual and intuitive mental modes [319,320]. The intellectual mode (systematic, methodical, analytical) can recognize that a problem exists and can analyze its nature. In the intuitive mode (erratic, inconstant, non-calculable) a sense of dissatisfaction can then arise, which triggers and motivates the mental interaction with the intellect to attempt to solve the problem. The oscillatory interplay between intuitive and systematic thinking modes and related procedures can result in creative proposals for solutions. Sufficient oscillatory tension must exist between the intellectual (scientific, methodical) and the intuitive (playful) mental modes for creativity to occur. Creativity (proposing novel solutions to a problem) has occurred as a direct result from using a systematic approach to searching for solutions, as has been demonstrated, e.g., in [141]. The likelihood of obtaining a "best" solution can be increased, but not guaranteed, by systematic work.

Neither a rigid systematic approach to designing, nor a fully intuitive mode of working are adequate by themselves to make it likely that good engineering products result from the processes of designing. Equally, neither approach alone can lead to effective learning of design. A judiciously chosen combination of system and freedom of action, adapted to the problem, and performed by knowledgeable and well-educated people using good tools in a suitable atmosphere (supervision) is likely to increase the chances of getting good results.

An insight reported by Akin [38] shows that creativity depends not only on mental ability, but predominantly on experience and technical knowledge (see also the quotation of Edison in Section 1.4.5). This has to be transformed by the individual person "from declarative knowledge into procedural knowledge;" the procedural manner has to be abstracted from experience and technical knowledge, and absorbed, but also the experience and the technical knowledge has to be built into the mental maps.

In order to be acknowledged and effective as *creative* in problem solving, and consequently in designing, persons must show simultaneously the following components (i.e. all of these at once, before you *can* be recognized as creative) [130,135]:

- an adequate *knowledge of objects and principles*, possessed or available (as information) to an individual or a group of problem solvers, including *tacit knowledge* contained in experiences, heuristics and gut feelings that have not been explicitly expressed;
- a *knowledge of processes*, especially a knowledge about design and problem solving processes, both of which should be intuitively and/or consciously available (with some practice) to the problem solvers; this includes knowledge about different ways of modeling objects (products and processes) and interacting with models;
- adequate *judgment*, a sense of what is reasonable to expect under the circumstances;
- an *open-minded attitude*, apparently the *essence of creativity*, a willingness to accept ideas and suggestions (self-generated, or generated or brought in by others) and to associate them with other knowledge; also a *sense of care and attention* for the subject to strive for excellence;
- sufficient *motivation*, including self-motivation (inspiration) and externally induced motivation (rewards);
- ability to *communicate* (verbally, graphically and symbolically) to make the generated proposals visible, to present the proposals in useful forms;
- level of *stress*, too much or too little stress, inadequate preparation, or effects of pressure of time or unsuitable management style and conditions can reduce creativity;
- recognition and *ownership* of *the existence of a problem*.

This composition of creativity, and the admitted variability among persons for each of these factors, probably account for the fact that psychological test instruments that try to measure a person's "creativity" and report one unitary result seem to be so inconclusive.

Most of these factors should be acquired *in parallel*, at the same time. Especially process knowledge must be practiced under guidance and supervision to generate a *stereotype* of action procedure. This list is *not* a prescription that "you must have all the object knowledge before you can start to learn to design." Learning is also iterative, knowledge of objects and knowledge of processes support each other.

Even with all of these, there is no guarantee that you *will* be creative, it depends on the circumstances. And if a creative event occurs, the result may not be useful. The levels of stress and of "ownership" of the problem seem to be mainly responsible for motivation, and these factors frequently include perceptions of security, hope, general economic and cultural well-being, and others.

Creativity can usually only be *recognized* after the event. It seems that the highest level of creativity usually appears in historic waves. Many times, these waves have unfortunately coincided with wars and other catastrophes.

Creativity is likely after good preparation of a problem. Its first evidence seems to be a feeling of urgency and sometimes fear, an initial rush of adrenalin. This changes to likely panic and/or despondency, which can lead to inaction, almost "freezing up." If incubation is allowed to happen, then a possible enlightenment (illumination) can take place. The results must in any case be checked, validated, verified.

Some methods and instructions for improving this mental ability are described in the literature. Osborne [330] has proposed "brainstorming." Gordon [173] has developed the system of "synectics." Adams [37] shows how stereotype modes of thinking can be overcome and mental obstacles to free thinking removed. Nadler and Hibino [317] have developed advice for "breakthrough thinking" to help in opening organizational and product ideas. Some investigators recommend contradiction-orientation, looking for contradictions in the search directions demanded by the given requirements to produce a "best" solution, to trigger an idea for a concept that can overcome the contradiction (developed from the proposals of Altschuller [42,43]). These works have made the creativity to some extent "learnable" (see Section 1.5), in that they have provided some heuristic *algorithms* (see also Section 1.4.6) as recommended consequence of operations, with the necessary conditions, which try to make a possible success more probable. Here, too, success is not guaranteed. Producing good and new thoughts is not the only measure for success, the ideas must also be made realizable. Creativity thus shows similar advantages and disadvantages to intuition, as explained in Section 1.4.1.

### 1.4.4 Designing and Innovation

The word *innovation* (sometimes used as a slogan) not only contains the implied positive objective, i.e. to create the new and original (analogous to creativity). It also contains a danger, if "the new" (also a relative term and content) should be elevated to the supreme goal of the designers. The goal of the designing must always and only be *to achieve an optimal solution* in the given conditions. That does not exclude the existing solution, it could still be the optimal. It is therefore more meaningful, to look for creative rather than innovative designers, although "creativity" has also become a fashion word.

Innovation is also used in a slightly different context, and is consequently almost wrong within designing. This word has particular meaning in the context of introducing or implementing a newer realized solution. This process of the "renewing introduction" can contain designing, but can also happen with already existing means. Only when a renewal is brought onto the market or is taken in service, should one speak of an *innovation*.

### 1.4.5 Designing and Inventing

If a new idea (that is not deducible from the current state of the art) is devised for completing a technical artifact, this is called an *invention*. Such inventions are normally patentable, i.e. they can be protected against copying under certain conditions through granting a patent to the inventors.

The combination of inventing with designing has a varied character. Designers should examine solutions with regard to the possibility of patent applications as inventions, if this would be advantageous for an enterprise because of license possibilities. Again we emphasize that invention should not be a primary objective for designers. Higher on the scale of striving should be the combination of an optimal solution and an invention. Whether the art of evasion of patents should practiced is an ethical and legal question.

The great inventor Thomas Alva Edison [126] maintained, "there is no substitute for hard work," and "genius (creator power) is one percent inspiration and ninety-nine percent perspiration." He explains (quoted in Runes [378]):

> *In trying to perfect a thing, I sometimes run straight up against a granite wall a hundred feet high. If, after trying and trying and trying again, I can't get over it, I turn to something else. Then, someday, it may be months or it may be years later, something is discovered either by myself or someone else, or something happens in*

*some parts of the world, which I recognize may help me to scale at least part of that wall. ..... We sometimes learn a lot from our failures if we have put into the effort the best thought and work we are capable of.*

Confirmation comes also from other sources, e.g.:

*Sir Alexander Fleming did not, as legend would have it, look at the mould on a piece of cheese and get the idea of penicillin there and then. He experimented with anti-bacterial substances for nine years before he made his discovery. Inventions and innovations almost always come out of laborious trial and error. Innovation is like hockey: Even the best players miss the net and have their shots blocked more frequently than they score. (Royal Bank Letter, March/April 1988, Robert Stewart, author/editor)*

Invention theory is a theory about the ability to invent, therefore has a certain parallel to design methodology. The algorithm of inventing by Altschuller [42,43] (based on works by Ostwald and Backovsky, see Section 3.1.9) assembled the triggers to processes by which, with high probability, problems in (mainly conceptual) design tasks will be overcome. In constructing this algorithm, many patents were analyzed to find principles by which the inventors came to their inventive thought. In this way several principles of procedure were found, and those principles which were applied most were statistically explored. It was shown that the solution was found with a relatively small number of working processes (principles) in the predominant majority of cases. These processes were assembled into the *algorithm of inventing*. They have been further developed into a contradiction-oriented procedure. Results from applying this algorithm are usually in the form of concepts, and realization depends on designing. Further remarks to this theme may be found in the Sections 3.1.8, 3.1.9 and 6.8.

### 1.4.6 Designing and Heuristics

The treatment of heuristics in this context already points to their role as a bridge to design methodology and as their most important basis. Opinions about the contents of heuristics diverge traditionally, but each class of opinions contains different aspects that are interesting for designers. For that reason will we enter more deeply into a discussion about heuristics.

As an adjective, the word *heuristic* refers to an instruction or advice which is not necessarily based on science. In this sense, an heuristic is simply a rule of thumb, derived by experience or myth, which can lead to an acceptable result with good probability.

Using this interpretation, Koen [254,255] proposed a hypothesis that in engineering all theories and instructions are to be regarded as heuristics. Three main reasons can be mentioned. Even the most strictly explored science serves in engineering only to formulate heuristics for the realization of technical systems (as directive to their design and manufacture). In formulating the engineering sciences, many secondary effects were consciously neglected. The form in which the results are usually expressed in engineering science is not directly suitable for the explorative ways of designing. Koen even maintains that the application of these heuristics is the only method used by engineers: "Everything is heuristic," including the method of application of heuristics, and also this statement itself. The interpretation by Koen is useful, on one hand for enhancing the humility of engineers, on the other hand as weapon against the conceited opinion of some advocates of the pure and applied sciences, that only they care for the progress of the mankind. As a directive or prescription about how one can accomplish designing more effectively and rationally, the proposition about heuristics by Koen is not really useful.

In the German language regions a stricter interpretation prevails. Rules, instructions, programs and other working means are provided under the generic term *heuristics* (in German, the expression is "*die Heuristik*," singular with a definite article). These collect, analyze, organize and apply proven productive methods necessary for intellectual working processes to accomplish research, design and development. This idea is not new, heuristics has a long history. A pupil of Euclid left notes about how his master proceeded when he looked for new mathematical proofs. The study of heuristics is linked with psychology (Bromme [84]).

Heuristics was defined (Klaus [249]) as the "science of the methods and rules of discovery and invention." The heuristic method is a particular form of the trial-and-error method. It differs from the deductive method among other things in that it works with conjectures, working hypothesizes, provisional models, etc. The heuristic method is not a strict method of proof, but only a process which helps in the search for proof. Heuristics studies the actually occurring cases of discoveries and inventions, and tries to derive general laws for discovering and inventing which are not dependent on the respective concrete task. In so far, it is an empirical science. It uses results and processes of experimental psychology, information theory, information psychology, and neuro-physiology. The heuristic methods can be simulated on digital electronic calculating machines (computers). Such "heuristic" machines and programs work like modern chess playing programs, i.e. they contain and use a set of general strategic principles and the situation of the case, in combination with the "trial-and-error" method.

Heuristics is an important element of dialectic logic. If a proof is found, or a task solved with the help of heuristic methods, so in general the proof can be presented strictly logically, and the original heuristic considerations have become superfluous.

The position of heuristic methods is also characterized by the relationships between object, method and theory, according to Klaus [248]. His considerations with respect to our theme are discussed in Section 4.6.

The best known proponent of modern heuristics is Polya [353,354]. He is regarded as the re-discoverer of heuristics. His views rarely reach beyond the area of mathematics. Within this area, Polya was versatile, exciting and original. He developed the heuristic working scheme, which is presented in figure 1--4.

The main pillar of the bridge between heuristics and design methodology is formed by systematic heuristics, which was proposed by J. Müller [308,309,310], H. Lohmann [280,281] and K. Steuer [413]. Some of the knowledge from this discipline is as follows.

Müller has drawn on several sources for his systematic heuristics [212], among them the available literature about design methods, observations of experienced designers, system analysis and heuristic methods. The following simple assumptions are made:

1. innovative designing is a problem solving activity;
2. each design problem can be divided in a finite number of sub-problems, which can be solved simultaneous (concurrently) or successively;
3. each sub-problem needs a different method, which must be prepared for this task;
4. design activity must be observed and analyzed continuously, so that methodological experience can be derived and stored from it;
5. each professional community has its own and peculiar experience. Systematic heuristics should be flexible enough to be applicable in different fields and methodological environments.

Systematic heuristics has three fundamental elements:

The heuristic working scheme according to Polya:

1. Understanding the problem and task
   -- what is unknown?
   -- what are the data (what is given)?
   -- what are the conditions, restrictions?

   } Introductory section

2. Devising a plan for finding a solution
   -- find the connection between the data and the unknowns

3. Carrying out (executing) the plan

   } Solution section

4. Looking back -- examining the solution obtained
   -- can the results be checked?
   -- can the result or the method be used for any other problem or task?

   } Closure section

Indications from the heuristic working scheme:
-- it is of very general validity, and yet important aspects can be clearly seen;
-- it allows approximate capture and comprehension of goals. 'You would not seek me, if you had not already found me.' (Blaise Pascal)
-- Aporia = doubt, perplexity, impass, blockage, obstruction in the way. Surmounting (overcoming) of aporia by means of 'method' (Greek ... following a path)

Schema of a situation in which a method can be applied to overcome an 'aporia':

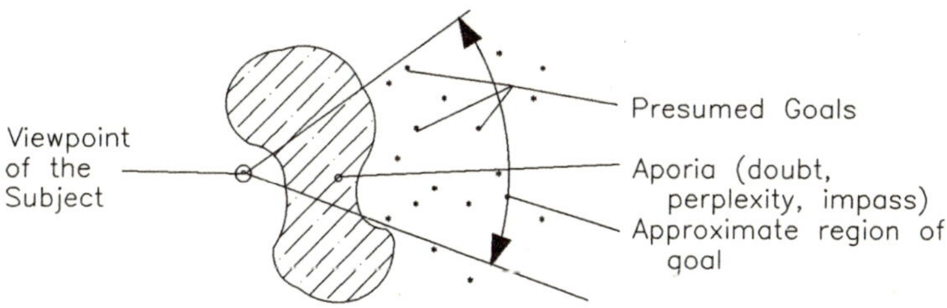


'Aporia' occurs in various forms:

-- subjective aporia (i.e. only for certain individual cases, for instance Chinese language for persons not capable of speaking Chinese);
-- methodical aporia (i.e. a method is not known);
-- aporia stemming from the object (i.e. the ontology of the object is at present obscure, for instance the 'object Hades' in the new astronomy).

Literature:

Polya, G., How To Solve It, Princeton, N.J.: Princeton U. P., 1945
Polya, G., Schule des Denkens -- vom Lösen mathematischer Probleme, Bern: Francke, 1980
Pascal, B., Pensées, transl. A.J. Krailsheimer (1966), Harmondsworth: Penguin Classics, 1970

| Figure 1--4 | Heuristic Working Scheme according to Polya |
|---|---|

- processes, systems of activities, which should be accomplished in a given sequence;
- a program library, an organized assembly of heuristic programs, which were developed for different innovative design purposes;
- a collection of additional instructions, which contains all those heuristic instructions which were not installed in the programs, but which complement them.

An *heuristic program* is understood as a finite organized quantity of heuristic methods, supported by further heuristic instructions, with which a process is to be executed, not uniquely, but as far as needed so that the corresponding staff members reach the anticipated goal with a reliability which is significantly higher than 0 (*zero*), but is smaller than 1 (*one*). This is one of the essential differences from a deterministic algorithm.

From the view of the respective subject, is the process executed stereotypically, planned or intuitively? The differences are not of necessity conditioned by objective reasons, but through the knowledge and ability of the examiner.

- A process proceeds according to *stereotype*, if the logic of the procedure is known, the information to be processed exists, and the sequence of steps to be executed has not only been learned, but practiced so often that it is not necessarily brought into consciousness to execute it. It suffices to grasp the intention, everything else runs without thinking.
- A process proceeds by *planning*, if logic and branch information are given as before, and the sequence of steps to be executed is known. The steps are however not among the routine items for that person, but are to be realized as a deliberate and planned course to achieve the purposes. The step sequence is not yet trained. It contains too many operations, to be able to store them routinely, or the routine has been lost because it is not constantly required.
- A procedure is done *intuitively*, if the examiners at that point in time do not know of any sufficient procedure which leads to the goal. The reasons can be objective (the logic is not yet clarified) and/or subjective (the examiners have not yet learned this method, or they consider these to be too costly in the given case).

With what certainty can we assume that, in the course of a thinking process, an event will enter which can be accepted as a partial solution? This question allows us to distinguish deterministic and heuristic operation sequences. This differentiation is objectively conditioned:

- assuming sufficient branch information is available, a *deterministic* operation sequence admits a sure prediction, because the logic is fully cleared up, the step sequence contains no gaps, and the circumstances of the course are irrelevant or with certainty controllable. Such operation sequences can in principle be simulated by computer techniques, they are algorithmizable;
- an *heuristic* operation or operation sequence admits of no sure forecast, whether the occurring event can count as a (partial-) solution.

In systematic heuristics, Müller [212] has created a tool which can produce a suitable working program (for human application) for each concrete case, under application of universally valid principles, and under re-use of existing methodical experience. This procedure is characteristic for systematic heuristics.

### 1.4.7 Closing Remarks

In any discussion about design methods and Design Science, we may neither forget, nor overestimate these partial areas. Huxley (1825-1895, in *The Progress of Science*) notes: "No delusion is greater that the notion that method and industry can make up for lack of motherwit, either in science or in practical life." On the other hand motherwit and creativity cannot compensate for defects in methodology, science and technical knowledge. Methodology and science can also lead to innovations and even inventions, as was shown by Ehrlenspiel [141], see also [220].

## 1.5 Is Designing Teachable/Learnable?

We are aware that some readers cannot easily find a connection between Design Science and the teachability of designing; we therefore consider it reasonable to discuss this question.

Let us consider first the teaching/learning process from the viewpoint of the engineering sciences (see also Section 2.4). This again refers to a transformation, a process, in which the state (condition) of the learning person is transformed with regard to knowledge (and the insights, concepts and relationships among items of knowledge), abilities, skills and attitudes. From the view of the design teaching, the desired condition should correspond to the professional picture of designers.

With this model we have arrived at the kernel of our question. If namely designing is regarded as a purely artistic activity and includes no explainable design (process) knowledge, then we must deny teachability, and even learnability, of designing, for only knowledge is learnable.

Although formulated design process knowledge was missing, designers were educated for decades. The time to maturity (and capability) of designers is the problematic parameter; for example, the maturity time for designing engineers was numbered at about ten post-school years. Such a long time span was needed (on average), to gain and to internalize (abstract and process) the knowledge lacking from their own experiences and own study. This design process knowledge is, however, not explained in a most suitable form for designers and teachers, who should in turn explain it and make it clear to their students.

Designing is also a skill, i.e. it needs both knowledge and abilities. Abilities are obtained through working, exercises, and training. But how quickly one attains abilities depends also on the presented or available knowledge. If no or only little relevant *and organized* knowledge is presented (branch knowledge, process knowledge, material knowledge, engineering knowledge, see Section 2.4.4), the learning time will be long -- education and training are inseparable. And especially long and unproductive can be a learning period under unfavorable conditions. Not each enterprise puts good instructors in charge of a new generation of designers, and the best designers are mostly not good instructors. But that is also decisive, for only that knowledge presented in advance, including the newest, objective and organized knowledge, brings positive results and allows also those who do not have so much perseverance and motivation, to master this learning period more successfully and quickly.

The learning goals should be revised in the light of Design Science, as is implied in figure 1--5.

It can therefore be postulated without further proof that the introduction of Design Science can fashion the teaching/learning process for engineering design more effectively and increases the chances for design education. This can be greatly enhanced, if in all design courses a harmonized instruction based on General Design Science (GDS) is introduced, and its applicability and (open-ended) problem solving modes of operation are also included in the "purer" engineering science subjects. See Chapter 8 about the derivation of Specialized Design Sciences (SDS) from the General Design Science.

Further knowledge and discussion about design education is contained in several books and articles by WDK-authors, see references quoted in the preface.

| Acquisition of: | Teaching and Learning Goals | |
|---|---|---|
| | TODAY (Design Science not used or incompletely used) | FUTURE (fully used Design Science) |
| Understanding and Knowledge | -- machine elements<br>-- individual families of machines<br>-- standards | -- technical system (theory)<br>-- branch knowledge about designing<br>-- theory of design processes<br>-- branch knowledge (machine elements and some TS–families) |
| Ability | -- representing with a feel for forms (shapes)<br>-- transfer from existing examples<br>-- application of relationships | -- methodical procedure<br>-- abstracting and combining<br>-- representation (modeling) at various levels of abstraction<br>-- transfer from existing examples<br>-- application of relationships |
| Skill | -- sketching<br>-- drawing<br>-- calculating | -- sketching<br>-- drawing by hand<br>-- drawing by computer<br>-- calculating, orders of magnitude<br>-- calculating with computers |

Figure 1--5 Teaching and Learning Goals With and Without Design Science

## 1.6 Summary and Outlook

This completes our initial survey in breadth about engineering design and its contexts. The survey should set the scene for the following chapters. The presentation so far has been largely narrative, therefore not really suitable to form a basis for Design Science. These factors, and many more, must be rearranged. The separate strands in this tangled web must be followed to show their relevance and to formulate a clear structure. The interrelationships must also be defined in a better way. With this complex picture in mind, we can now proceed to explore, develop and suitably arrange the forms of recorded knowledge about engineering design in more depth -- Design Science.

Clearer and better can only be achieved if we follow the principles of science. Chapter 2 outlines the essence and implications of these principles. The further chapters show the paths leading to Design Science and its definition -- the purpose of this book.

# 2 KNOWLEDGE SYSTEMS (SCIENCE) AND DESIGNING

*Knowledge is power*
***Francis Bacon***

*Wenn man wissen will, WOHIN man geht, sollte man zuerst feststellen, WOHER man kommt. (If you wish to know WHERE you are going, you should first establish FROM WHERE you came)*
*Vizebürgermeister Franz Jochum, Klagenfurt (Vice-Mayor of Klagenfurt)*

Question:
What is a science?
Of what does it consist?
How can it be recognized?
Which purposes does it pursue?

With the introduction of Design Science we can achieve neither spectacular and rapid successes, nor double-digit percentages of savings. With certainty we can say that Design Science is a decisive step in the direction of progress and will bring a long-term improvement of designing, both in the procedures of the designers and with respect to the quality of the manufactured products. But, as for any science or theory, it is not expected that it can be directly applied to the real problems of designing engineers. Improvements must on one hand be derived from science and adapted to the practice, and on the other hand where improvements originate from the practice, they must be inserted and absorbed into the science. Technical knowledge presented in a new order, totality and form then becomes a striking and productive tool for designers. Essential positive effects are to be expected in a relatively short time in engineering education, and through suitable derivations and adjustments of this science for particular purposes.

The range of knowledge required by designers can now be defined in a new classification scheme, with better completeness and form, and can provide a powerful and accessible tool for designers.

## 2.1 Knowledge and Science -- General Considerations

In all areas of human life, not only in science and technology, knowledge determines power. Therefore relevant knowledge and experiences from the individual areas is collected and organized, and is thereby formed into individual sciences, and abstracted to general rules, models and laws.

It is unfortunate that to date there is no uniform, logically summarized and classified *system* of sciences. Instead, we must make do with a multitude of individual sciences, with boundaries that are only ill-defined, porous and variable, with frequent overlaps to other knowledge areas. New disciplines have emerged and still emerge through opening up new object and process areas (especially phenomena). The present division of science is determined mostly by tradition. The system of science tends to reveal the historic development of knowledge, rather than a homogeneous whole set up according to a set of guiding principles. Newly emerging disciplines become added to those parts which in earlier periods "logically" belonged together. Similar to a city, in which new quarters are more or less organically connected to old parts, so (in a simplified view) the science systems collected around the old kernel of philosophy, mathematics, theology and other knowledge areas. Such "extensions" to knowledge are not necessarily inferior; the new planning of Paris by its architect under Napoleon III (Emperor of France, 1852-71) is an example in city construction. Similar exceptions are also found in science, in areas in which logical structures were recognized through scientific methods, as for example in biology.

New knowledge often caused a change in the recognized divisions (taxonomies), structures and elements. The title "biology," as science which deals with phenomena and forms of living systems (the objects of this science), appeared as a collective term around the year 1800 with Burdach and other scientists. Biology collected the then existing areas of anthropology (object of study: the person), zoology (object: animal), botany (object: plant) and paleontology (object: fossils) into a unit. This was then a logical and consistent step, in which known objects of the world (humans, animals, plants and fossils) were combined into one "living" kingdom. The transition from the singular statements to generalized statements (the range of validity is broadened) characterizes this development of science. The overall structure of knowledge is simplified and made clearer.

### 2.1.1 Science and Designing

One important role for science is collecting, defining, categorizing and classifying knowledge. This includes finding relationships, structuring, and systematizing. An accepted model for scientific procedure is: (a) ask an appropriate question, (b) propose a model and a hypothesis, (c) collect data, (d) analyze the data, (e) formulate an answer, (f) accept the new knowledge and revert to (a). The hypothesis may reach the detail of a proposed theory, if the prior knowledge and its logic is sufficiently well established. The proposed model may try to describe a phenomenon, it may try to simulate the information, it may be the basis for a computer implementation, or have some other purpose. If this sequence of actions is interrupted at any point, the result is a lack of knowledge -- this cannot ever be construed as disproving the hypothesis.

Knowledge is needed for the dual purposes of (a) obtaining consensus on its nature and making it accessible for retrieval and reference, and (b) putting it to use. Putting knowledge to use relates to social, artistic, economic, political and other activities, including designing. The structuring of knowledge should usually be different for these two purposes.

Normally, any phenomenon to be studied is considered as an inseparable whole, a *holistic* viewpoint. There is a fundamental philosophical objection to artificially dividing *theory* (logic) and *practice* (experience) [362]. The separation lacks a sufficient epistemological foundation. But to understand a phenomenon in its context, we must separate, decompose, and describe, not only for individual parts, but also with respect to theory and practice. If our investigations stay only at this level, we can be accused of being *reductionistic* and *deconstructionistic*. On the basis of knowledge of the parts, we can then more effectively abstract, conjecture, discover relationships, and *synthesize* into a holistic description, but also validate, verify, and test.

These processes use intuition and logic, they proceed by alternately deductive and inductive means, using methods that are creative and imaginative, but also careful and systematic. Popper [355], in presenting one view of the form and history of scientific progress, pointed out the supreme importance of intuition, imagination and speculative hypothesis. Progress was also discussed by Kuhn [266,267], and divided into normal accretion of knowledge within a disciplinary matrix (paradigm) for that region of knowledge, and scientific revolution when the disciplinary matrix is subject to revision.

There is an unfortunate tendency in human thought to consider any phenomenon as a dualistic polarity [389,404] to see everything in isolated and simple (simplistic) dualistic opposites, in "black and white." In almost all cases, the behavior of a phenomenon tends to be of spectral continuity, and should be seen in "shades of gray."

Natural science, according to [85], is based on experimentation and observations. Hypotheses are proposed to describe the expected repetition of a chain of occurrences, and the internal and external conditions under which this chain progresses. The observations cover the happening of the activities of occurrence, not the occurrences themselves. We cannot therefore establish the *reasons* for the occurrences, we can only establish, and possibly measure, what our observation instruments can reveal or detect. The hypothesis can then declare that "if *a*, then *b*," establishing the observed effects (teleologically) from the prior conditions and causes. Unless *b* happens differently, the hypothesis is not disproved (a deliberate double negative) -- but the hypothesis can never be proved.

The experimental situation is a reasonable methodical abstraction that considers the conditions and causes, and observes the resulting effects, i.e. delivers repeatable and valid results within established limits by excluding certain factors. Repeatability does not refer to the happening itself (or necessarily to the resulting state of nature), but only to its active progression. The phenomenon or the experiment may in fact destroy part of its own evidence, and therefore not be strictly repeatable, e.g., an archaeological dig. The result of interpreting such experimentation and observation is a model that describes causes and effects with reasonably achievable accuracy, and knowledge about the expectation of phenomena. The aim of the experiment is to stabilize the progression, such that its constancy can be guaranteed. This leads to the proper category of scientific objectivity, its complete *verifiability* (Nachprüfbarkeit).

All laws of *correctness* must be adhered to and acknowledged to obtain valid knowledge. The experiments and observations must also be *repeatable* (nachvollziehbar), *validated* and *verifiable* (überprüfbar). The results must be prepared in such a way that they are freely *available* (verfügbar) for us to interpret and act on. Natural science in this sense cannot deal with reasons or purposes. Either for starting the search for knowledge, or for using the results, we need a particular *category of thought*, namely "because *a*, therefore *b*." This is a form of *Anschauung* (view, opinion, perception, idea, conception, intuition, point of view), a human consideration and valuation that is not found in the happening. There are many other human needs, to explain, predict, create, design, confer, heal, communicate, and so on -- and most of these must be accomplished before all action relationships are clear.

With the need for more rational and effective learning, design research becomes necessary. This involves production of conjectures and hypotheses, observations, developments, gathering and interpreting data, digesting and generalizing, etc. All this knowledge, suitably structured should add to the scope and understanding of Design Science, and outlining its structure is the main purpose of this book. Design research has a short history, as outlined in Chapter 3, dating from the 1850's or 1950's according to interpretation.

According to Cross [99], three forms of design research are useful:

- research *into* design, by various kinds of observation, e.g., protocols;
- research *for* design, to create tools (especially computer-resident), design methods, forms of modeling;
- research *through* design, e.g., abstraction from self-observation and other observations during designing, hypothesizing and testing.

For validity, this research should be:

- purposive -- based on identification of an issue or problem worthy and capable of investigation;
- inquisitive -- seeking to acquire new knowledge (or new relationships among knowledge elements);
- informed -- conducted from an awareness of previous, related research;
- methodical -- planned and carried out in an efficient and disciplined manner;
- communicable -- generating and reporting results which are testable and accessible by others.

Gregory [176] states that the maturity of science about a body of knowledge (for instance the knowledge about designing) is characterized by a gradation of behavior which may be expected, and which causes that knowledge to be reckoned as scientific. Clearly, this relates to the *epistemology* of the knowledge, the theory of the methods or grounds of knowledge, the ways of knowing. This behavior and the state of organization of the body of knowledge may range through:

(1) description of phenomena (natural history phase);
(2) categorization in terms of apparently significant concepts;
(3) ordered categorization whose pattern may be deemed a model (the evolutionary taxonomy or periodic table phase);
(4) isolation and test of phenomena, with implied reproducibility by independent observers;
(5) quantification (classical physics phase).

Even the first of these is commonly acknowledged as "science" when applied to a traditional area of study. Many areas of science cannot reach the final "quantification" stage in which mathematical relationships are formulated. Engineering design is obviously such an area, but this observation must be qualified by the connections to other knowledge. Some of the knowledge we use and interpret for engineering design exists at the highest scientific levels -- it may be useful for analysis of proposed solutions, but the choices and values are necessarily human.

Some directions that characterize the progress of finding new and improved design methods and theories, each of which has good and bad points, are (Eder [130]):

(a) self-observation whilst designing and directing a design team, or participating third-party observation [181], and abstracting from this experience; active industrial supervision and tutoring can be useful, but interpreting the observations can also be biassed by circumstances;
(b) third-party observing, protocol analysis, etc. is apparently scientific, independent and relatively objective, but can only detect the obviously observable;
(c) formalizing, theorizing, proposing models, etc. is capable of logical coherence, but may

result in invalid generalizations, especially if the observer is not intimately familiar with the phenomenon;

(d) developing procedures and methods, etc. can be directly applicable, but may produce more problems than it solves;

(e) developing tools, aids, computer programs, etc. can be very useful, but can also lead to elegance without substance, relevant applicability, or friendliness to the potential user.

These directions (in both Gregory's and Eder's forms) are almost always used in combination with each other. In continental Europe the progression from (1) through (3) has mainly been from (a) to (c), with some attention to (d) and (e) -- Professors of engineering design were called from industry with typically 20 years experience of design and project leadership, and then published their main works, e.g., [140,257,333,334,376]. Japan is characterized mainly by (d) and (e). Design publications in the English-speaking regions were based on (a) in the 1950's and 60's, but more recent design research tends to emphasize (b) in the hope of being able to achieve (c), with (d) and (e) taking place in parallel efforts, but significantly by almost omitting (2) and trying directly to achieve (4).

This book (and its German predecessor [229]) has been developed mainly from (a) and (c) to elaborate (1), (2) and (3). We have nevertheless been careful to relate many published results from the other directions to our considerations. In this way we have tried to validate our models and to collect Design Science as a basis for systematic designing and redesigning of objects as systems.

Research, together with existing knowledge, must also be interpreted for use in engineering design practice. Some examples are [47,48,49].

## 2.2 Technical (Branch, Domain) Knowledge

Technical knowledge is knowledge about artificial objects (existing in a branch of industry or a domain of activity) which have been created and produced to accomplish certain goals. In today's world we find many objects which serve human needs, especially to improve human living conditions. Humans owe their standard of living to an immense quantity and variety of devices, machines, tools, equipments, etc. These show the inclusive influence of *technology* (in its broader interpretation). The constantly growing quantity of technical products penetrate from industry into the personal environment of today's people -- domestic appliances, household machines, fitness devices, motor vehicles, telephones, telefax, television, information "highways" and so on. These objects characterize the modern world at least as intensively as plants and animals stamped the previous world scene. We do not want to discuss in this connection whether this condition is right and desirable, such a question leads to moral and political judgments -- for us it is simply the given reality.

A vast amount of knowledge exists about technical objects ("hardware"), much of it in written form. How is this written and printed (recorded) knowledge about technical objects organized now? This question should lead to suggestions for possible improvements. Various of the individual objects or object families (for example motor vehicles, bridges or television) are treated in general, or in specific detail, depending on who is the intended recipient of the knowledge -- whether for designers (for instance the connection dimensions), for buyers (performance statements), for consumers (instructions for use), or for students (working principles, completion examples, theoretical substructure). According to the addressees and their specific needs, different aspects are emphasized.

As is well known, the written and printed (recorded) *knowledge* forms only a part of the entire wealth of knowledge. People also possess a range of *experience* knowledge --

"know-how" -- which considerably influences their proficiency in using or creating technical objects. Such knowledge tends to be *internalized*, absorbed into an individual's mental maps. It can then only be explained with difficulty, and is called "intuitive" or "tacit" (silent). In addition, peoples' *abilities* and *skills* are involved in achieving the total result. These factors reflect the long development of a person from novice, through apprentice to master.

The presented knowledge refers to a differing breadth of objects, according to the kind of the statement (singular *vs.* general): A statement treats either only a certain object (or a small group or type series as in case of "instructions for use"), or the knowledge is valid for a smaller or larger family of objects. One can deal with broader areas such as bridges or piston machines (see figure 2--1), or narrower areas such as concrete bridges or railroad bridges (evidently two overlapping classes), or with spark-ignition motors or piston pumps (same principle, opposite effects or actions).

The picture about ways of structuring the knowledge would be not complete, if we did not mention a separate treatment of common properties (characteristics) of technical objects. If the knowledge about a technical object is to be complete, as described above, we would have to repeat for each family a certain range of special knowledge, for example about strength, raw material, energy transfer or manufacturing technology. To avoid such duplication, these areas were collected into special disciplines -- not systematically, but according to needs. An analogy with biology shows that morphology, anatomy, histology or cytology act as bridging disciplines across the common themes of human, animal and plant life.

The reasons for developing such structuring as a two- or multidimensional matrix are obvious. Consider just which level of knowledge (e.g., about strength or other areas) would have been reached, if the relevant considerations proceeded only within individual product

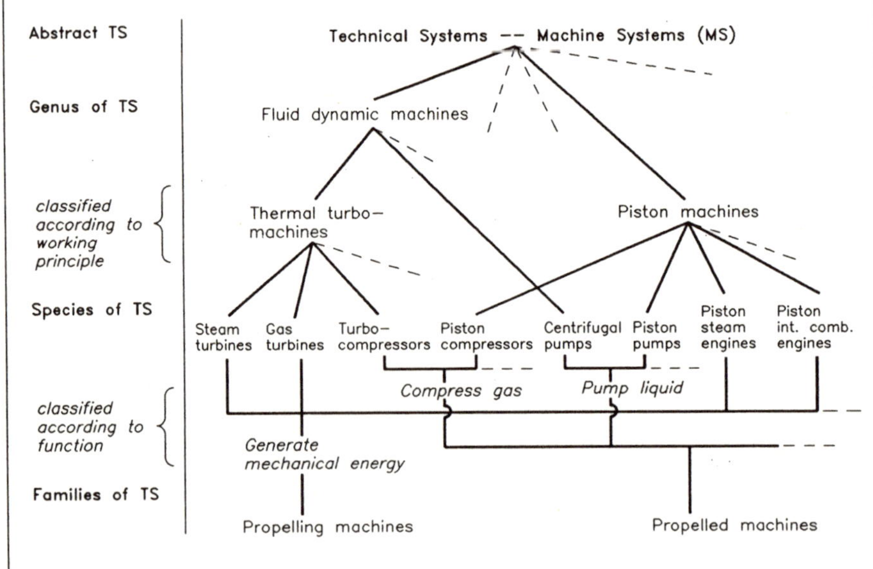


Figure 2--1 Regions of Technical Knowledge -- Partial Hierarchy

families (branches), and not in the bridging area of a self-contained (engineering) science. The superior performance of a specialized science with general statements for a broad area of validity is evident (specialization effect).

Distinguishing and separating the problems common to several knowledge areas and formulating a new discipline is not as easy as it may at first appear. This is especially difficult if the relevant knowledge areas are on the same plane or in a narrow hierarchical area (object family *vs*. object). Only a higher position in the hierarchy allows a survey, and common problems can be recognized. This is the main reason why hierarchical knowledge structures with appropriate levels are formed, not only in science, but in all fields where orientation and order should exist. They are particularly useful in organizations (management) and general system technology. The spirit of this principle emerged in biology, as it recognized and defined "living objects" as a higher (inclusive, collective) classes containing humans, animals and plants.

Hierarchical structures of knowledge in science can also be described using the ideas of set theory (classification theory), i.e. a hierarchy of sets (quantities) -- sub-sets -- elements.

## 2.3 Objects Hierarchy in Technological (Engineering) Science

For good orientation and order, a clear and complete hierarchy of technical objects should be set up. On the lowest plane we can place the familiar real technical objects. The higher planes are hierarchically occupied by object groups, object families, etc. Each of the collective (inclusive) terms used implies that this level contains one or more sub-group of objects in the next lower plane. Many of these collective terms are already known and familiar. For instance, the collective term "lathe" (turning machine) includes all machines suitable for processing rotational parts by chip-removing cutting operations with stationary tool and rotating workpiece. Conversely, a lathe is an element (of the set) of machine tools; to this collective term belong milling, drilling, grinding and many other machines. Such a clear *taxonomy* cannot emerge for every kind of technical objects, or it remains incomplete, or poorly structured.

Which objects and/or collective terms occupy the uppermost planes of this hierarchy-pyramid? We know that up to now no uniformly and exactly defined term (*terminus technicus*) has been generally accepted for the top level. If we wish to include all technical objects (those that existed in the past, exist now or will exist in future) within a *terminus technicus*, we would have to use one or several the following collective terms: technical object, technical product, artifact, technical work, technical creation, and so on.

The term (*terminus technicus*) "artifact," although extensively used, is too broad for our application in engineering design, it also includes non-engineering designed objects, wrongly interpreted observations (e.g., in a microscope or in chemical analyses), and archaeological finds which display indications of human intervention. "Product" contains everything which is submitted by a natural, artificial or intellectual process as output: material, energy and information; for consumption, application (as a tool), or other uses. A designed product has almost always a purpose for certain persons. The designed product, as treated in this book, should find applications as means (tools) in a desired transformation, in which available material, energy and/or information from a source is transformed into other forms which are better suited for that person (as part of mankind). Such means, products of engineering (thought and instructions) and technical manufacture, we call technical systems.

For the area of designing in engineering, the term *technical system* (TS) was introduced to emphasize the most important characteristic (property) of technical objects, namely belonging to a system. Otherwise the terms "technical object" or "technical work" would also be useful. We can therefore define "technical systems" as artificial (implemented or manufactured by humans) process objects and material (real) objects. The technical system is the *object of study for the technological (engineering) sciences*.

Searching for the next lower level we can explore the present arrangements and classifications of technology (again in its broader definition). Engineering has been divided, according to the purpose for which the means (especially tools) were applied, or according to their historic developments, into sections such as mining, smelting, mechanical (power and processing machines), electrical (power, communications, electronics), instrumentation technology and metrology, transportation (land, water and air traffic), civil (building construction, roads and hydraulic works), machine tools and materials processing (metal, wood, plastics, textiles), chemical and metallurgical (plastics and other materials), etc. It is evident that this division delivers no clear classification and that other schemes are possible.

We can recognize immediately that this view is heterogeneous, both with respect to the choice and arrangement of hierarchical levels, and to their contents. Some areas are arranged according to objects, others oriented towards processing systems (procedure, process). Obvious difficulties also arise if one would like to see the subordinate object families clearly arranged. Do the machine tools belong to the processing machines and thereby to mechanical engineering, or do they belong to the materials processing machines? Should a chain bucket excavator for lignite cutting be placed among mining technology, or into mechanical engineering? -- It is clearly designed and manufactured in mechanical engineering. And where do computer controlled electric spark erosion machines fit?

For pragmatic purposes, such as for example the classification of industry branches within commerce and the (inter-)national economies, such a division can be useful and necessary; for scientific consideration however not.

A further well-known suggestion appears attractive, namely to arrange technical objects (systems) according to whether they process materials, energy or information (signals). Even this division is not unique or clear, some parts of existing machines process predominantly one of these three components, other parts of the same machine process predominantly another, and generally all three must always be present. The following chapters will show other possible arrangements for classifications of technical systems. At this stage we will depart from the arrangements of objects, and turn to technological (engineering) sciences.

## 2.4 Technological (Engineering) Sciences

### 2.4.1 Task and Object

The technological sciences should organize the technical knowledge as expediently as possible, and strive for a complete and suitable form.

Frequently technology (in its broad sense) is quoted as the object of engineering sciences. However this term is associated with different contents. Generally three basic elements must be treated in engineering sciences:

1. the things (objects, including systems of things);
2. processes; and
3. materials.

They can be explored independently of each other; but they form an inseparable unit.

### 2.4.2 Importance and Position

The high value of engineering sciences for humanity lies in their social dimension. They help to prepare better living conditions for humans and help to satisfy their needs. Society determines the goals and means, and technology mediates the knowledge from other knowledge areas to reach the goals. This fact leads to conjectures that doubt the independence of engineering sciences, they are then considered merely as application areas of conventional sciences. A simplistic and misleading statement that "engineering is applied science" is frequently heard in this context.

Even if this polemic argument is of secondary nature for us, such a position corresponds to the truth. It is correct that humans change nature by using the laws of nature, and give the revised nature those forms and characteristics which correspond to human goals. This technical-natural side (i.e. functioning) is however inseparably joined with the *social nature* of technical objects. The technical system must fulfill a whole range of characteristics (requirements for properties) to satisfy the social, economic, esthetic, psychological, ecological, and even political demands which the technology must meet.

Another fact bears witness to the importance and position of engineering sciences: With the rising number of scientific specialties the proportion of technical disciplines increases continually. Using the example of the previous Soviet Union, figure 2--2 illustrates this statement from the numbers of staff members in individual science branches. In 1972, the number of scientific disciplines was quoted as 514, of which 232 were engineering sciences.

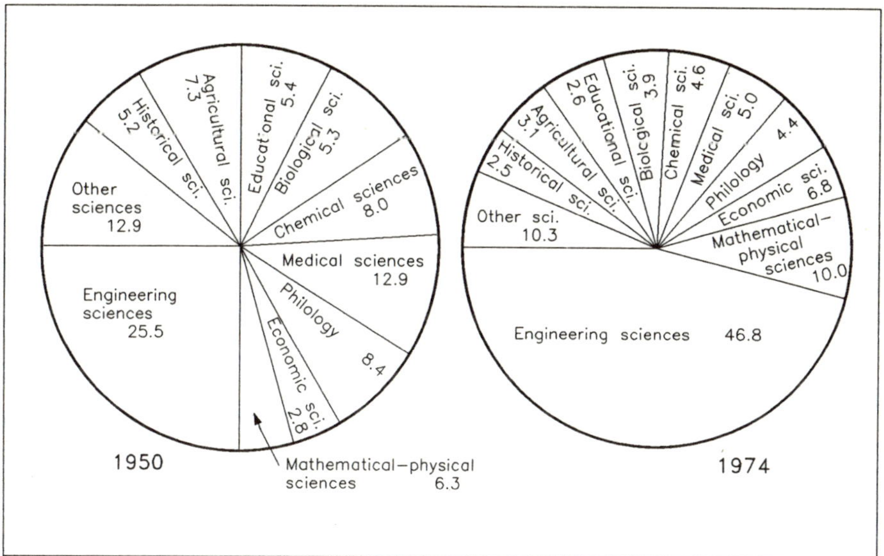


| Figure 2--2 | Distribution of Scientific Colleagues according to Knowledge Areas in the previous USSR |
|---|---|

### 2.4.3 The Problem Treated by Engineering Sciences

Technical knowledge must provide sufficient guarantees that the objects, processes and materials (and their complexes) that society needs are also planned, can be realized (made, manufactured, produced, implemented), can be used, eliminated (disposed of), and are available in an expedient quality, including expediency with respect to the environment, disturbances, outputs, etc.

In the area of real objects (things -- we adopt the technical term "real objects" from Ropohl [374]) and process objects (processes), primarily the desired function must be achieved.

Thus the relationships between function (behavior) and structure of a real object or process form the real problem of the technical sciences. The technical knowledge describes the correlations among five aspects:

- the technical properties of the object (real/material or process);
- the structure of the object;
- the processes, which progress according to the laws of nature within the object;
- the external processes and objects which are influenced by the object under consideration; and
- the influences of external processes and objects onto the object being considered.

The knowledge from the natural sciences is used (among other knowledge) by the engineering sciences towards the solution of engineering tasks. Technical sciences therefore receive from the natural sciences information about processes, materials, structures among others. This can be called object knowledge or branch (domain) knowledge.

The transformations which are realized through different processes demand a different kind of knowledge (process knowledge), particularly the technological knowledge about the conditions under which a process can be realized. This contains the knowledge about which transformations exist, which are achievable, and within what boundary conditions, etc.

Apart from the problem as outlined, persons active in technology (engineering) must be equipped with methods, techniques and instruments (tools) which make their work possible and easier. In the course of technical work, things must be calculated, modeled, presented, simulated, experiments performed, etc. Of the knowledge mentioned above, engineering knowledge therefore also contains such knowledge derived from and coordinated with the entire natural sciences, particularly physics, chemistry and many others.

### 2.4.4 Types of Engineering Sciences

The analysis of the problem in the previous paragraph delivered some meaningful main classes of technological sciences:

- object sciences,
- process sciences,
- material sciences and
- engineering sciences.

#### 2.4.4.1 Technical Object Sciences

The knowledge about things (real objects as first basic element) is probably the most important knowledge of technology. This class contains individual technical systems, as well as their families, as we discussed in the object hierarchy (Section 2.3). The *Theory of Technical Systems* [214,219] should act as the fundamental discipline that determines the problems and structure. All hierarchically lower object theories, which concretize their contents, can be derived from and/or should be related back to the general theory. More about the content of the object sciences follows in Chapter 7.

#### 2.4.4.2 Technical Process Sciences

This class mediates technical knowledge about individual processes and types of processes. The basis for this knowledge is offered by the model of the transformation systems (see figure 7--2), from which all other problem ranges can be derived. This model also incorporates the technical system into the transformation system. Designing is one of these processes. The unity between processes and real objects has already been mentioned in Section 2.4.1.

#### 2.4.4.3 Technical Material Sciences

The technical knowledge about the area "material" has gained importance with the available and newly developed materials and technologies. If the material is defined as the starting piece for a component (the raw part, therefore not only rolled steel but also semi-finished goods, e.g., castings, forgings or weldments), then many possibilities are offered for rational manufacture of components. The material also has to be designed. For example, fiber reinforced plastics technology opens new and very progressive possibilities to adjust the characteristics the "material" (composite) to the demands of each individual case.

The existing order according to material types in this area is largely applicable for engineering designers with little further adjustment.

#### 2.4.4.4 Engineering Sciences

The designation of this class as engineering sciences perhaps oversteps the average understanding of this term, which limits itself normally to technical mathematics, mechanics, thermo-fluid dynamics, electrics and similar areas.

As already defined, we are dealing with knowledge which makes possible the treatment of the technical objects and processes in all of their life phases (planning, designing, manufacturing, distributing, using, eliminating) so that design work is facilitated.

The essential professional groups that work with technical objects and use special branch information are related to the individual life phases of the technical system. Figure 2--3 shows some of the life phases and the professional groups associated them.

Each group of technical experts (designers, process planners, production and industrial engineers) who work with a certain technical object (real) system or process, requires a different kind of technical knowledge (branch or domain information). This technical knowledge is composed of two parts.

The first part is the technical knowledge which is relevant to work in that branch, namely

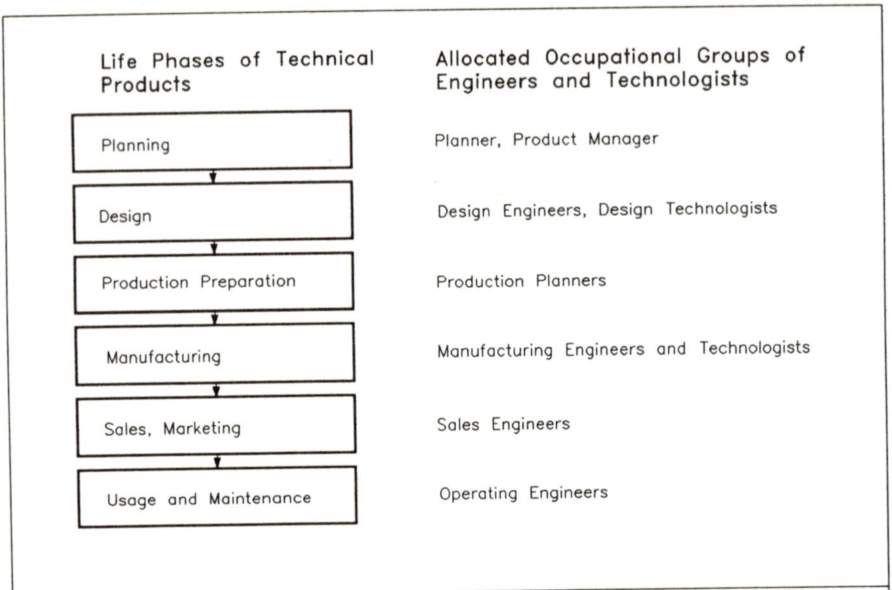


Figure 2--3 Life Phases of a Technical System and Task Allocation to Occupational Groups

the more general knowledge (relating to activities) about designing or planning, manufacturing, assembling, selling, cost calculating, packaging, operating, maintaining, and so on. The second refers to the more specific knowledge about the corresponding real object system, the technical system, on which work is being done.

In the total knowledge of individual branch experts, the specialized profession knowledge is usually dominant. A turner turns a shaft on a lathe, without needing to know whether it is to be used in an electric motor or a pump. Engineering designers cannot change their product family so simply. If they have to design an electric motor instead of a pump, they would take a much longer time to work up to competence and speed. A "learning curve" operates, because designers' work depends far more on the activity-, branch- and domain-related and branch-referenced knowledge than does the work of turners. In addition, the content and the form the object knowledge must be formulated differently for individual professions. The statements range from a simple "branch expertise" for turners, up to a set of theories with complicated laws for engineering designers.

The knowledge for each professional group should theoretically form the contents of an independent science, together with a further alignment towards a certain object (real system or process) and/or an object group. Thereby several professional sciences could emerge on each plane of the hierarchy (see figure 2--4).

In the next chapters and paragraphs we turn to Design Science, whereby the main attention is directed towards the highest abstraction level of the hierarchy, namely that of *technical systems* (TS).

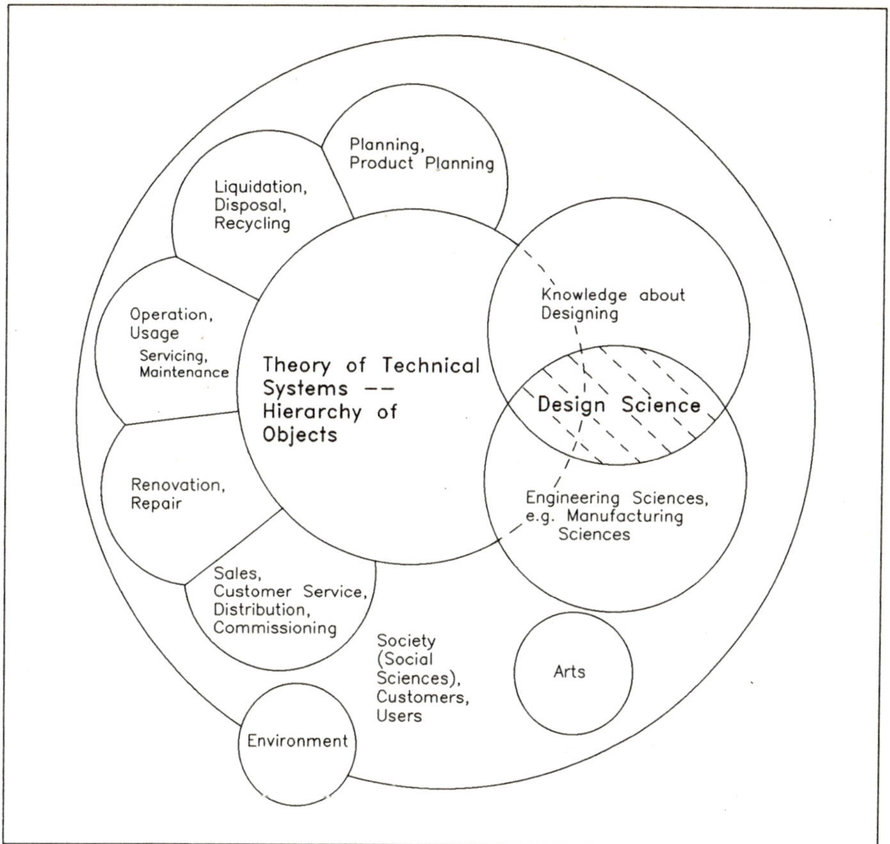


| Figure 2--4 | Branch (Domain) Knowledge in Hierarchical Planes |
|---|---|

# 3 HISTORIC DEVELOPMENT OF KNOWLEDGE TO DESIGN SCIENCE

*He who cannot draw on three thousand years is living from hand to mouth.*
***Johann Wolfgang Goethe***

## 3.1 Sketch of the Development of Ideas about Rationalizing of the Area of Designing

In this sketch we will concentrate:

- with regard to content -- only on designing in technology (therefore especially "engineering design"),
- with respect to time -- on the period since approximately 1940.

We do not wish to discuss "design" in its total breadth. The term "industrial design," in the sense as it is used in England (with emphasis on ergonomics and esthetics), is only taken into consideration if the properties relating to this characteristic of the technical system are addressed. The similarly sounding term "industrial engineering" comes also only conditionally in our considerations; it is used in North America for engineering-related activities in the preparation and rationalization of manufacture and assembly.

We emphasize that the goal to establish a complete treatment cannot be reached, but we wish to establish some facts which contribute to understanding our new viewpoints.

> *To expect anything like complete success at stating in a completely rigorous and formal way what the correct procedures are in interpretation would obviously be utopian. This may not, in itself, seem like a terribly important point. After all, complete success at stating the laws of physics may well be a utopian project; certainly complete success at describing the functional organization of a human brain is a utopian project. But physics is possible, even though complete success eludes us, because the laws of physics can be successively approximated; and functional psychology is possible because (we hope) the functional organization of the human brain can be partly described and approximated. Similarly, one might hope that one could obtain partial success in describing the practices and procedures of interpretation.*
> ***Hilary Putnam*** *[360]*

And similarly, our aim is to establish Design Science in this book. We start by exploring elements of its historic development in this chapter.

### 3.1.1 Requirements for Efforts Towards Rationalizing

Before the task of rationalizing the work of engineering design can begin, certain *opinions* (prejudices) must be *completely refuted*, namely:

- designing has been said to be an art and only especially talented persons can execute it;
- designing has been said not to be a general (or generalizable) activity, but it is always bound to the particular object to be designed, e.g., "designing machine tools," but not "designing" generally.

As *premise* for entry to the new direction, two theses must therefore be established:

**Thesis 1: Designing is a rational (cognitive) activity, which can be decomposed into smaller (design) steps, stages and/or phases** (see Chapter 1);

**Thesis 2: The design process (the procedure) is, of course, dependent on the object to be designed, but can be studied and presented in a very general form. The more concrete the object class, the more completely and detailed can the design process be defined.**

These two theses *do not imply* that "the design process must be followed linearly (without iterations or recursions)," nor "exactly" (without deviations or alterations), neither do they imply that "no intuitive steps are allowed." On the contrary, all methods must be adapted to the subjects and objects. Design methods must be adjusted for the objects to be designed, and for the designers and their environment.

Design education is closely related with the first thesis. Contrary to earlier insights a further important thesis can now be formulated:

**Thesis 3: Designing is teachable, conditioned by the existence of the theory (i.e. Design Science), and the right educational methods and media.**

This change of opinion (even if it has entered only unconsciously) has opened the gate for improvement efforts concerning the situation in designing. The first evidence of change originates from the period of the Second World War and from the following reconstruction and construction period.

Which were the particular *features of these situations* which have caused the need for improvements? On one hand it was an unusual pressure towards performance in a highly developed industry, especially new and very demanding needs. A further influence was the availability of newly developed means, e.g., computers. On the other hand, a large shortage of means existed, for example shortage of capable people and raw materials, with which one should attain the performances. Time pressure also contributed.

Only rarely did all these conditions occur simultaneously; and so the solutions were found at different times. Up to the year 1967 we could only find some widely scattered and isolated groups or individual experts who proposed certain solution for improvement of design work.

The next period, after about 1967 until today and especially in the seventies, can be labeled as the prime time for the initial development of Design Science. Increasing numbers of research contracts and the founding of Institutes for Design Technology (especially at Technical Universities in Germany) provide for the enlarged capacity to solve the design problems. In addition, the exchange of opinions has been expanded considerably through international conferences (first Prague Conference 1967, the International Conference on

Engineering Design -- ICED series since 1981), and has increased quality.

The development of Design Science displays characteristic features in the area of individual countries: communication and agreement are understandably more intense on a national level than internationally. Therefore we will develop our sketch of the entire development as partial sketches of selected countries. The selection of representatives is necessarily incomplete, and was made according to originality of the views (so that as many directions as possible are recorded) and also from the importance of the national movement for international development.

### 3.1.2 Development in the German-Language Area

#### 3.1.2.1 Federal Republic of Germany and Switzerland

In a detailed study, J. Müller [310] (1990) traced the historic development of design methodology in the German language region. Our outline recapitulates both the most important stages of this evolution and the expansion of these themes into Design Science.

In Germany and Switzerland, important personalities such as F. Redtenbacher (1809-1869), F. Reuleaux (1829-1905), C. Bach (1847-1931), A. Riedler (1850-1936) provided the basis for support, and others could build on their suggestions for improvements. Especially F. Reuleaux recognized that a Design Science was necessary (see also Section 1.3).

Independent from the design area, Polya [353,354] (Switzerland) developed general instructions for solving problems in mathematics. Of particular importance was the approach to morphology, which Goethe (1749-1832) had already used. This approach was formulated by Zwicky [465] (Switzerland) for modern science; his attempt to gather and record the whole knowledge of the world in clearly viewable and retrievable form has since then been recognized as impossible.

The pressure of external and internal conditions was especially strong in Germany. Two important works emerged there in the 1940's. The one originates from H. Wögerbauer (1943) [461], who had begun to set up a design methodology. The other work originates from F. Kesselring [245] (1943, Switzerland).

The next wave of rationalizing came after a long interval. The treatments of designing by H. Tschochner [426] (1954), R. Matousek [288] (1957) and A. Leyer [274] (1963-68, Switzerland) contained several new questions, which are not considered as direct contribution to Design Science, but as an important inspiration for some of the characteristics of technical systems, especially design for manufacture.

The period of intensive research commenced only around 1965, with the establishment the first University chair and the Institute for Design Technology at the Technical University of Munich (W.G. Rodenacker [369]). Measured on the "pulse" of the journal *Konstruktion* (Engineering Design), the peak in the treatment of the design problem was reached in the years 1972-75.

The convinced pioneer of Design Science, F. Kesselring, continued working on this problem. With his book *Kompositionslehre* (Study of Composition) [245] (1954) he not only developed an analysis of design work, but also expressed a warm relationship to his profession. Later he took part in these efforts as committee leader within the Verein Deutscher Ingenieure (VDI -- Society of German Engineers) and was decisively involved in originating the VDI-guidelines for the design area (VDI-R 2222, 2225) [23,24,26].

A "bottleneck design" was detected in 1965 (see Beitz [67]), and later a further "bottleneck designers" was described. The VDI participated intensively in the developments and in the unification of existing opinions and ideas.

After 1965 several new Institutes of Design Technology emerged (including aspects of computer-processing) at the Technical and other Universities. Their professors (W. Beitz, K. Ehrlenspiel, R. Koller, G. Pahl, K. Roth, H. Seifert) produced new and original approaches and new publications. Numerous dissertations about designing have been written. The thematically broadest view can be found in the book of Pahl and Beitz [333] *Konstruktionslehre* (Study of Design), which aligns closely with the view of Design Science. Also R. Koller expanded the theme, as he changed the title of his book from *Konstruktionsmethode* (Design Method) [256] to *Konstruktionslehre* (Study of Design) [257]. *Designing with Catalogs* by K. Roth [376] contains valuable references to processing the object knowledge in the engineering sciences.

V. Hubka continued the work which he began in the 1960's in Czechoslovakia (see Section 3.1.9 -- Czechoslovakia). His book *Theorie der Maschinensysteme* (Theory of Machine Systems) [199] expands the horizon of design knowledge by generalization and recognition of object knowledge. The problem of designing viewed as a process was debated comprehensively in *Theorie der Konstruktionsprozesse* (Theory of Design Processes) [202] and transformed into a general procedural model [204]. He also dealt with design education [203,225]. He started considering Design Science in the essay "Konstruktionswissenschaft" (Design Science) [200].

The characteristics of the socio-technical system, especially the interaction of technology with the social and economic relationships, were explored by Ropohl [374] and scientifically processed to a theory (system theory of technology -- general technology).

Slowly a "second generation" of scholars and researchers is emerging, in which the students of the previously mentioned professors appear (for example, H.J. Franke, G.W. Diekhöner, H. Birkhofer, and others), and there are approaches from other specialties, such as Schregenberger [386] (Switzerland) about problem solving in building construction (civil engineering).

Under the leadership of W. Beitz, the VDI-guideline 2221 [21,22] was issued in 1985, which strove for a "universally valid, branch-independent basis of methodical developing and designing," and expresses the tendency towards unification.

Research also progressed in the area of computer support. Not only was a theoretical basis (a design logic) worked out, but also CAD-systems were developed (e.g., PROREN by J. Seifert [390]).

Design management, planning, representation and other themes were developed parallel and independently.

In 1992, the German edition of *Konstruktionswissenschaft* [229] was published. A new edition of the 1982 English publication *Principles of Engineering Design* (reprinted as [215]) was issued by Heurista under the shortened title of *Engineering Design* [228].

Appraisals of the developments in the discussed areas was undertaken repeatedly, and discussed especially at ICED 81 and 83. This conference series "International Conference on Engineering Design" -- ICED, lead by V. Hubka (Switzerland), M.M. Andreasen (Denmark) and W.E. Eder (Canada), has occurred regularly since 1981. The Proceedings [29,125,207,208,209,213,215,221,226,227,230,372] contain more than 1200 contributions of many prominent scientists representing about 50 countries. Several workshops have been held to clarify particular themes, among others in Zürich and Rigi Kaltbad (Switzerland), Pilsen and Prague (Czech Republic), Copenhagen (Denmark), Rome (Italy), and New Orleans and San Luis Obispo (USA). For additional information about these activities, see the preface.

#### 3.1.2.2 The Previous German Democratic Republic (GDR)

During the time period considered here, this part of Germany was practically cut off from the developments in other countries. Here we report on these independent developments.

The first conference of designers in the GDR, in which "the characteristics of design systematics were presented" by F. Hansen, was held in 1954 in Leipzig (with involvement from H. Wögerbauer and W.G. Rodenacker). Many other scholars in the GDR developed their work from that of F. Hansen and his frequently quoted books *Konstruktionssystematik* (Design Systematics) (1965) [186] and *Konstruktionswissenschaft* (Design Science) [187] (1974). These researchers originate both from engineering practice and from the universities (A. Bock, G. Höhne, J. Müller, W. Heinrich, J. Rugenstein and others). Müller [308,309, 310,311] has contributed particularly from the view of a philosopher, and appears with own relevant works, especially *Systematische Heuristik* (Systematic Heuristics, see Section 1.4.6).

Interesting approaches for the methodology of problem solving emerge also in supporting the innovators-movement. This movement aims at renewal of the modes of thought for the purpose of advancing innovation and invention of workers, see also Section 1.4.5. This stems mainly from the Russian ideas (see Section 3.1.8).

Engineering pedagogics has been intensely developed in the GDR for instructional purposes. The strong personality was H. Lohmann from the Technical University Dresden, who expressed the fundamental ideas in his work *Die Technik und ihre Lehre* (Technology and its Study) 1955. A synthesis of the two directions (Hansen and Lohmann) was accomplished by K. Steuer in 1968 in his *Theorie des Konstruierens in der Ingenieurausbildung* (Theory of Designing in Engineering Education).

### 3.1.3 Great Britain

As was already stated at the Great World Exhibition in 1851, the originator country of the industrial revolution was slowly descending in the worldwide ranking of technical producers. After the second World War one of the recognized defects was that the offered products were technically and visually antiquated. In consequence the discipline of "Industrial Design" was founded at different schools to counteract one of the deficiencies, and has strongly increased in consulting practice. The most important works in this area originate from Ashford [63] and Mayall [291]. Later Mayall confirmed [292] that designing is of great importance, with particular consideration of quality, operational properties, esthetics and ergonomics.

Some early attempts at scientific investigation of the design process originate from England. Based his own design experience, Wallace [439] proposed as model of designing a cyclical series of steps under the mnemonic ATDM -- analyze, theorize, delineate, modify.

A consistent series of investigations began around 1960 at the University of Cambridge. Marples [285] followed new paths, by evaluating some design projects in industry through observation, a forerunner of protocol studies. He recognized that design work can be represented by the model of a decision tree. Others have drawn the conclusion that designers can or should use this "family tree" model as instruction for a part of designing (Jones [238,239]) or at least to record the decisions that were made (Eder and Gosling [120], Eder [121,123]). Shortly thereafter, Booker [76] wrote about the use of principles and precedents as guidelines and impulses in developing newly designed systems.

The work of Gosling [174] was directed especially towards electronic systems, however it contains sufficient general insights to a theory of technical systems. Among others, Norris [325] has evaluated and reported about morphology.

The entrance of Great Britain into Design Science occurred through the particularly appropriate analysis of the situation in the design area in the *Feilden Report*. The first "Royal Commission" under the direction of G.B.R. Feilden [149] submitted their report on engineering design, and proposed many innovations, especially for design education and for the respect of the engineer professions in society.

After the period of relative isolation from the continent, some German books were translated (for example, R. Matousek, German 1957, English 1963 [289]; A. Leyer, German 1963-68, English 1974 [275]; G. Pahl and W. Beitz, German 1977, English 1984 [334]; and V. Hubka, German 1980, 1981 and 1984, English 1982 and 1988 [217,219,220]).

Archer [51] published an approach to a systematic method of designing. Also further works by Archer [52,53,54,55] suggest interesting ideas for design methodology. Another attempt, which developed the earlier work of Gosling [174] for mechanical engineering, came from Eder and Gosling [120]. Checkland [88] contributed to system thinking.

An important conference was organized in Birmingham (Gregory [175]), which presented the state of the art in design methodology in Great Britain -- without international participation. This conference is important for our survey, primarily for the number and breadth of important contributions (among others from G.H. Broadbent, C.H. Buck, C.T. Corney, A.L. Davies, W.E. Eder, J. Farradane, S.A. Gregory, R.J. McCrory, P. McMullen, E. Matchett, W.H. Mayall, A.M. Penney, I.M. Ross, B. Shackel, A.F. Stobart, B.T. Turner and R.D. Watts) and also that Gregory already defined very well the scope of Design Science as goal of design research [176], see Section 2.1.1. In addition, for the first time attention was paid to definitions (Eder [121,122]).

A series of important books marks the further development: Cross [96], Ellinger [143] (from his own experience of designing), French [159,160] (conceptualizing, and evaluating using approximate and appropriate mathematical relationships), Glegg [168,169,170] (philosophy of design), Morrison [306] (decision theory), Pitts [349].

Also some contributions are dedicated to creativity: DeBono [102,103,104] published his volumes about "lateral thinking" for stimulating creativity.

The journal *Design Studies* (Butterworths, now Butterworth-Heinemann) with emphasis on Design Science and design in other fields was founded 1979 under the editorial supervision of the Design Research Society (DRS).

French [161] discussed analogies between artificial (human) constructions and natural formations and living thing. He shows that the natural structures are a good approximation to the theoretically optimal, and how artificial formations can be conceptually optimized. A similar project was pursued by Maunder [290], however with particular relation to movement and mechanisms.

The *Journal of Engineering Design* (Carfax), founded 1990, aims at bridging between design research and industrial application.

Important conferences were held and the Proceedings published in book form, among others directed by Booker [77,78], Cross [95], DeSimone [108], Gregory [175,177], IMechE [17], Jones and Thornley [237], Langdon [270], Loughborough University [19] and Pitts [350].

Not only design methodology, but also other areas of Design Science are practiced, for example, the management of design (B.T. Turner), information systems for the designer (G. Pitts), managing the design process (Hollins and Pugh [194], Leech [272], Wearne [447]) among others.

Newer works on engineering design include those by Cross [98], and Pugh [359].

A peculiarity of the development in Great Britain is the *institutionalization* of the efforts towards improvement. Already 1944, the "Council for Industrial Design" was founded, with emphasis on appearance and human operability (ergonomics) of products. In 1981 it was renamed "The Design Council," to give more emphasis to engineering. Also in 1981 an "Engineering Council" was founded with the task of improving the situation, particularly with regard on the splintered nature of the engineer organizations, consisting among others of "The Institution of Mechanical Engineers" (IMechE) and "The Institution of Engineering Designers" (IED), each of which have also provided their positive contributions.

An occasional series of reports was demanded and prepared for government offices, but also by other interested organizations. The most important were Corfield [92], Department of Trade and Industry [10], Feilden [149], Fellowship of Engineering [5], Finniston [156], Lickley [276], Mellor [296] and Moulton [307]. In the late 1980's the government has encouraged the causes of design for the purpose of reconstructing the industry. Some centers for research and dissemination of knowledge about design and design management have emerged, and have seized this initiative, e.g., the group around Glasgow and Strathclyde (Engineering Design Research Centre), and the Universities of Cambridge, Lancaster and Newcastle, with City University (Engineering Design Centre). Various publications originated from this institutionalizing, e.g., from Abbott [31,32], the teaching aids [3,4,185,358] from the organization SEED (Sharing Experiences in Engineering Design, see also Kimber [247]).

### 3.1.4 France

The French literature related to scientific designing is quite varied. Already in the seventies the *Méthodologie de la construction mécanique* appears from J. Chabal, R. De Preester and R. Ducel (1973). Methodical approaches are also available in all later books with the theme of design study (for example A. Chevalier, P. Poignon). Likewise the processing of technical knowledge (for example *Technologie de construction mécanique* from M. Norbert *et al*, 1969) shows a very systematic approach which in form come close to the design catalogs (compare the works of Roth [376] and Koller [257]).

The particular interest in these books is that they were published only as textbooks and all authors are without exception professors at technical intermediate schools (technical colleges). No attempt was undertaken to our knowledge to expand this bare alignment on teaching, or to present the available experiences in an international discussion and to compare them with the systems in other countries.

### 3.1.5 Italy

The research and teaching in the area of design knowledge have been taken up in Italy at two universities: in Rome (U. Pighini) and in Milan (G. Biggioggero, E. Rovida), where the problems of representation and modeling has been assigned much research capacity.

### 3.1.6 Scandinavia

In the industrially highly developed Scandinavia the solutions the problems of design began in the sixties under pressure of the market.

In Sweden, individual approaches and guidelines as part of the Sveriges Mekaförbund have been prepared (e.g., Ko7 -- quality of design, 1958). Also in design instruction (teaching) we report about progressive courses (F. Olsson: Compendium, TU Lund, 1966).

In Denmark, in the Laboratory for Design under leadership of V.A. Jeppesen, a progressive design teaching system was set up. It was carried by M. Myrup Andreasen, and realized several new approaches especially to the themes of "design for assembly" and "integrated product development" [47,48,49,50] in a series of books and lectures. Andreasen was already named as one of the leading persons of the ICED-conferences. His cooperation with E. Tjalve [423,424] has enriched the theme of "representing-modeling" with new ideas.

Also in Norway and Finland interesting design themes were initiated at the universities.

### 3.1.7 USA and Canada

Attempts to outline the development of design knowledge together with efforts for improvement, as we have done for other political areas, is not appropriate for the USA and Canada. An early blossoming of design thought occurred almost simultaneously with the trends in Great Britain. The *Design Methods Group* (DMG) that was founded in California in those years still exists, but after a short burst of high general activity it seems to have lost its influence, perhaps because of its emphasis on architectural design, and was overtaken by other events. The impulses for these events lie in other directions, and so we must describe certain partial areas as sources for the present knowledge potential. As in the earlier development stages, the knowledge of these sources is accepted into the area of designing and serves often rather unconsciously as inspiration.

Efforts in this vast geographic region seem to have been patchy. Specific areas of design were addressed, and many different schools of thought grew up around regional centers. There seems to have been little in the way of collecting, coordinating and synthesizing into a more general abstraction.

#### 3.1.7.1 Bases for Design Knowledge

***-- Problem Solving***

The work about problem solving methods for mathematics by Polya [353,354] was taken up and developed by Wickelgren [455]. Newell and Simon [318,397,398] treated problem solving thinking. Problem solving was generalized by Wales [437,438]. Relationships to computer techniques were established by Starfield *et al* [408]. A more direct relationship to design was not produced, but the application is recognized by the originators of this direction as important.

***-- Systems Theory***

Systems theory and systems thinking were developed in the USA since the 1950's, for instance represented by Bertalanffy [69] and Hall [183]. Further progress and concretizations can be recorded by Churchman [90], and Klir [250,251,252], both in system theory, and in their application to analysis and problem solving. The goals of system theory were disclosed through an announcement in the journal *Philosophy of Science* (Vol. 22, 1955, p. 331):

1. to explore the isomorphy of the concepts, laws and models in different areas and to support useful transfers from one area to another;

2. to encourage the development of suitable (or sufficient) theoretical models in areas, where they are missing;
3. to avoid the duplication of theoretical efforts in different areas;
4. to advance the uniformity of science through improvement of communication between specialists.

### – *Decision and Management Bases*

A structure of human decisions as one of the important steps of designing was developed by Miller [300]. Nadler [314,316] has developed a method of planning from the point of view of company organization, which he considers also suitable for designing.

From company management, the works of Ackoff [33,34,35], Argyris [57,58], Drucker [117,118] and Schön [385] should be mentioned. Combinations between system thinking and management occur in Ackoff [36] and Churchman [90].

Starr [409] defined design as an almost pure decision process. His work is broadly laid out in the mathematical processes, which cohere with decision theory. This line was later extended and popularized by Suh [415], unfortunately under the claim of a design method.

In order to make engineering design accountable for its actions and decisions, two trends have been derived from legal actions regarding product liability [400,401,402,403]. One is towards improved definitions and knowledge about human factors (ergonomics and capabilities) for engineering application [65,100,138,379,399,460]. The second led to methods and procedures for independently validating, verifying, checking, reviewing and auditing the intermediate and/or final results of design work [382].

Concern for "satisfying the customers" has led to introduction of various management techniques with links to designing, e.g., TQM, QFD, Taguchi experimentation (see also Section 1.2.5).

### – *Error Events, Error and Failure Elimination*

One of the many reactions to error events came from Warfield [442,443]. His thesis is that appropriate efforts of political and operational supervision are needed, to keep error events within acceptable boundaries. This can occur only by using suitable methodical measures on the part of the management. A systematic method of design management is proposed and called "generic design."

Abstracting from historic reports of failures that occurred, mainly in civil and structural engineering, Petroski proposed a model for how development of systems over time takes place [343,344,345,346]. He pointed out several paradigmatic failure causes and tried to relate these to available knowledge and social conditions.

### – *Knowledge about Creativity*

Advancement of creativity *per se* was the goal of works by Jewkes [235], von Fange [434,435], Weisburg [449] and Whiting [452]. Gordon [173] proposed and developed the system "synectics" as a formal problem processing method for a group of participants, and this method is considered especially useful for organizational problems. Osborne [330] developed "brainstorming" as a further group method to enhance creativity. A partial support for these methods through research in psychology followed only five years later through the work of Guilford [178].

Another problem of creativity, namely how one can free the human brain of prejudiced and fixed ideas, is clarified by Adams [37], presumably as a basis on which creativity can be enhanced. Similar explanations in reference to intuition and its advancement stem from Goldberg [171]. In a further development, Nadler and Hibino [317] proposed a philosophic system for renewal of management approaches, especially regarding products.

### 3.1.7.2 Design Knowledge

Examining the USA literature regarding engineering design shows two typical characteristics for the improvement efforts. Firstly the number of works peaks clearly in the period 1960-1970, secondly the absolute majority of the works deals with or is based on creativity.

Psychological insights were incorporated consciously and unconsciously in the instructions for designing, especially for advancement of creativity, as with Adams [37], Crawford [93], Dixon [111], Gordon [173], Nadler [317], Osborne [330], Schön [385], von Fange [434,435], Whiting [451] and others. Already the first important book in the engineering area, authored by D.S. Pearson, meshes into these themes and carries the title *Creativeness for Engineers* (1959). Also the much used book of H.B. Buhl *Creative Engineering Design* (1960 [86]) and many others, including the "Prentice-Hall Series in Engineering Design," belong to this class.

To explain engineering design and general engineering for students is the main purpose for Krick [261,262,263], whereby creativity is emphasized and the description of the design processes occurs almost only as a secondary matter.

Further introductions of similar kind were published by Gibson [167], Middendorf [298], and Vidosic [433]. Assemblies of projects were compiled by Spotts [405] and Vidosic [432], as exercises for design. Wilson [457] shows the development of a product from an idea up to a working model.

In this sense, but with more relevance to designing, works of Harrisberger [188] and Woodson [459] are noted. Likewise a large number of these works which deal with "design instruction" (*Introduction to Engineering Design*), emphasize creativity or invention (I.E. Edel 1967; J.R. Dixon [111] 1966 and partly also J.P. Vidosic [433] 1969).

Another direction is opened by the works of M. Asimow [64] (*Introduction to Design*, 1962), R.J. McCrory [293] (*The Design Method - A Scientific Approach to Valid Design*, 1964), and G.N. Sandor [380] (*The Seven Stages of Engineering Design*, 1964). In this more discursive direction are also the books about "design instruction" by E. Pare *et al* [336], T.T. Woodson (1966), R.E. Parr (1970), I.R. Wilson (1970) and others. Alger [41] places large value on the creative processes, but also describes some mathematical methods, especially for evaluating the proposed solutions. Roe *et al* (Canada) [370] have tried to develop a rational model of designing. Also in Canada, Love [282] opened a consulting office and developed courses for systematic design, based mainly on methods and approaches known in the English language region.

In the following years, treatments of "Engineering Design" are relatively rare. In contrast, more general works about methods (therefore with extended validity and directed towards management) accumulate, as for example from G. Nadler [314,315,316]. The object field of the methods approaches the one in Great Britain.

A new situation came about in the USA during the eighties. The need for new knowledge about and for designing with computers developed more rapidly in the computer area than in design practice.

The scene in the USA was subjected to a rapid development and a quantity of research money was made available through the NSF (National Science Foundation) Initiative on

Design Theory and Methodology. The appeal [15] was answered by a report published by Rabins [361]. The recipients of research money have held an annual seminar, and the papers usually published as proceedings volumes (e.g., Newsome [321]).

Following ICED 87 Boston, a conference series was started in 1989 under the title of "ASME Design Theory and Methodology Conference," and has been run annually since then (e.g., Rinderle [367]). Although this conference is directly sponsored by the ASME (American Society of Mechanical Engineers), its scope has included other engineering disciplines, architecture, and even clothing manufacture.

A newly founded journal, *Research in Engineering Design* (Springer-Verlag, New York), aims at dissemination of research results, with particular emphasis on observation and protocol methods, and on progress in the computer application.

The theme of observations and protocols has received much attention in recent times in the USA. For instance, Ullman [429] has belatedly discovered the importance of the sketching for conceptualization through observation and protocol research. A similar quest as Tjalve [423,424] about the importance of sketching, but from the point of view of psychology, is treated by McKim [294].

From his own observations of the processes of designing of chemical industry plant, Westerberg [450,451] set up a theory of the design process. Some of these approaches are similar to the Theory of Technical Systems [214,219] (see Section 7.1).

Further significant works, especially related to embodiment, layout and detail design, have been published by Ertas [145] and Ullman [430]. Dym [119] produced a noteworthy attempt at synthesis of the existing (mainly North American) views, with particular emphasis on the use of computers in design.

*-- Artificial Intelligence and Computer Applications*

If design problems should be supported by applications of computer systems, the predominant approaches are developed from artificial intelligence (AI), especially the method of knowledge-based systems (expert systems). The latter are suitable on one hand for diagnosis, on the other hand they are being developed within design research as advisers for evaluation, to coordinate the activities of design groups ("design teams") and their management, and for other activities. Examples of such program attempts originate from Eisenberger [142], Papalambros [335], Rinderle [366], Subramanian [414], Talukdar (and collaborators) [418,419,420] and Westerberg [450,451].

A more general application of CAD, coupled with AI within the design processes (see Section 7.5), is the goal for the IFIP (International Federation for Information Processing), especially through conference series of different study groups, e.g., Yoshikawa and Warman [464]. From this, work and applying the theories of Yoshikawa [425,462,463] (see Japan, Section 3.1.11), new approaches for CAD-programming in object-oriented languages are undertaken (e.g., Warman [444]), which is better adjusted to the stages of layout, drafting and detailing ("embodiment design") for the known modes of operation of designers.

Newer developments include the use of *genetic algorithms* within computer-assisted design processes [87], and of *fuzzy set theory*, *ambiguity*, *nonspecificity*, and *strife* [253].

### 3.1.8 Russia -- Previous USSR

The first known work in scientific design originates from P.I. Orlov: *Basis of Designing* [329]. The subtitle "technical knowledge and methodology" points to the methodical aspect.

Otherwise the references to methodical designing in the literature are fairly scarce. It appears however, that the application of the computer increases the pressure towards the investigation of the design processes and objects, and has brought new points of view and results (compare Klimov, Lebedeva in WDK 10 [213]).

Completely different is the situation in the area of the support of innovators and inventors. One direction consists of general references, e.g., I.N. Sereda, who works at the Peoples University for Technical Creation in Riga, and whose book *Worker - Inventor* [392] belongs among the very widespread books in the USSR. Another direction is represented by Altschuller: *Inventions - (Not) A Problem* [42] (see also Section 1.4.5).

Based on a detailed investigation of patent submissions, Altschuller has found [42,43] that the majority of the inventions were prepared using a small number of known methods. He proposed five kinds of problems and their solution.

The methods that may be utilized were summarized in a system recommended for inventing, an invention algorithm (for example ARIZ 59, improved as ARIZ 61 and ARIZ 80), to indicate to inventors the question about the way to a solution (invention). This is a system of organized methods and planned activities which are based on logical rules and instructions. The algorithm has been computerized by the Invention Machine Laboratory, Minsk, Republic of Belarus [427], and exported to subsidiary companies in Germany, the USA and other places. In some respect this algorithm is similar to systematic heuristics after Müller [212] (see Section 1.5.4).

Two further study groups under the direction of Odrin [326,327,328] and Powilejko [356,357] are interesting here. They have developed innovative design methods as expansions of morphological analysis (as reported by Arciszewski [56]).

Another widespread research direction in the USSR aims at organizational-economic questions of development. J.S. Sapiro belongs to this circle. His book *Organization and Effectiveness of Technical Development* (1980) [381] treats this problem.

As in other areas, the situation in Design Science in the previous USSR is fairly unknown, because the available literature does not necessarily reflect the status of knowledge.

### 3.1.9 Previous Czechoslovakia

In Czechoslovakia in the sixties, efforts for design improvement proceeded in three directions: first the invention method of K. Backovsky (1963) in continuation from W. Ostwald (1932) (the prior timing is noteworthy), second the movement of innovators and inventors who looked for a solution methodology (compare Sections 3.1.8 -- USSR and 3.1.2.2 -- GDR), and finally some designers in engineering practice who were trying to improve their work and have formulated relevant measures (V. Hubka, J. Smilauer, S. Vit). This phenomenon of the participation of designers from engineering practice in design research is singular in the history of Design Science.

These three directions were combined (at least partially) in the years 1962 and its members formed the Design Committee of the Scientific-Technical Society of Czechoslovakia. A series of conferences and seminars was organized by this committee, and cared for the transmission of knowledge, especially in design methodology. The conference in Prague 1967 is particularly important for the development in the world, because there a first international exchange of opinions occurred: Great Britain, the Federal Republic of Germany, the GDR, Switzerland, Poland and Czechoslovakia were represented by delegates.

In the following years attention was dedicated to the working means (tools, equipment, information sources, etc.) of engineering designers, including computer applications.

The particular development of Design Science shown in this book started with preparatory work in Prague, as Hubka reported in [198]. This work was transferred to Denmark as a consequence of circumstances, and later to Switzerland (see Section 3.1.2.1). The basis is a comprehensive Theory of Technical Systems, and its development through works by Hubka [199,214], Andreasen [46], and Hubka and Eder [219,228,229]. Derived from this work is a theory of design processes [202], a procedural model with instructions for implementation of the design process [204,217], and a set of case examples to illustrate the use of the procedural model for conceptualizing [220].

### 3.1.10 Poland

The very intense research in the whole area of Design Science in Poland today is based on the work of J. Dietrych. Already in the year 1967 he enriched the Prague conference around a new, holistic view of design. As professor at the Technical University of Glivice, he extended the study of machine elements to design in general. His efforts are directed especially towards the practical work of the designer. Dietrych [210] has produced a comprehensive theory based on a series of definitions for key-words and *termini technici*.

Beside that a very general direction of designing and project work has been developed. Building on the work of Kotarbinski [258,259], a group in Warsaw has tried to set up a science of sciences. From this, Gasparski [164] has developed an explanation and philosophy for design, which is known as Praxiology, and is also connected with the name A. Sielecki. Goralski [172] has taken the concepts of morphology into this work direction.

### 3.1.11 Japan

An attempt to apply formal logic with some axioms was undertaken by Yoshikawa [425,462,463] (see also Section 3.1.7) with the goal to generate a complete algorithm of designing on digital computers (especially for CAD-application). Kaoru Hongo presented thoughts about design education as part of the ICED-conferences.

### 3.1.12 Other International Developments

Apart from the ICED conference series, several other efforts at international cooperation should be acknowledged. A series of conferences and seminars on applications of computers to architectural design have taken place, starting around 1986, mainly under the leadership of J. Gerö (Australia). The methods and validity of protocol studies of designers in action have come under scrutiny in a recent seminar.

A significant book on engineering design was published by Lewis and Samuel [273], emphasizing methods in the context of detail design of machine elements.

### 3.1.13 Summary

Figure 3--1 presents an attempt to survey the development. The names of the most important design scholars as authors of books form a chronological series in four columns. This provides a comparison of the development tendencies in four geographical areas.

**Survey of the 'Evolution' of Contributions to Design Science**

Legend:

| | | |
|---|---|---|
| T ... Translation | G ... Germany (united) | S ... Sweden |
| C ... Conference Proceedings | F ... France | AU ... Australia |
| R ... Report | CZ ... Czech Republic | J ... Japan |
| G ... Guideline | PL ... Poland | SU ... USSR (previous) |
| J ... Journal | DK ... Denmark | |

69 ... refers also to later contributions -- some authors appear several times (with abbreviation of book title)

| | USA, CDN | D, CH | GB | Other countries |
|---|---|---|---|---|
| | | 1850 Redtenbacher | | |
| | | 1853 Reuleaux | | |
| | | 1919 Riedler | | 1929 Kotarbinski (PL) |
| 1940 | | | | |
| | | 43 Kesselring | | |
| | | 43 Wögerbauer | | |
| 45 | 45 Polya | | | |
| | 47 Miller | | | |
| | 48 Zwicky | 48 Konstruktion (J) | | |
| 1950 | | | | |
| | | 52 Bischof–Hansen | 52 Wallace | |
| | 53 Bross | | | |
| | 54 Crawford | 54 Tschochner | | |
| | 54 von Fange | | | |
| 55 | | 55 Lohmann | | |
| | | 57 Matousek | | |
| | | 57 Brandenburger | | |
| | 58 Jewkes | | | 58 Richtl. Ko7 (S) |
| | 58 Whiting | | | |
| | 59 Pearson | | | |
| 1960 | 60 Simon | | 60 Marples | |
| | 60 Buhl | | | |
| | 61 Gordon | | | 61 Sereda (SU) |
| | 62 Ackoff | | 62 Gosling | 62 Goranski (SU) |
| | 62 Asimov | | | |
| | 62 Hall | | | |
| | 63 Starr | 63 Leyer | 63 Feilden (R) | 63 Altschuller (SU) |
| | 63 Norris | | 63 Matousek (T) | 63 Hubka (CZ) |
| | 63 Pare | | 63 Jones (C) | 63 Smilauer (CZ) |
| | 63 Osborne | | | 63 Backovsky (CZ) |
| | 64 Alger | 64 VDI–R 2225 (G) | 64 Archer | |
| | 64 Drucker | | | |
| | 64 McCrory | | | |
| | 64 Sandor | | | |
| 65 | 65 Krick | 65 Engpass–Konstr. | 65 Eder/Gosling | |
| | | 65 Lehrstuhl München | | |
| | | 65 Hansen (KoSyst) | | |
| | 66 Woodson | | 66 Gregory (C) | 66 Olsson (S) |
| | 66 Harrisberger | | | |
| | 66 Dixon | | | |
| | 67 Tech–Innov (C) | 67 Müller (OpVerf) | 67 Mayall | 67 Dietrych (PL) |
| | 67 Miller | 67 Hansen (C) | | 67 Prag (C–CZ) |
| | 67 Nadler | | | |
| | 67 Roe | | | |
| | 68 Bertalanffy | 68 Steuer | 68 DeSimone (C) | |
| | 68 Gibson | | 68 Ellinger | |
| | | | 68 Morrison | |
| | 69 Churchman | | 69 Ashford | 69 Norbert (F) |
| | 69 Klir | | 69 Glegg | |
| | 69 Middendorf | | | |
| | 69 Vidosic | | | |
| 1970 | 70 Parr | 70 Rodenacker | 70 Jones | 70 Geminard (F) |
| | 70 Wilson | | | |
| | | | 71 French | 71 Vidal (F) |
| | | | 72 Gregory (C) | 72 Powilejko (SU) |
| | | | 72 Cross (C) | |

Figure 3--1 Part 1 of 2 — Historical Developments in the Literature

Legend:

| | | |
|---|---|---|
| T ... Translation | G ... Germany (united) | S ... Sweden |
| C ... Conference Proceedings | F ... France | AU ... Australia |
| R ... Report | CZ ... Czech Republic | J ... Japan |
| G ... Guideline | PL ... Poland | SU ... USSR (previous) |
| J ... Journal | DK ... Denmark | |

69 ... refers also to later contributions -- some authors appear several times (with abbreviation of book title)

| | USA, CDN | D, CH | GB | Other countries |
|---|---|---|---|---|
| | 73 Love | 73 VDI-R 2222 (G) | 73 Pitts | 73 Odrin (SU) |
| | 73 Holloway | 73 Schw.M.Mkt. (J) | 73 de Bono | 73 Chabal (F) |
| | 73 Miles | | | |
| | 74 Wickelgren | 74 Hubka (TMS) | 74 Leyer (T) | |
| | 74 Newell | 74 Hansen (KoWiss) | 74 Pitts (C) | 74 Svensson (AU) |
| 75 | | | | |
| | 76 Warfield | 76 Hubka (TKoP) | 76 Moulton (R) | |
| | | 76 Franke | | |
| | | 76 Steinwachs | | |
| | | 76 Koller | | |
| | | 77 Pahl/Beitz | | 77 Orlov (SU) |
| | 78 Argyris | | | 78 Yoshikawa (J) |
| | | 79 Ropohl | 79 Loughborough (C) | 79 Tjalve (DK) |
| | | | 79 Corfield (R) | |
| | | | 79 Design Studies (J) | |
| 1980 | 80 Adams | | 80 Finniston (R) | 80 Andreasen (DK) |
| | 80 McKim | | | 80 Sapiro (SR) |
| | | | | 81 ICED 81 Roma |
| | | 82 Roth | 82 Hubka/Eder (WDK1) | |
| | | 82 Schregenberger | | |
| | 83 Goldberg | | 83 Lickley (R) | 83 ICED 83 Kopenhagen |
| | 83 Schön | | | |
| | | | 84 Cross | 84 Gasparski (PL) |
| | | | 84 Langdon (C) | |
| | | | 84 Pahl/Beitz (T) | |
| 85 | 85 NSF | 85 Ehrenspiel | | |
| | | 85 ICED 85 Hamburg | | |
| | | 85 VDI-R 2221 (G) | | |
| | 86 Wales | 86 Seifert | | |
| | 86 Rabins (B) | | | |
| | 87 ICED 87 Boston | | | |
| | 88 Hubka/Eder (TTS) | | 88 Trade & Ind. (R) | 88 Lewis/Samuel (AU) |
| | | | 88 Hubka/Andr./Eder | 88 ICED 88 Budapest |
| | 89 Newsome (C) | | 89 ICED 89 Harrogate | 89 Andreasen/Hein |
| | 89 Westerberg | | | |
| | 89 Suh | | | |
| 1990 | 90 Res.E.Design (J) | 90 Müller (AMeth) | 90 J. Eng. Design (J) | 90 ICED 90 Dubrovnik |
| | 90 Starfield | | | |
| | | 91 ICED 91 Zürich | 91 J. Des. & Prod.(J) | |
| | | 91 Hubka/Eder (KoW) | 91 Pugh | |
| | 92 Ullman | 92 Hubka/Eder (ED) | | |
| | | 93 Pahl/Beitz (3rd) | | 93 ICED 93 The Hague |
| | 94 Dym | | 94 Cross | |
| 95 | | 95 Roth (2nd) | 95 Redford | 95 ICED 95 Prague |
| | | 95 Hubka/Eder (DSci) (current volume) | | |

Figure 3--1 Part 2 of 2 Historical Development of Design Science Literature

## 3.2 Description of the Development of Design Knowledge

### 3.2.1 Elements of the Development

The sketch of development of design knowledge has shown clearly the different rates and stages of progress. To the question about elements of these processes, only some can be hypothesized at present: the degree of industrial development, the levels of education, the organization and extent of research, the culture and tradition in the individual areas, the size of the area (country). The connections and interactions of these and other elements would show an even more complicated picture.

- The level of industrialization plays a large role, because first a need for rationalizing the design process must be recognized, so that solutions can be looked for. This happened certainly in the most of the described cases. This thesis is confirmed by the situation in other countries, e.g., in China, and India. There the research problems have been taken over as a task in some universities, but despite strong interest, the results remain currently of low significance, because the problems in those countries are not urgent. This thesis of the degree of industrializing appears to be problematic for the USA or Japan. It is surprising how sparsely and under which unfavorable circumstances the research projects for design knowledge began to work and -- this counts especially for the USA -- how weak the interest in the already existing knowledge was. A preliminary explanation to this could be (as an additional sub-thesis) that the degree of industrializing is changed through the economic power of the area. In areas, where this economic power is apparently strong, a persistence and inertia exists, which decreases research for the purpose of improvement of the modes of operation. To some extent, the prevailing cultural outlook and linguistic factors interact with economic power.
- The dependence of the intensity of improvement efforts on the degree of education has not been recorded in our historic sketch, because it would be necessary to expand the investigation to individual typical persons. It is, however, clear that university graduate engineers are more open for these questions than graduates of the lower engineering schools or industrially trained designers. It may be that university engineers are less represented in the design process, and they especially process the more abstract conceptual and analytic tasks in which the newer researches in design knowledge could be useful. In comparison, the graduates of the engineering technical colleges (e.g., two-year colleges) and trained designers dedicate themselves predominantly to the more concrete tasks, where results from the older existing engineering sciences appear more important. In comparison with the universities much less pressure exists for teaching personnel in the lower engineering schools to participate in research. These facts agree with the number of Institutes for Design Methodology (titles vary from place to place) at universities, if one compares them with similar institutes at the engineering technical colleges.
- Design research began mostly at university institutes -- and is promoted there -- either as part of the general research tasks or in funded research projects. Most scholars of the first generation (sixties and seventies) have chosen the range of problems from their own experience in design, having in part recognized the defects. A minority (especially in the USA, and more recently in the UK, Holland and other places) only discovered the research possibility with the availability of research money, and mostly without own design experience in industry. Motivation to research and to engineering design consequently appears as a relevant (human) element, and its influence can be seen also

in different functions and successes. A unique way to the solution has been shown in this respect by one group of designers from engineering practice in Czechoslovakia, as reported above.

- Generally considered, only relatively little research money has been invested in this problem until recently. The more theoretical and reflective tasks can be accomplished with less support. Relatively much of the available research money has flowed into the more concrete tasks, especially those of computer application.
- The cultural traditions also present a decisive element. The understanding about goals and means in the countries of the European continent is different from that in Great Britain or the USA. Concerning the situation in Japan, much more can be learned from the culture tradition here than from other points of view.
- The size of the country (and with it also the financial power) appears not to play as large a role as one would at first suppose. The results in Germany or in the Scandinavian countries towards Design Science is incomparable in this regard with those in the USA, the Soviet Union or Italy. In comparison, the output of results from computer tasks from the USA far outweigh any others.

Our considerations must remain in the hypothetical stage, because we do not intend to further explore these questions in any scientific way.

### 3.2.2 Development on Individual Planes

To obtain a more precise picture of the situation (a snap shot for a certain time), we would have to bring together on a time axis the current status and development, which runs on several planes, i.e. one would have to observe minimally the following partial areas:

- the development (and/or the situation) in research;
- the development (and/or the situation) in engineering design practice;
- the development (and/or the situation) in design education.

We have not done this systematically in our sketch, because the available material does not suffice for us to dare to make a founded statement. The biggest part of the information in our description affect the results of research, insofar as no other statements accompany a certain information.

### 3.2.3 Progress

To the sketch of the development in individual countries one can add also a description of the course of development:

- In the beginning in an organization or in an area, the rather isolated encounters have gradually changed with the time into a broader international movement, and the single problems have fused to a total problem of Design Science. Still the impression of fragmentation can emerge from individual work areas.

## 3.3 The Present State of Design Knowledge

In broad brush terms the state of design knowledge can be described (end of 1994) as follows:

1. Much knowledge has been accumulated, mainly however as "islands of knowledge," because too little synthesis was pursued. Many elements of design knowledge have striven for and approached the goal of the totality. Relationships between these elements have been inadequately explored. In the drive towards unification, only insignificant successes have been reached.
2. Within the design knowledge the available knowledge has not been prepared evenly.

   The individual areas were -- and are -- not evenly explored, which is, of course, a disadvantage, since the interrelationships do not emerge in the system of the knowledge. For that reason one must often return to make additional corrections to the existing conventional knowledge.

   Most frequently care was taken of the methodology of designing, because in the beginning it was a white spot on the map of general and design knowledge. There the achievements have been substantial. Some part problems crystallized either into partial tasks or even to principles for design methodology, for example:

   - Task to obtain a clear and actionable description of design procedures (the results of which can then be demanded by management);
   - Task to obtain a finest possible structuring of the process with clear separation of individual activities, especially the ones with special character, for example:
     -- searching for and finding solutions;
     -- evaluating;
     -- information activities;
     -- representing, etc.;
   - Principle of working out as many alternatives as possible and their optimization in the earliest tasks of the design process;
   - Task of adjustment of the general procedure to different elements, for example:
     -- for work in groups (team work);
     -- for the application of computers, etc.;
3. The quality of knowledge is not uniform, and extends from experience knowledge up to precise statements.
4. Because of linguistic and conceptual barriers (especially between cultural and language regions), understanding and agreement is difficult and has not reached a satisfactory degree.
5. The way into engineering practice has not yet been found.

These efforts are in large part little known or acknowledged, or their importance is underestimated. Symptoms for this are the typically human outlooks, in English usage occasionally summarized as NIH ("not invented here," thus allegedly not useful) and NIMBY ("not in my back yard," because even the thought appears often too dangerous).

*Part II.*

# Model of Design Science

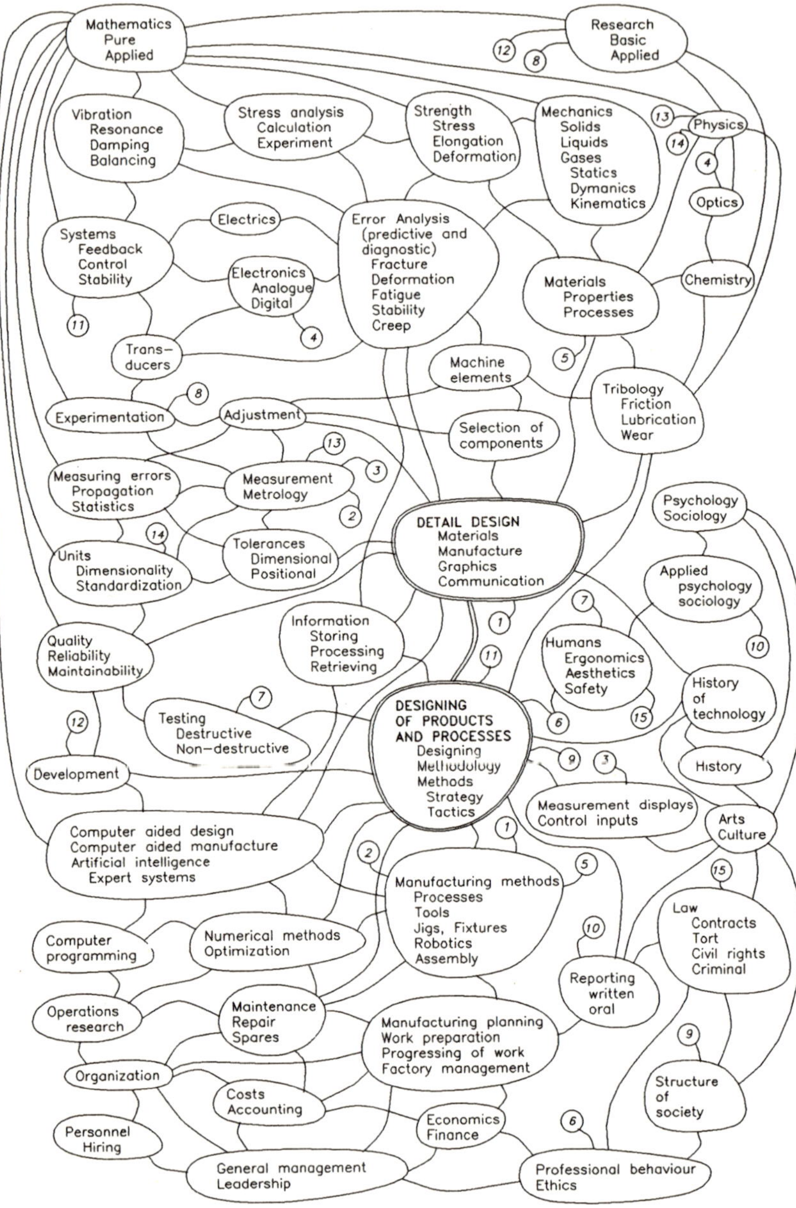
Mathematics Pure Applied
Research Basic Applied
Vibration Resonance Damping Balancing
Stress analysis Calculation Experiment
Strength Stress Elongation Deformation
Mechanics Solids Liquids Gases Statics Dymanics Kinematics
Physics
Optics
Chemistry
Systems Feedback Control Stability
Electrics
Electronics Analogue Digital
Error Analysis (predictive and diagnostic) Fracture Deformation Fatigue Stability Creep
Materials Properties Processes
Trans-ducers
Machine elements
Tribology Friction Lubrication Wear
Experimentation
Adjustment
Selection of components
Measuring errors Propagation Statistics
Measurement Metrology
Tolerances Dimensional Positional
Units Dimensionality Standardization
DETAIL DESIGN Materials Manufacture Graphics Communication
Psychology Sociology
Applied psychology sociology
Information Storing Processing Retrieving
Quality Reliability Maintainability
Humans Ergonomics Aesthetics Safety
History of technology
Testing Destructive Non-destructive
DESIGNING OF PRODUCTS AND PROCESSES Designing Methodology Methods Strategy Tactics
Development
History
Measurement displays Control inputs
Arts Culture
Computer aided design Computer aided manufacture Artificial intelligence Expert systems
Manufacturing methods Processes Tools Jigs, Fixtures Robotics Assembly
Law Contracts Tort Civil rights Criminal
Computer programming
Numerical methods Optimization
Reporting written oral
Operations research
Maintenance Repair Spares
Manufacturing planning Work preparation Progressing of work Factory management
Organization
Costs Accounting
Structure of society
Economics Finance
Personnel Hiring
General management Leadership
Professional behaviour Ethics

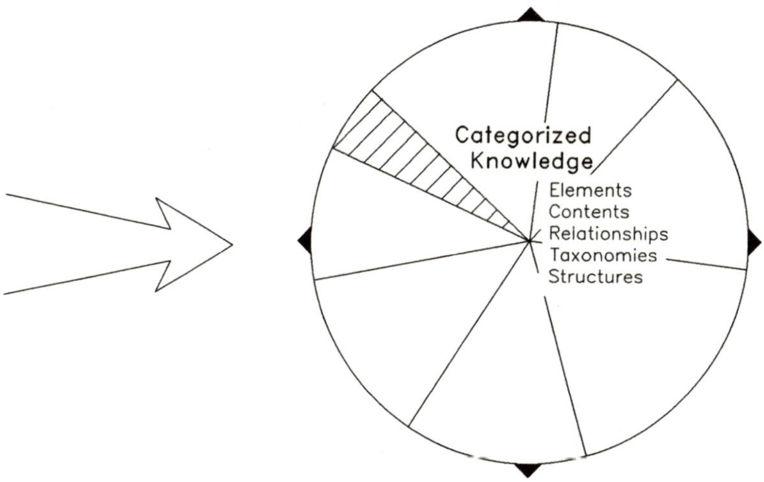


*Transition from scattered to categorized knowledge*

# 4 DESIGN SCIENCE AND ITS GOALS

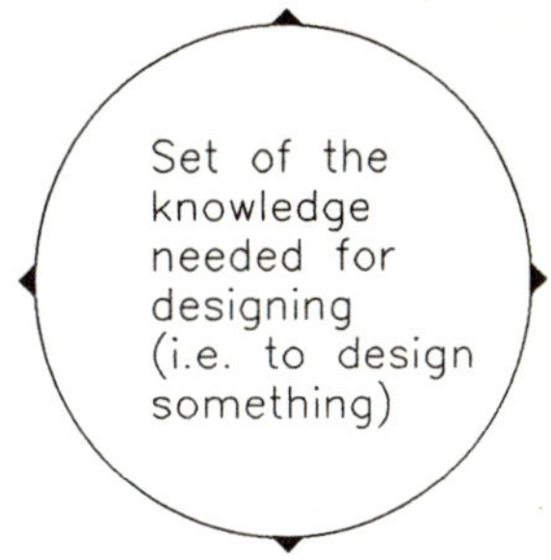


*Scientists don't ask questions very loudly until they can see the beginnings of an answer.*
***Peter Medawar***

One of the declared purposes for this book, as illustrated in the title figure to this second part, is to develop and present a proposal for a coherent and comprehensive view of knowledge about engineering design.

Several schemes could be used to interpret the knowledge about engineering design, one of these was presented as an informal narrative in Chapter 1. This knowledge should help to explain what designers do. It should also suggest and develop ways in which the knowledge about engineering design (the design process and the objects being designed) can help to make the practice of designing more rational, and to improve the products resulting from designing, both in the quality of the objects and in more rapid and rational procedures.

The aim is to get useful groupings (taxonomies), an overview of the information -- a structure. Defining a useful structure is a first step in preparing a scientific rationale for a field, in this case the field of engineering design. A structure should consist of its proposed elements (the groupings of knowledge), an outline of contents (necessary characteristics), relationships among the elements (internal to the knowledge base) and to the outside (to related knowledge) -- it should reveal a system. A system, in turn, should allow some investigation and prediction of behavior. This applies to the behavior of the system itself, and of the (human, technical and natural) participants involved with the system.

It is worth noting that no phenomenon or its behavior is completely predictable, no branch of science is ever exact, complete and comprehensive (refer to the quotation in Section 3.1). Some phenomena and behaviors, and their branches of science, are more predictable than others ([131,360]).

The "unordered" side of the title figure to this part already presents one possible form for collecting and classifying some of the typical information needed by engineering designers to perform their work of designing a product. Designing is at the center, and many interrelated and relevant topics surround that set of tasks. These topics could be further collected into larger groupings (a more abstract level in the hierarchy), which may typically be labeled pure science, engineering science, deliverables, information, measurement and quality assurance, computational and computer, manufacturing, management, social, cultural, human interaction, psychological (and several other) systems. Most of these sub-systems refer to partial knowledge with respect to the product and its material/physical context, the object. Procedures, methods and design processes are apparently embedded into designing.

We propose that there are advantages in using a different set of classifications, as developed in the following chapters. This involves differentiating designing (the procedures, methods, methodologies, activities, tools, etc.) as one aspect, from the product (its requirements and constraints, properties, life cycle, behavior, development, etc.) as a different aspect. We claim a major advantage by being able to relate various processes of designing, on one hand with various kinds of models of the product being designed, on the other hand with progressive development and refinement of models of the product itself.

## 4.1 What is Design Science?

In Chapter 1 we analyzed designing as the sum of activities of the designer. In this analysis of the activities, the task of the *design process* itself has crystallized as a core. This process transforms the information from the initial state of needs, demands, requirements, and constraints into the final state, a description of the designed technical system (compare figure 1--1). If "the design process" is used instead of "the design activity" (designing), an important reason is that these terms are not identical, they must be consciously differentiated. The activity of the designer -- designing -- is only one of the possibilities for realizing the transformation of designing in the design process. Other possibilities exist -- at least besides designers also computers should be considered as transforming systems.

For designing (transformation of information), an inclusive and abundant quantity of information and knowledge is needed, in varying depths of understanding and penetration and theoretical underpinning. Some of it must be known in concepts and in analytical depth, other parts can exist merely as awareness. We cannot try to completely enumerate this knowledge, but only to imply the extent. The information system exists particularly of:

- knowledge (basic knowledge) in different extent and for different operations of designing, taken from the individual engineering sciences like strength, materials, and manufacturing technology (and certainly also mathematics, physics, and others);
- knowledge from the branch and specialty of the relevant product family with regard to functions, mode of operation, organization, maintenance, etc., and especially guideline (heuristic) values for these parts;
- knowledge of typical components (constructional and anatomic elements), of which the more complex units are or can be built;
- knowledge of the production possibilities, and delivery of raw materials, semi-finished goods, purchased items from other manufacturers, and OEM (original equipment manufacturers') supplies;
- knowledge of procedures, methods and techniques, which can facilitate the design work;
- knowledge of representation techniques and models (and their relationships and possibilities for transformation among them), which make possible the communication and description of the ideas;
- knowledge of organization, management and administration techniques;
- knowledge of the possible working means, which support or take over fulfilling the design tasks, including libraries, handbooks, calculators, drafting equipment, computers and workstations, up to partial or full automation;
- knowledge of standards, regulations, patents;
- etc.

In vain we look in the current literature for a single source to obtain all this information. The most inclusive answer can be found in an engineering handbook, for example Kent's, Machinery, Mark's, Perry's, Dubbel, Hütte, etc., which however always present only a fraction of the required information. They practically omit discussions of design processes.

In addition many knowledge disciplines must be consulted.

Another problem is that only a part of this knowledge exists in written form. A further part exists in experience, in mostly unprocessed and less accessible forms, e.g., internalized by a designer. Individual designers must under these conditions continue to set up and maintain their own information systems and include them in their design knowledge, abilities and skills (Eder [126]).

That this design knowledge is decisive for the quality of the resulting products (Hubka [223]), is a separate question. It surprises perhaps that the collection and evolution of such knowledge was left to each individual designer. In education and the subsequent practice, only some elements of all we have enumerated here were (and are) transmitted to designers, without presenting them the total, organized into a knowledge system. The reason for this was simply that such an entire knowledge system did not yet exist.

Only in the last decades have the efforts for the solution of this problem and the construction of such a knowledge system become visible.

**The term *Design Science* is to be understood as a system of logically related knowledge, which should contain and organize the complete knowledge about and for designing.**

For science, the tasks have increased, especially in comparison with the earlier, usage-related, occasional collection of knowledge (empiricism). In contrast to empiricism (disorderly knowledge experience) science sets itself the *goal* to attain the totality of knowledge, to organize it according to a set of principles (Kant), and to recognize the reasons and establish the causes of the things (Aristotle) [383,384].

Design Science must therefore explain the causal connections and laws of the area (in this book the area "designing of technical systems") in its whole breadth. The knowledge system must be fixed in the forms of its terminology, classes (taxonomy), relationships (including inputs, throughputs and outputs), determination of measures, laws, theories and hypotheses, so that it can serve as basis for consciously guided design activity.

The contents and the boundaries of Design Science should be well considered, so that it forms favorable combinations with the already existing engineering sciences as the necessary system. In each case Design Science should explain the complete structure and the theories on which it is based. With Design Science the connections of the individual knowledge elements is made clear. Further discussion follows in Chapter 5.

## 4.2 Is Design Science Necessary Today?

The critical question about whether Design Science is indispensable can be regarded as justified from the historic context.

The reasons why Design Science is today so current and necessary can be found especially in the new situation of the marketplace:

- needs increase constantly with time, not only with respect to quality and quantity, but also with demands on shortened time from ordering to delivery of technical systems;
- the pressure of competition rises tremendously, and will continue to grow in future in the connection with the political developments (e.g., reduction of trade barriers, internationalization);
- a new situation emerges also from legislation, for example regulations with respect to quality, quality assurance, safety, environment (and danger of pollution) and especially

legal liability for freedom from danger and liability for quality, which also demand that sufficient design records are kept;
- the raw material situation also changes and can generate particular demands.

These changes force the enterprise, especially however the designers, to utilize new methods and modes of operation in order to remain competitive. The computer is a useful tool towards this aim, but it is only one tool among many.

## 4.3 To the Term Design Science

Such a knowledge system is not yet uniformly named. Besides Design Science we find terms like “philosophy of design” (design philosophy, similar to natural philosophy), “design theory” (as used in the USA, e.g., by the National Science Foundation), “scientific design” (Hansen), “design instructions” (Pahl and Beitz), or “design studies” (Design Research Society, U.K.). These names are not suitable to indicate the total knowledge of designing. In the first place they were chosen to designate only a part of the design knowledge. Secondly the intent was not to replace “science” with any of these terms. Science needs not only systematic descriptions (declarative knowledge, descriptive statements which belong in the area of the theory), it also needs methodology, instructions for the practical activity (procedural or prescriptive knowledge), and/or (deterministic and flexible) algorithms and techniques for part processes and operations.

## 4.4 Goals of Design Science

The way towards Design Science did not historically lead straight to an entirety, but first to different uncoordinated elements, see Chapter 3. The common superior goal and ideal which experts pursued was, however, clear:

*The situation in the design area is to be improved, and the existing problems are to be eliminated.*

This could be reached either directly through certain measures in the practice or indirectly through better education. Three areas of objectives therefore crystallized out, by which the direct effect can be reached:

1. *Goal area **practice***: direct improvement of the situation in engineering practice, i.e. in a company or an organization group, or in a project, with two addressees:
   - directly to designers;
   - indirectly in the enterprises (as the superior system);
2. *Goal area **science***: Answering of scientific questions, as for example in a research project or a dissertation;
3. *Goal area **education***: Improvement of design education in the schools.

The results in these areas have turned out very differently, but in their final effect they were all aimed at improvement of engineering practice (the long-term goal is the first goal area). Certainly in this way no compact Design Science could emerge and be fashioned. Developments always take place simultaneously in two directions: from practice towards abstractions, and from theory towards application. A further research goal had to be set to bring these directions to a useful convergence. This goal is to create an organized knowledge system for the design area within the area of science (second goal area).

We must recognize as a premise, that the task of Design Science itself cannot be limited only to the delivery of technical-scientific information for achieving the most important technical properties (as was the predominant case in the engineering sciences). Design Science should also offer complete information about the design process, i.e. information about models, procedures and methods with which all properties of the technical system can be established and realized in optimal quality, and also how they can be implemented to perform the design process economically and quickly.

The following partial goals can be derived from the above:

- generation of optimal quality of the technical systems to be designed, with respect to all of their properties [140,223];
- reduction of the design times; decrease in the costs of the design process; control of the responsibilities and down-stream commitments [50];
- reduction of the risks for designers and the enterprise;
- decrease in the proportion of human routine work in designing and thereby formation of space for creative work;
- increasing the attractiveness of design work for talented engineers;
- shortening the complete education (maturity time to competence) of designers;
- creation of the knowledge base for computer application in designing;
- capturing and recording of all available information (including experience) from the enterprise and formulation of a specialized information system;
- the design process should become scientifically transparent, and the bases for designing technical systems, be it without or with computer support, should be improved;
- the support of learnability and teachability of designing is an explicit goal in the education area, and should provide a suitable foundation for learning the further knowledge necessary for designing (including motivation).

Also the demands of the theory of science must be fulfilled, particularly:

- the science may contain only statements (and systems of statements) which are objectifiable (expressible in linguistic forms) and can be understood by the experts;
- these statements must also be examined continuously through the practice, whereby their quality can be constantly increased.

## 4.5 Acceptance of Design Science

Whether and how far the Design Science presented here is accepted depends on many circumstances. Prerequisites on the part of the recipient are a certain self-motivation, openness to newer outlooks and insights, and sufficient and suitable prior knowledge.

This Design Science has up to now only been explored for internal regularity and consistency through cautious processing. Temporarily no other test has occurred for the full Design Science, therefore no experience exists, the pattern of this book is the result of this first processing. Some part areas were already introduced some time ago, and they have withstood examination as separate parts.

Design Science should deliver a suitable taxonomy, and good references to further processing, for research in the areas of design theory, methodology, ways of thinking, establishing of working means (e.g., computer support), etc. For education some resources are already available, but the existing experience is limited. The designer in engineering practice can be helped first, if this theory is expanded by modifying the knowledge and providing better instruction. Even then, the largest usage is foreseen in two of the conventional (management-related) processes of designing -- in *clarifying the problem*, formulating a design specification, and in the design stages of *conceptualizing* where innovations should be generated in the task field. It is during these stages that the most important innovative, creative and holistic thoughts can bring the greatest benefits for the product to be designed. The more closely the task comes to changes in the detail of the solution, the smaller is the contribution which Design Science is likely to deliver.

It must also be clear that many psychological barriers are present, which work against acceptance of the newer research and design methods in engineer practice [311]. Modification of the knowledge and provision of instructions and education is only one of the necessary requirements for acceptance, others must also be fulfilled.

# 5 CONCEPTION OF DESIGN SCIENCE AND ITS METHODS

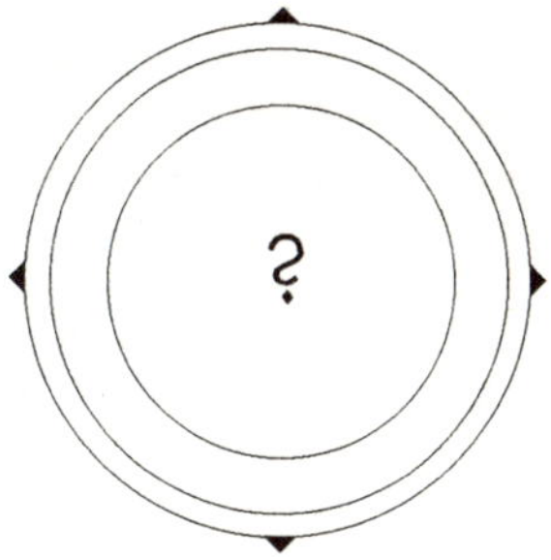


*I am aware that as long as anything remains in the area of magic, it cannot fruitfully be used.*

***Virginia M. Satir**, p. 269, **Satir Step by Step**, Virginia Satir & Michele Baldwin, Palo Alto CA: Science and Behavior Books, 1983*

Design Science has set itself the goal to explore the design process as generally as possible, and to organize, store and reference all knowledge for and about designing. To create a possible and near-optimal system, two important decisions must be made, namely:

1. about contents, elements, terminology, etc. and thereby about the boundaries of the system; and
2. about its structure, internal and trans-boundary relationships, taxonomy, etc.

## 5.1 To the Conception of Design Science Regarding Contents

As has been shown, extensive knowledge exists which can be designated as *traditional design knowledge*. Its extent cannot be exactly defined. Especially unclear is whether individual knowledge classes belong to designing, or not. Designers know also that for designing they need only certain parts of the existing knowledge, such as from strength of materials, manufacturing technology, and other areas. Also, these parts are often only obtainable in a form which is not directly suitable for designing. The existing knowledge must be sorted out and revised.

Beside that, much knowledge about technical system and designing is missing, we generally call these *completions*. They deal especially with design knowledge, as was shown in Chapter 3.

From these considerations, four possible fundamental concepts for the contents of Design Science emerge, each with completely different extent:

1. traditional knowledge plus completions;
2. selections from the traditional knowledge plus completions;
3. revised selections from the traditional knowledge plus completions;
4. completions only.

The first concept includes all engineering sciences into Design Science, which would be neither meaningful nor expedient. It would also be excessively extensive and tangled.

Design Science does not want to be a competitive discipline, but wants to deliver a framework discipline which allows and encourages inclusion, transfer or reference of the relevant knowledge from other areas.

In the second concept the extent improves, if only common knowledge is selected. Even so, the form of the knowledge does not yet correspond to the questions asked by designers.

Only the third concept can fulfill the objective of delivering relevant knowledge for designing in suitable form.

The last concept, in which only the completions form the content of Design Science, is attractive, because no contradictions emerge for the existing order. In addition, research of the design process appears to be the right research task. In this sense, F. Hansen [187] conceived his Design Science. This view would be possible only if designers are assumed as the sole executing powers of the design process, i.e. only "design activity". No complete basis can emerge in this way to describe the general transformations of information as they occur in the design process, including those performed on computers.

Our model, as it is described in the following pages, rests on the third concept. That means: the contents of Design Science consist on the one hand of that knowledge selected and revised from the existing global knowledge, on the other hand from necessary "completions" according to the chosen structure (see Section 5.2). This knowledge system must form a logical and organized entirety, in which the individual elements mutually and synergistically fertilize each other and their effects.

This solution has the advantage that the order of existing science is not altered, yet a viable discipline which can fulfill the stated mission still emerges. An important consequence of this choice (compare also Chapter 6) is, that numerous and close relationships emerge to many other disciplines, especially to engineering science. We can speak also of an enrichment of engineering science through Design Science.

## 5.2 To Content and Structure of Design Science

Each system, and therefore also each knowledge system, can be structured in several ways. The chosen structure should offer the users in each situation a well-defined orientation, so that they can find the desired information quickly and reliably, comprising both facts (including theories and hypotheses) and explanations. The elements of the structure define at the same time the contents of the system.

For structuring Design Science, the morphological order of the Design Science statements (information) as shown in figure 5--1 can be used. The aspects (features) and their states of embodiment as contained in the morphological matrix, together with the accompanying text of the figure, delivers the explanations for the structure. The individual statements can be described with the help of the states of embodiment, and larger statement complexes (e.g., books) can at least be profiled. This orientation facilitates the work of the searcher, because works about designing often bring much juxtaposition, intermeshing and interpenetration of statements with completely different epistemological status (for instance unconsidered mixture and confusion of descriptive theory-oriented statements with prescriptive statements), and under circumstances therefore awake wrong expectations for the recipients.

For constructing the fundamental structure of Design Science, we choose the *methodological class of statements* as first feature, and the *object character* as the second feature. Thus we arrive at four fundamental classes of statements to reach the set goal. The essential parts of the contents are formed by two elements from the object character:

- knowledge about the design process and its operators (the design system);
- knowledge about the designed objects (process and real systems).

Each of these parts is divided according to the methodological class into descriptive (theory) and prescriptive knowledge (Note: please refer to the comments on terminology in the preface, especially concerning the word "descriptive").

Figure 5--2 contains the derived hierarchical structure and the resulting "map" of Design Science, as well as a comprehensive description of the individual classes. The assumed structural arrangement is certainly not the only one possible, as the variants we have already listed have shown. The other features of the morphological matrix (figure 5--1) can be used for additional structuring of statements and knowledge. For instance, aspect 2 -- empirical support -- can become the ordering feature for "scientific quality," and allows arrangement of the statements into pre-scientific, scientific and statistically-scientific statements. Such an analysis will form the emphasis in Chapter 10.

## 5.3 Kinds of Design Science

The other features of Design Scientific statement, as specified in the morphological matrix (figure 5--1) with their states of embodiment, contribute to the determination of contents and the structure of each Design Scientific essay or work. Fundamental orientation is thereby eased for the user (see branch, domain and professional knowledge, Section 9.2).

Two aspects are of particular importance for the purpose of the users: the recipient (feature 3) and the object extent (feature 5).

According to which recipient circle the work addresses, three fundamental kinds of Design Scientific statement systems emerge:

1. *design educational instruction*, destined for instruction of designing, design education;
2. *Design Scientific essays*, addressed mainly at scholars and teachers;
3. *design manuals and handbooks*, which should serve design practice.

From the content, from additional knowledge, and from other aspects, very different works can emerge which are adapted to the different recipient circles. For instance, design educational instruction assumes different contents, forms, and pedagogic and didactic order, depending on whether it is intended for a technical university, technical college or lower technical school. And a handbook or manual intended for design engineers should differ considerably from that for detail designers.

The extent of the object (e.g., technical system and its degree of complexity and family) decisively influences the contents of the work. We will distinguish two basic kinds of Design Science:

1. *General Design Science* (GDS), which deals generically with designing of technical systems (and/or machines or building systems). Naturally the statements are very general and abstract, but they have a very broad validity, as presented in this book. Scholars, researchers and teachers should have particular interest for this kind;
2. *Specialized Design Science* (SDS), which treat a concrete product family as designed object. The groups of technical systems (TS-groups) form a hierarchical order, which is also taken over for the Specialized Design Sciences (compare Section 2.3). The recipient circle of this group is composed of staff members in engineering practice.

| Characteristics of Statements of Design Science | | State of Embodiment A | B | C |
|---|---|---|---|---|
| 1 | Methodological Category of Statement | Primarily descriptive (d–statement) | Primarily prescriptive (p–statement) | Normative (n–statement) |
| 2 | Empirical Support for Statement | Pre–scientific (practice – experiences) | Scientific: singular 'understanding' | Scientific: inductive –– statistical |
| 3 | Recipient of Statement | Typically: novice, student | Typically: teacher, researcher | Typically: practitioner, designer |
| 4 | Aspects of Designing | The technical system | The process of designing | |
| 5 | Range of Objects as subject of Statement | Universal: all artificial real and process systems | Technical real systems = TS | Branch–specific reference objects of engineering: machine systems, building systems, etc. |
| 6 | Experience and Status of Author | Branch or discipline | Position in organization | Primary activity: –– practice –– research –– education |
| 7 | Declared Aims of Researcher | 'Automation' of (all or part of) a process of designing | Better empirical foundation for a methodology or method | Others |

Morphology of Statements of Design Science (Hubka & Schregenberger 1987)

(Any given statement of design science may be characterized by combining the appropriate states of embodiment from each characteristic. The 'states of embodiment' describe discrete or continuous scales for the individual characteristics. They should be interpreted as alternative solutions to the problem stated by the 'characteristic'; the format of the columns labelled (A), (B) and (C) is only convenient for referring to these states of embodiment, it has no other significance.)

The contents of the matrix can be characterized as follows (Hubka & Schregenberger 1987):

**Characteristic 1: Methodological (Epistemological) Category of Statement (Compare Stegmüller 1973a)**

(A) Descriptive statements establish certain circumstances (facts) as they appear, and/or explain and substantiate them. They express what exists, how it behaves and why it is or appears as it does. The real circumstances are often not accurately known, only the apparent symptoms of a phenomenon can be specified. Likewise, the causal reasons for a phenomenon can be obscure. Explanations can refer to past, present or future occurrences. In those systems that are largely non–deterministic (including for instance all socio–technical systems) the past cannot be uniquely explained, nor the future predicted. Every theory is a set (system) of descriptive statements (including some that take the form of 'laws') which govern a wider range of objects.

(B) Prescriptive statements declare (and give directions) how a conceived circumstance can (but not must) be actually realized (know–how). They are thus non–compulsory instructions for actions or for realizing the desired structures and functions by using appropriate means. A general 'program for action' is usually termed a method. A methodology for a certain branch of engineering contains its methods (as elements) and instructions for relevance and use of these methods (as relationships in the system of methodology), and implies thereby a number of d–statements about the object to be formulated. Some statements may be interpreted as being both descriptive and prescriptive, for instance the 'heuristic principles' (Bromme & Hömberg 1977), or principles (guidelines) for form–giving.
(C) By means of the normative statements the guidelines are complemented by enforcable rules that are prescribed for attention, including the laws of the land, and where applicable the standards and codes of practice.

| Figure 5––1 Part 1 of 2 | Design Science: Structure of Knowledge about Engineering Design |
|---|---|

## Characteristic 2: Empirical Support for Statement (Stegmüller 1973b)

(A) Pre–scientific is the term used for those statements that appear to be less systematic and consistent, or that cannot be (or to date have not been) tested or verified. Many declarations from the practice for the practice fall into this category of statements.
(B) Scientific: singular understanding is the term used for the social–scientific approach. It leads to acceptable practice–based hypotheses by means of case descriptions and in–depth analyses (typically for historical science and leadership theories). The action–research concept is a more recent method in this area.
(C) Scientific: inductive –– statistical is the term applied to the method of substantiation, and the research concept, used in the empirical (hard) sciences. According to the most recent scientific understanding, all knowledge is hypothetical in character, and must be continually re–examined and substantiated, especially for application (Koen 1986, Schurrman 1977). The statistical (inductive) support of practice–based hypotheses is widely used, but cannot deliver 'proofs' for the 'truth' of statements (Stegmüller 1973c).

## Characteristic 3: Recipient of Statement

We distinguish statements that are primarily aimed at interest groups such as:
(A) inexperienced persons, novices, (B) teachers/researchers, or (C) practitioners.

## Characteristic 4: Aspects of Designing

Statements about 'designing' (especially in engineering) can refer to two different phenomena:
(A) to the technical system in its various states of existence (e.g. from specifications, to concepts, plans, and hardware);
(B) to the design process in which the proposals for the technical system pass through successively developed states (forms), in stages that may be referred to as (a) generating a design specification, (b) conceptual design, (c) laying out, or embodiment design, and (d) detail design, or elaboration.

## Characteristic 5: Range of Objects as Subject of Statement

The statements can refer to larger or smaller classes of objects:
(A) socio–technical systems –– artificial concrete (real or hardware) or process systems;
(B) technical systems in general;
(C) branch–specific systems –– reference objects of engineering disciplines (mechanical, electrical, constructional systems, etc.) –– and product families, system component sub–groupings of varying complexity, difficulty of designing, etc. –– machine, machine element, sub–assembly, part, feature of a part.

## Characteristic 6: Author

All statements are substantially influenced by the experience and status of the author(s):
(A) in which branch or discipline they are active,
(B) which position they hold within their organization,
(C) whether they are primarily active in engineering practice, research, or education.
The author's profile is important when interpreting statements of design science.

## Characteristic 7: Declared Aim of Researcher

The general direction of aims that design scientists follow in their teaching/learning and research activities determine the view and treatment accorded to design–scientific phenomena. For instance, authors can aim (A) to automate a (small or substantial) part of the design process, (B) to provide an improved empirical foundation for a method or a methodology, (C) others –– e.g. to improve educational experiences of designing. It is therefore of supreme importance to consider the aims of the authors when interpreting their statements about engineering design and its scientific explanation.

Literature:


Bromme, R. & Hömberg, E., Psychologie und Heuristik (Psychology and Heuristics), Darmstadt: Steinkopf, 1977

Hubka, V. & Schregenberger, J.W., 'Eine Ordnung Konstruktionswissenschaftlicher Aussagen' (An Arrangement for Classifying Design–Scientific Statements), VDI–Z vol. 131 (1989) no. 3.

Hubka, V. & Schregenberger, J.W., 'Eine neue Systematik konstruktionswissenschaftlicher Aussagen –– Ihre Struktur und Funktion' (A New Systematic Order for Design–Scientific Statements –– its Structure and Function), in V. Hubka, J. Baratossy & U. Pighini (eds), WDK 16: Proceedings of ICED 88, Budapest: GTE u. Zürich: Heurista, 1988, Vol. 1, p. 103–117

Hubka, V. & Schregenberger, J.W., 'Paths Towards Design Science', in W.E. Eder (ed), WDK 13: Proceedings of the 1987 International Conference on Engineering Design, New York: ASME, 1987, p. 3–14

Koen, B.V., Definition of the Engineering Method, Washington, D.C.: ASEE, 1985

Schurrman, E., Reflections on the Technological Society, Toronto: Wedge Publ., 1977

Stegmüller, W., Wissenschaftliche Erklärung und Begründung (Scientific Explanation and Substantiation), Berlin: Springer, 1973a

Stegmüller, W., Theorie und Erfahrung (Theory and Experience), Berlin: Springer, 1973b

Stegmüller, W., Personelle und statistische Wahrscheinlichkeit (Personal and Statistical Probability), Berlin: Springer, 1973c


| Figure 5––1<br>Part 2 of 2 | Design Science: Structure of Knowledge about Engineering Design |
|---|---|

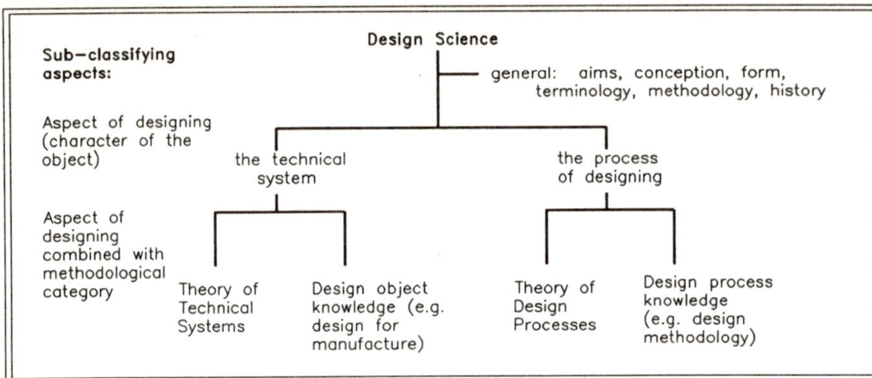


Diagram A -- Sub-Regions of Design Science

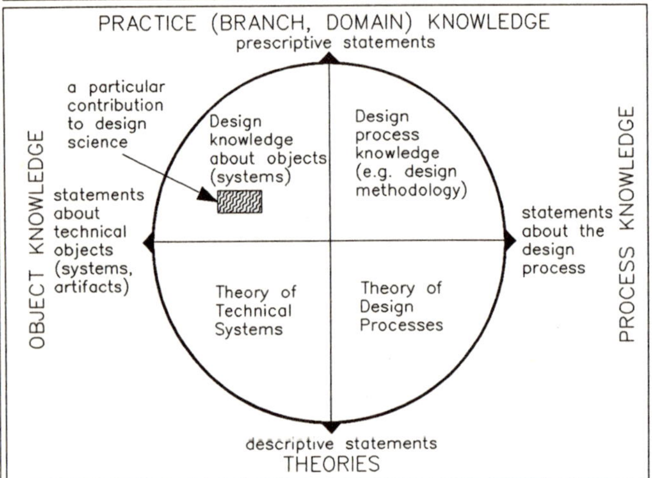


Diagram B -- Main Categories of Design Science Statements (Hubka & Schregenberger 1988)

The hierarchical structure of design science as shown in Diagram A is based on the two most important characteristics (numbers 4 and 1) from figure 5--1. In order to create a better and more memorable over-view, Diagram B shows a 'map' of those statements contained in design science (Hubka & Schregenberger 1988). The four sectors contain statements that can be classi-fied into the following main categories (Hubka & Schregenberger 1987):

-- <u>Descriptive statements about technical systems</u>: The 'Theory of Technical Systems' describes, explains, establishes and substantiates the structures, their elements, properties, modes of action, functions (etc.) of technical systems.

-- <u>Prescriptive statements about technical system</u>: The 'branch-related design know-ledge' (domain knowledge in its narrower sense) contains the know-how regarding ways of satisfying (realizing, achieving) the functions in a realized technical system, i.e. the knowledge about the <u>ways and means</u> in which technical products can and must be laid out and detailed in concrete forms to fulfill the required functions. The statements find their theoretical foundation in the engineering sciences, in the collected heuristic knowledge from practical experience, and by a further step in the pure sciences which provide the connections to the descriptive statements about technical systems.

-- <u>Descriptive statements about the design process</u>: The 'Theory of Design Processes' describes, explains, establishes and substantiates the elements, properties, sequences and effects (actions, results, successes) of actually observed design and possible processes in their socio-technical context, including all company-related, operational, organizational and leadership aspects.

-- <u>Prescriptive statements about the design process</u>: Design process knowledge con-tains indications about all operators of the design process. Of particular importance is the 'methodology of designing' (design methodology) which shows ways (methods,

Figure 5--2 Part 1 of 2 — Design Science: Structure

procedures, strategies, tactics) for successfully performing and managing design processes in an industrial context. This contains in its widest sense all the formal and ideal means of assistance (including methods) which designers can use in their tasks of thinking out (conceptualizing, inventing), representing (modeling), calculating, analyzing and evaluating designed systems (products and processes).

Literature:

Hubka, V. & Schregenberger, J.W., 'Eine Ordnung Konstruktionswissenschaftlicher Aussagen' (An Arrangement for Classifying Design-Scientific Statements), VDI-Z, vol. 131 (1989) no. 3.

Hubka, V. & Schregenberger, J.W., 'Eine neue Systematik konstruktionswissenschaftlicher Aussagen -- Ihre Struktur und Funktion' (A New Systematic Order for Design-Scientific Statements -- its Structure and Function), in V. Hubka, J. Baratossy & U. Pighini (eds), WDK 16: Proceedings of ICED 88, Budapest: GTE u. Zürich: Heurista, 1988, Vol. 1, p. 103-117

Hubka, V. & Schregenberger, J.W., 'Paths Towards Design Science', in W.E. Eder (ed), WDK 13: Proceedings of the 1987 International Conference on Engineering Design, New York: ASME, 1987, p. 3-14

Hubka, V & Ropohl, G., 'Was ist ein technisches System? Zur Grundlegung der Konstruktions wissenschaft' (What is a Technical System? To the Foundation of Design Science), VDI-Z vol. 128 (1986), no. 22, p. 864-874.

Stegmüller, W., Wissenschaftliche Erklärung und Begründung (Scientific Explanation and Substantiation), Berlin: Springer, 1973.

Schregenberger, J.W., 'Erfolgreicher konstruieren - aber wie?' (Designing More Successfully -- But How?), Schweizer Maschinenmarkt (1986), no. 23, p. 46-49.

| Figure 5--2 Part 2 of 2 | Design Science: Structure |
|---|---|

All these considerations which affect the kinds of Design Science are of interest for Design Science itself, and will therefore be treated in more detail in Chapters 8, 9 and 10.

In our book we want to present and discuss mainly the General Design Science as well as its influence on the specialized disciplines. Despite the central position of the technical system as general design object (the object treated in the design process, the object process and/or system being designed) with validity in all branches, the emphasis of the cited examples lies on mechanical engineering, the specialization branch of both authors of this book. These examples should not divert the reader from the generality of the statements, but to excite them towards application in their own field [242,386].

## 5.4 Principles for the Construction of Design Science

Beside the application of scientific methods, the quality of Design Scientific statements can be increased by consistent application of some principles. Thereby the contents of both the statements and the totality of Design Science can become qualitatively more homogeneous.

In particular, the following principles are needed, and can be considered as a certain procedural tactic (the hierarchical relationships among the principles are not considered):

1. acknowledged scientific principles (and methods) should find consistent application (especially those from the areas of science);

   Example: the thinking principles of Descartes can be called a scientific methodology, and can be consistently used (see figure 5--3).
2. the knowledge contained in all knowledge areas should be used to the maximum, including transfer of experience and adjustment of the required knowledge (in novelty and suitable order) even from relatively remote areas;
3. the principle of totality, and of systems science and technology should be utilized;
4. graphic models, as suitable language of the engineer, should find maximum application,

'And just as violations of laws can often be excused because of the large number of laws in force, so that a state is far better regulated if it only has a very few laws, but these are strictly followed, so instead of the large number of rules of which logic consists I believed that I could find enough in the following four rules, under the condition that I make the firm and persistent resolution to always follow them.

The first was: never to assume a thing as (axiomatically *) true that I would not recognize as true with certainty and conviction, i.e. to carefully avoid any undue haste and prejudice and to only understand so much in my assessments as represents itself clearly and distinctly in my mind, so that I would have absolutely no possibility to doubt it.

The second: to decompose every difficulty that I would investigate into so many parts as would be possible and desirable to better achieve a solution.

The third: to organize my thoughts: to begin with the simplest and easiest to comprehend objects and to advance gradually and in easy steps up to the recognition of the most complex, and even to hold those things as somehow organized of which by nature one does not precede the other.

And the last: everywhere to produce such complete enumerations and such comprehensive summaries that I would be certain not to have omitted anything.'

* ... addition by the authors

Literature:

René Descartes, Discours de la méthode pour bien conduire sa raison et chercher la vérité dans les sciences, Verlag Girardet
René Descartes, A Discourse on Method (transl. J. Veitch), London: Dent Books, 1912

| Figure 5--3 | Principles of Thinking by Descartes (1596–1650) |
|---|---|

along with verbal explanations and (where possible) with suitable mathematical-symbolic relationships;

5. a broad variety of solutions for individual problems should be offered;
6. thought should be consistently directed to human science and working techniques, and maximum application of technical means including computers, especially in the adjustment of technical means to the person and in exploiting human adaptability;
7. the order and form of the presented knowledge should correspond to the ways of thinking in designing, and should at any time be visible and surveyable (transparent);
8. the knowledge should be presented in clear and explicit forms, and therefore should contain no, or as little as possible, paradigm or pattern knowledge. The form should allow also deductive derivation for hierarchically lower levels of technical systems (as a Specialized Design Science);

   Example: If students should learn designing of machine elements ($ME_x$), then one can:

   - show how $ME_y$ is designed, and as task have them design $ME_x$. The transfer of knowledge between $x$ and $y$ can be shorter or longer, or:
   - explain how $ME$ are designed in general, then give data for the class $y$ and ask them to design $ME_y$.

   The second way of operating is the one which we recommend.
9. at all places shown, hierarchical ordering should be accomplished, and quantities, sub-quantities and hierarchical structures should be formed.

## 5.5 Methods of Design Science for Obtaining New Knowledge

The way of gaining knowledge and insights, starting from a hypothesis, which scholars usually pursue, is called a *method* in the areas of science. Hypotheses are scientific assumptions, which tend to complete and deepen the fragmentary empirical knowledge at a certain place, or join different empirical knowledge to a totality. "The essential function of a hypothesis exists therein, that it leads to new observations and attempts, whereby our conjectures are confirmed, refuted or modified; in a nutshell, whereby our experience is expanded," as E. Mach appropriately wrote in 1926.

As in other branches of science, here also scientific progress is achieved by a combination of:

1. proceeding from the whole to the parts (*analysis*),
2. progressing from the parts to the wholes (*synthesis*),
3. going from experiences and observations towards definitions (of terms) and causes, from the particular to the general (*induction*),
4. moving from the general to the particular (*deduction*),
5. transposing from a more concrete to a more abstract formulation (*abstraction*), and/or
6. establishing a more concrete reality from a more abstract formulation (*concretizing*).

always examining the one by comparing with the others. Thus the systematic advancement into the reality of designing (in breadth and depth) will have a totality and quality of knowledge as its consequence. These six procedures are of equal importance, none has precedence, neither in research, nor in application or education.

The actual historic way to Design Science was surely not fashioned so completely according to such schoolbookish plans, because their goal was set relatively late. The more likely process was that a start was made in the first phase from the knowledge in some areas, and only later was the model of General Design Science defined (induction), and at that time also the whole built up from available parts (synthesis). The examination and validation (verification) of the model occurred (and occurs) by means of analysis and deduction. Renewed induction and synthesis caused modifications to be made, which tried to eliminate the defects that were found.

The second phase was (and is) filling the model with specific knowledge, which is not available in the desired form at this time (compare Section 5.1 -- completions). On the other hand a broad knowledge basis exists as source, and is used as experience and "know-how" in engineering practice (see selection and revision of knowledge, Section 7.2.2.6).

One way in this second phase is presented by the *method of derivation* (deduction) of suitable prescriptive knowledge from the available theory. The requirement here is the existence of a theory. This rarely occurs, because usually first the practical, prescriptive statements are defined and used. Prescriptions are frequently formulated on the basis of experience, i.e. with little scientific foundation -- this happens in all but the newest technologies. Such prescriptive statements can indeed also be used as starting point for forming a theory, and serve by abstraction especially as basis for a hypothesis (induction).

The other way of filling the structural parts with knowledge starts from engineering science, by *transformation* of the available knowledge into a more useful form (form change, especially concretizing). If demands are formulated clearly, this method should find no serious problems. Other critical questions concern the quality and totality of the available knowledge.

A particular research area (also from the methodical point of view) is *design methodology*. The employment of special methods and hypotheses is demanded, especially when algorithms should be formed for computer application in the area of designing (therefore without directly utilized human thinking processes).

In the first phase, several attempts to define methods emerged on the basis of self-observations and experience (compare Chapter 3). The multiple layers of the problem and the originality of the object of the investigation (i.e. designing by persons), as well as the application in expanded breadth (industry branches) have led to the current large extent of methodology, and employment of special empirical methods. Only a few attempts have been made to synthesize the current knowledge into larger methodological units.

We must not forget in our judgments about the results, that most of the research was not conducted by specialized scientists, but by people who tried to process their own experiences (in very different specialties) by means of self-observation and different attempts to formulate a knowledge system. Thereby the methodological aspect was not given much attention.

Only in the later years, attempts were made to prove these hypotheses through strict scientific experiments, and especially to obtain higher quality by comparisons (e.g., ICED-Conferences, and later the NSF Initiative). One can speak of a second period, in which without doubt a further goal should exist for Design Science, namely to arrive at uniform methods, and to acquire and use also further methods (such as mathematical logic, semantics, model theory, psychology, artificial intelligence, and so on) more intensely in research on the design process (designing).

## 5.6 Relationship Between Method, Object and Theory

According to G. Klaus [249], close relationships exist between methods, the objects under consideration, and the basic theory, as figure 5--4 shows. The theory should describe and provide a foundation for both the behavior of the object (with adequate and sufficient precision), and the utilized methods. The method should also be sufficiently well adapted to the object. These three phenomena are of equivalent status to each other, none has fundamental precedence in time or direction of development.

A mutual interplay between object (and phenomenon), theory and method, one refined and examined on the other, characterizes the normal human and social development and progress, whose scale is the state of technology.

The current relationships among objects, methods and theories are extremely significant for the situation of heuristic methods. Quoting from Klaus:

*'Both method and theory emerge from the phenomenon of the object.'*

If the theory of an object–region is mature, then theory and method are definitely in such a relationship that the method is founded in the theory. The theory declares what is in reality the case, the method describes, on the basis of the declared facts, how the scientific and practical activities and behaviors of the humans has to take place.

Where no comprehensive theory is available, the relationships are of a different kind. Methods to deal with objects can be proposed even where the structure of the objects or their behavior is not completely known (cybernetic, newer interpretation). The method can conceivably have the character of an input–output–relationship ('black–box' principle, first formulated by Ashby in 1956). We know that corresponding results will be generated when we act on a system in a certain fashion. The theory will then –– and often with lengthy delay –– give an explanation of why this is so (to some extent an interpretation of the input–output relationship).

Schematic representation:

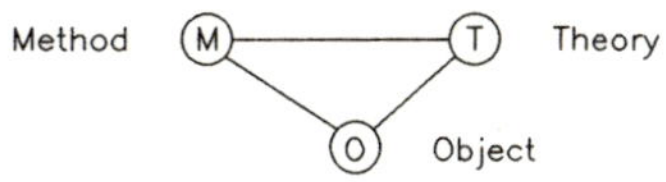


Relationships:
R(T,M)
S(T,O)
U(M,O)

For many currently interesting problem groups we lack an appropriate theory which can conclusively explain the method for its treatment and solution. In such problem situations, the method (frequently an heuristic method) must first serve to open up the problem field and disclose the structure of the problem.

This kind of problem situation is increasingly found in recent research efforts, and therefore the interest in heuristic methods is rising.

Literature:

Klaus, G., Kybernetik in philosophischer Sicht (Cybernetics in Philosophical View) 4th ed., Berlin: Dietz Verlag, 1965
Ashby, W.R., An Introduction to Cybernetics, London: Methuen Univ. Paperbacks, 1968

Figure 5––4 Relationships Among Method, Object and Theory

# 6 SOURCES OF THE KNOWLEDGE AND IMPULSES FOR DESIGN SCIENCE

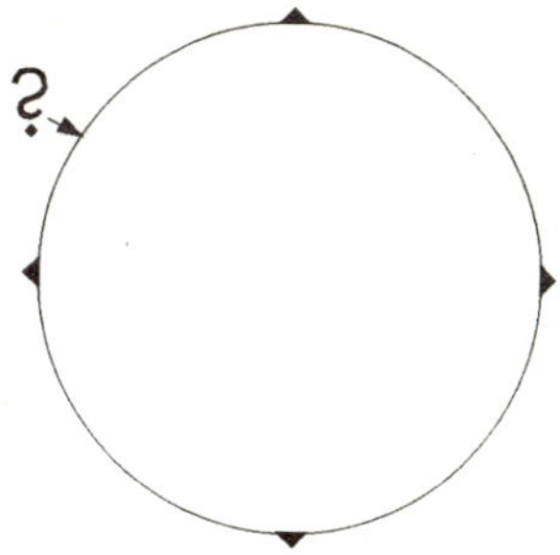


*All fact collectors, who have no aim beyond their facts, are one-story men. Two-story men compare, reason, generalize, using the labors of the fact-collectors as well as their own. Three-story men idealize, imagine, predict; their best illumination comes from above through the skylight.*
***Oliver Wendell Holmes, Sr.**, in **The Poet at the Breakfast Table**, 1872, p. 2*

*Who dares to teach must never cease to learn.*
***John Cotton Dana, Libraries: Addresses and Essays**, fac ed LC 67-22088, New York: Irvington, 1982*

Design Science has many pertinent relationships to other branches of science. It accepts and adapts much knowledge from other disciplines. The viewpoints to the solution of the design problem region do not emerge isolated from other scientific areas. Design Science is therefore not autonomous in its development, it and the other sciences mutually influence each other. Wc will select some important knowledge areas and mention their influences. References may be found in the bibliography, many of the reference numbers are quoted in other sections of this book.

## 6.1 Philosophy

As part of philosophy, the science of knowledge and heuristics has especially enriched Design Science. For instance, as a philosopher, J. Müller has used knowledge of these areas (compare "systematic heuristics" in Section 2.4.6). Ethical philosophy has also served as source for engineering codes of ethics.

## 6.2 Psychology, Sociology

Up to the present day, the human has practically always solved design tasks with the help of "purely" mechanical working means which usually have no decision abilities of their own. It does not astonish therefore if the insights from the psychology of thinking are used as basis for the development of design education, especially in the area of the solution methods. For the work in a group, the knowledge delivered by investigations of team work and group dynamics is applicable. Several works of psychologists are repeatedly quoted by design scholars and researchers, for example A.F. Osborne, H.R. Gordon, A. Newell, H.A.Simon, or

R.P. Crawford, see also Sections 1.4.3 and 3.1.7. Figure 6--1 shows a short survey of some such attempts.

Cognitive psychology has recognized several types of knowledge (compare [324]):

- declarative knowledge [44,45] -- factual information, *knowing what*;
- procedural knowledge [44,45] -- how to use declarative knowledge, *knowing how*;
- situation knowledge -- understanding where and when to access declarative and procedural knowledge;
- strategic knowledge -- knowledge processes that are planned and used to facilitate the use of knowledge;
- tacit knowledge [352] -- knowledge existing (in a possibly recallable form) in the mind.

The first category, declarative knowledge, is almost coincident with *object knowledge* as defined in figure 5--2. The next three fit well into *process knowledge* as we understand it. Procedural and situation knowledge could be combined as *tactical* knowledge.

The limitations of human capacity for mental processing of information must be considered for engineering design work [301,302,303,304]. This indicates that information is "chunked" by humans, and that an absolute maximum of nine chunks can be processed at any time in short-term memory.

An important partial area is educational psychology (the psychology of learning), represent in the literature by names such as J. Piaget, R. Gagné, J.S. Bruner, R.F. Mager, B.S. Bloom, D.R. Krathwohl and A. Melezinek. However, they have rarely referred directly to designing. Within Design Science it is the teachers of designing who deal above all with educational psychology.

Further exciting areas for design knowledge are the psychology of motivation research and of human interaction (team work, team building), which can deliver relevant knowledge for design managers and for design education. They also try to explain the partly justified resistance of designers in engineering practice against acceptance of systematic and structured methods [311].

## 6.3 Work Science

The knowledge of the study of personal "working technology" or the "technology of mental work" form an important source of supply for design researchers and scholars. The different general working principles, such as being methodical or the principles of validation and verification, can be completely taken over into designing [228,295].

## 6.4 Mathematics

G. Polya with his considerations about *How to solve it* has brought important impulses for scholars and researchers in Design Science, see also Section 3.1.2.1. The search for general laws in engineering sciences, but also in the engineering design field, have always considered mathematics as its most important tool. The increasing demands made on decisions and representations have always led to more intensive exploitation of the "old" as well as developing "new" methods of mathematics. Let us name here mathematical logic, set theory, combinatorics, statistics, graph theory, fuzzy quantities (fuzzy sets), cluster analysis, matrix methods, iterative approximation methods, and many others. Lately other

## Scientific Method

The classical analysis of the scientific method according to Dewey is explained in the following five steps:

1) Emergence of a difficulty;
2) Definition of that difficulty;
3) Emergence of a conjectured explanation and a possible solution;
4) Rational elaboration and development of an idea;
5) Confirmation of an idea and formulation of a pursuasive conviction.

Steps 2, 4 and 5 can obviously be regarded as rational and can therefore be executed by a computer. According to Dewey, step 3 is irrational and needs a different kind of 'black box'. This is exactly the kind of way in which a computer is different from a human brain, and why the human mind is the most appropriate tool for this third step.

## Incubation

According to Wallas (and attributed to H. Poincaré and J. Hadamard) problem solving can be supported by a procedure consisting of four stages:

1) preparation
2) incubation
3) illumination
4) verification.

In the preparation stage, the brain is programmed with all the facts relevant (appropriate, applicable) to the problem. Once the facts have been assimilated, the problem is dismissed from the forefront of the mind, so that incubation can take place at a subconscious level. A variety of other mental and physical activities should be undertaken during this time. The creative 'leap' may occur unexpectedly at any time. Illumination, in fact, is likely to strike at an unguarded moment, when mind and body are relaxed. After the leap the conjectured solution must, of course, be subjected to rigorous analysis and verification, if possible by experimental test.

## Hierarchy of Thinking Strategies

According to Wales, Nardi and Stager, the steps of problem-solving are:

1) establish (define) the situation (similar in form to a theatrical play):
   -- who is involved? (actors, players)
   -- what things are involved? (props, auxiliary pieces, scenery)
   -- what happened? (actions)
   -- when did it happen? (scenes)
   -- where did it happen? (locations)
   -- why did it happen? (causes)
   -- how serious is it? (effects)
2) state the goal
3) generate ideas
4) prepare a plan (or plans)
5) take action

The procedural plan sketched here is similar to the basic operations in figure 7--12. This sequence represents the lowest level of a hierarchy of thinking strategies, the higher lèvels are:

| | | |
|---|---|---|
| Purposes: | internal and external communication | --> serve each of the thinking skills |
| Skills: | creative and critical thinking | --> serve the thinking modes |
| Modes: | analysis, synthesis, evaluation | --> serve the thinking operations |
| Operations: | establish the situation, state goals, generate ideas, prepare plans, take action | --> help effective persons during problem solving |

Literature:

Gregory, S.A. (ed.), The Design Method, London: Butterworth, 1966
Wallas, G., The Art of Thought, London: Cape, 1926 (reprint 1931)
Wales, C., Nardi, A, & Stager, R., Thinking Skills: Making a Choice, Morgantown, WV: Center for Guided Design, 1986
Hadamard, J.S., Essai sur la psychologie de l'invention dans le domaine mathematique, Paris: Gauthier-Villers, 1975
Poincaré, H., The Foundations of Science, Washington 1982

Figure 6--1 Problem Solving in Psychology and Pedagogics

regions have been included within mathematics, as for example decision theory, optimization, operations research, chaos theory, computer algorithm developments, artificial intelligence and others, which also find application in Design Science.

## 6.5 Cybernetics

This area was introduced by Norbert Wiener's: *Human Use of Human Beings* [456]. Building on this basis, "cybernetic thinking" (particularly systems thinking) and knowledge from further areas (for example from the general system theory of G.J. Klir, 1969) bring impulses for many areas in Design Science, among other things also for modeling (for example G.J. Klir: *Cybernetic Modeling*, 1965). Another kind of needs can be fulfilled through information systems and data processing (making a connection to mathematics).

Also for design education, "cybernetic pedagogics" has importance (H. Frank, 1969). Cybernetics can also be associated with other applied disciplines, as for example "system engineering" (A.D. Hall, 1962, S.M. Shinners, 1967, among others). In connection with influences of technology and society, G. Ropohl (1979) developed the theory of socio-technical systems, which overlaps in some ways with sociology, see Section 3.1.2.1.

## 6.6 Information Science

The new discipline of Design Science confronts design researchers and scholars with new questions, especially when they try to expand the logic for "stupid" computers. The experiences with the computer applications benefit not only scholars, but also designers at the practical level. New ideas like "artificial intelligence" (AI) and "expert systems" (knowledge-based systems) have already taken root. Several developments are being commercially implemented (solid modeling, parametric modeling, attachment of other product data to computer models, common data bases for design and manufacture including management, etc.), or are under investigation with a view to using them for parts of design procedures, e.g., genetic algorithms, decision theory, and so on.

## 6.7 Management

The design process must be organized and directed in its progress. Management methods and insights (e.g. [447]) can contribute substantially to obtaining the right solutions. Management questions were discussed for example by D.J. Leech: *Management of Engineering Design* [272] (1972). Management was often discussed at the ICED-conferences, especially in relation to engineering design management and leadership. The discussions covered management and development of the products ("design" as a noun), and managing the design processes and their information ("design" as a verb). The methods QFD (Quality Function Deployment), TQM (Total Quality Management), simultaneous or concurrent engineering, product development strategies, etc. are also counted among the boundary region between management and design.

Several areas overlap between psychology and management. Team work and team building [396] is an important area for current management techniques and philosophy, and draws on psychology and sociology. The issues include control of conflict, obtaining mutual recognition and support, interchanging and stimulating ideas (e.g., by free association), and so on. The effects of individual personalities, psychological types (e.g., MBTI -- Myers-Briggs Type Indicator [83]), and learning styles also affects team operation. The danger must also be recognized that a "silly" idea can be accepted by a group because nobody wants to create a conflict. The effects of human behavior in general, especially their group actions, and resulting failures (e.g., "turf wars" and other power gambits), has been the subject of many studies, for instance in [68,338,339,340,341,342].

## 6.8 Invention Theory

Invention theory pursues very similar goals to some aspects of design methodology. The most important representatives were already mentioned in the historic survey (Sections 3.1.8 and 3.1.9), W. Ostwald, *The Theory of Inventing*, 1932, K. Backovsky (1963), and Altschuller (1969, 1984).

## 6.9 Other Design Fields

Several themes have been regarded as issues for research and development in fields of design not directly involved in engineering. Such fields contain architecture, industrial design, typography and graphic design, packaging and advertising design, but also Computer Integrated Manufacturing (CIM) as a management concept, design for easy and economic manufacture, and others. The titles obviously indicate some overlap of the subjects. A typical list of issues with sparse comments where a characterization is readily available to us, arbitrarily in alphabetical order, consists of:

- Agent-based Reasoning;
- Case-based Reasoning -- adapting the lessons learned from a previous design problem into a (computerized) knowledge base for guiding a new problem;
- Cooperative, Participatory or Collaborative Design -- allowing several designers to interact (even across long distances) with a design data base to modify sections that are their responsibility, including encouraging "conversations" in graphical and verbal forms; customers and users should also be enabled to take active part in the processes;
- Construction Robotics;
- Decision Support Systems -- mainly computer-based systems to help in evaluating and recording decisions;
- Design Kinematics and Animation -- allowing a time-scale dependent change of the graphic computer output to visualize the actions of a mechanism;
- Electronic Design Media -- using computers and their displays in similar ways to paper and other artistic media;
- Environmental Simulation -- modeling an environment to discover its behavior;
- Generative Systems;
- Geographic Information Systems -- generating and transmitting information of geographic nature, e.g., to advise motorists about travel routes, or to produce visualizations related to the exploration of the sea;
- Human-computer Interaction -- the ergonomic, procedural (understanding and learning) and psychological problems of computer interfaces to enable and enhance human work with computers, e.g., the values and disadvantages of "intuitive" interfaces with "point

and click" actions, compared to keyboard interactions;
- Information Delivery Systems for Design -- not just information storage and retrieval, but also interpreting and presenting in design-suitable forms;
- Knowledge-based Design Tools and AI -- capturing knowledge, especially human-resident and unstructured experiences and data;
- Model-based Reasoning;
- Modeling and simulation -- e.g., of marine ecosystems, or of behavioral scenes and artificial living structures;
- Multi-media -- storing information in verbal, graphical and oral modes, for display and play-back (including animation) by computers, and permitting retrieving by random access of selections made by the user;
- Neural Networks -- computer "learning" by deriving common patterns from case input, practically without abstraction or definition of any rules that may be discoverable;
- Object Oriented Integration between Databases;
- Optimization, especially mathematical and computer techniques;
- Precedence and Prototypes;
- Prediction and Evaluation;
- Rapid Prototyping -- computer-driven manufacturing of a solid object by depositing or hardening a material (usually a plastic) layer by layer, but other expressions are used, see the preface;
- Rule-based logic (e.g., "if--then");
- Scene and Product Modeling;
- Shape Grammar -- formulating a computer model to contain a class of shapes that can be modified by adding varying dimensions, and other properties;
- Shape Recognition and Emergence -- computer-automatic matching of a perceived (camera-formed) image with an existing stored image file;
- Simulation -- computational investigations of the behavior of systems, e.g., fluid dynamics, plastic material flow, heat transfer;
- Space Planning and Facilities Management -- computer-resident;
- Surveillance -- e.g., for marine or outer space environment, monitoring, remote sensing, etc.;
- Virtual Reality -- currently the state of the art of animation seems to exist as "virtual reality", where a person can visually "walk" through an environment (e.g., a building) that has only been built as a computer model;
- Visualization of data -- e.g., for observing marine data, pressure distributions, etc.

## 6.10 Summary

Figure 6--2 shows on one hand many of the impulse sources, on the other hand the role of Design Science as condenser, transformer and developer of information for education and engineering design practice, as well as for the areas in which the competence of designers ultimately is rooted. We consider in this diagram only information flows in one direction, from source to recipients, and neglect all the admittedly important interactions and couplings (especially also the feedback coupling) in this knowledge system.

Two important directions of information transformation are clear from the figure:

- Knowledge from some disciplines is concretized in Design Science: mathematics, philosophy, biology, sociology, physics, geometry, chemistry, cybernetics, psychology,

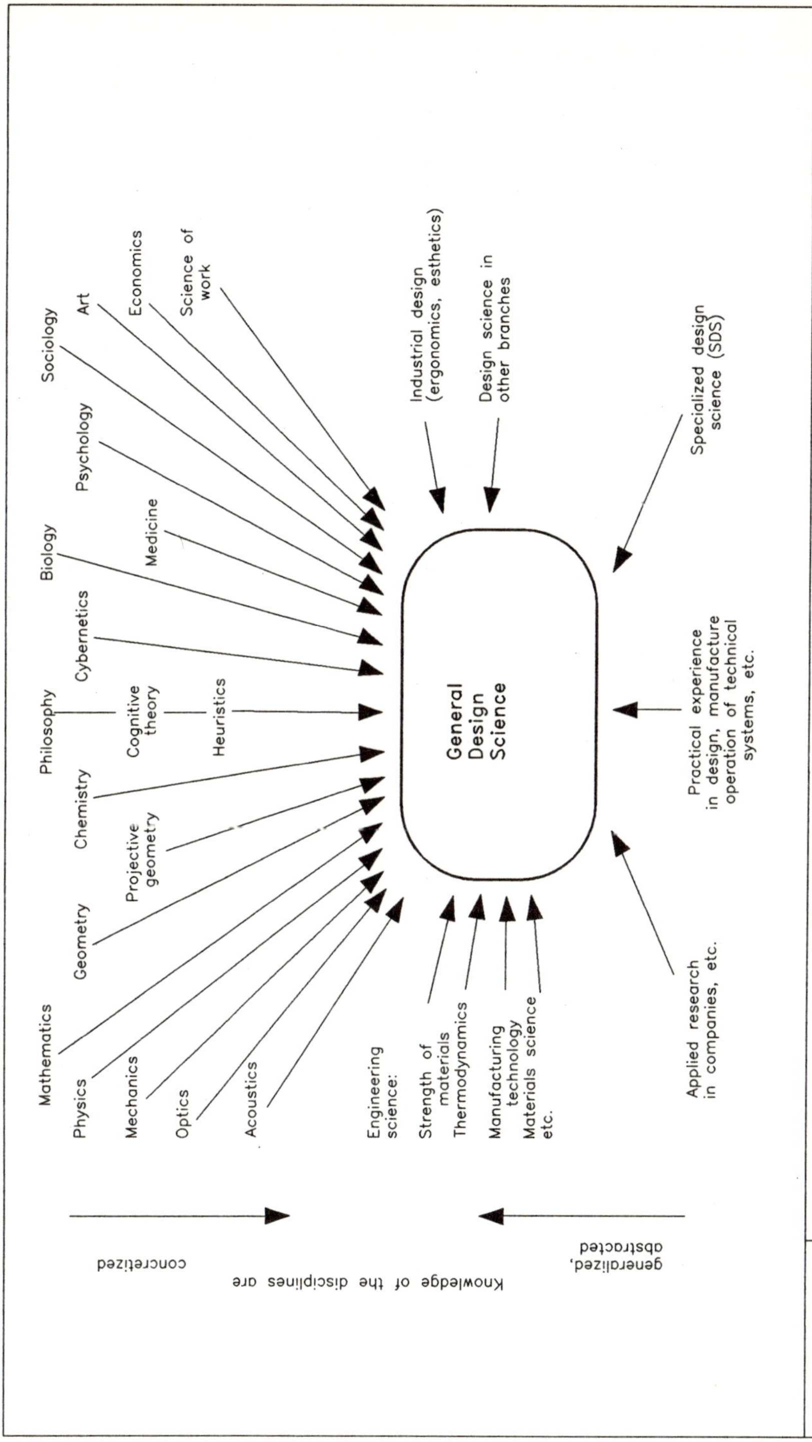


Figure 6--2 Information Flux: Sources for Design Science

art, mechanics, representational (descriptive) geometry, knowledge theory, medicine, economics, optics, heuristics, work science, acoustics, and many more;

- Knowledge from other disciplines is generalized: engineering science: strength of materials, thermodynamics, manufacturing technology, material science, etc.; but also industrial design, Design Sciences in other areas, applied research in service and practice, practical experience in design, production, operation of machine and other systems, Specialized Design Sciences; General Design Science in (mechanical) engineering.

# 7 CONSTITUENT AREAS OF DESIGN SCIENCE

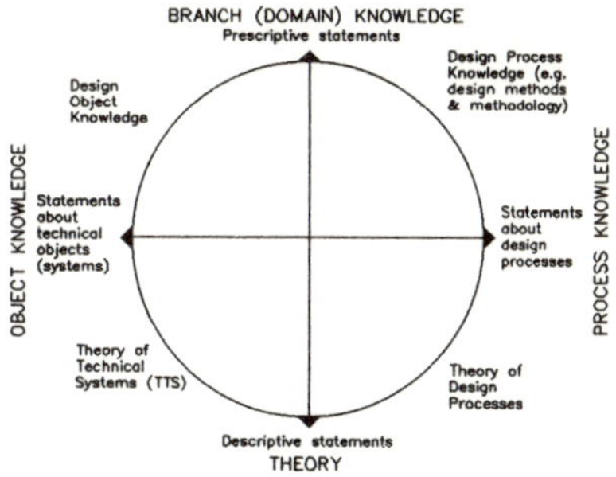


*According to our times and our experience we represent the natural and human world by a great set of images. To this set of images we apply, as a template, a system of hypotheses which seems to us coherent. The difficulty in scientific advance arises when some new experience necessitates a reassembling of the pattern of our images.*

*__Carl Frederick Abel Pantin__, in __The Relation between the Sciences__, London: Cambridge U.P., 1968*

*Theory and Practice are not antagonistic, as is so often tacitly assumed. Theory is not necessarily unpractical, nor Practice unscientific, although both of these things may occur ..... It is not a matter of merely setting forth in a new form and order that which is already known ..... On the contrary, if the new theory is to lay claim to general interest, it must be capable of producing something new; it must make problems solvable which before could not be solved in a systematic way.*

*Reuleaux, F., __Theoretische Kinematik__, Braunschweig: Vieweg, 1875*

*Reuleaux, F., __The Kinematics of Machinery__, London: Macmillan, 1876 and New York: Dover, 1963 (translated by B.W. Kennedy), Introduction*

In Chapter 5, we presented the structure of Design Science. Its elements were shown in figure 5--2 as the four partial areas of Design Science, which were derived from two selected features of Design Scientific statements from figure 5--1. In this chapter we will explore these partial areas, and formulate an accurate position of questions which should be answered by the corresponding area, so that the goals of Design Science can be reached. In addition the possible forms of representation are considered and the most favorable variants selected and justified. Thus the structures of the partial areas in which the technical knowledge is categorized and ordered can be established. The relationships among these areas and to the outside of the system will also be explored, where this is likely to enhance understanding.

We want to discuss only the fundamentals for these partial areas. The detailed representation of each partial area needs separate treatment. A part of this treatment on the next more concrete plane already exists, e.g., in [202,219,228].

## 7.1 Theory of Technical Systems (TTS)

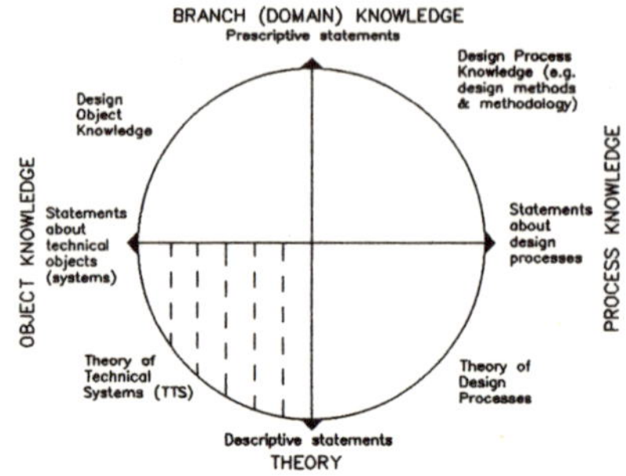


A partial area of Design Science -- TTS [214,219] -- should describe, explain and justify the object of designing (the technical system to be designed), from all points of view important for designing. These *descriptive statements* (i.e. statements of theory) affect primarily the transformation and the effects of technical systems on the operands, the mode of action (internal mode of operation of the technical system), component (anatomic) structures and constructional elements (components), different possibilities of modeling of technical systems, their characteristics, but also their origination and development. The meaning and importance of these terms is indicated in the guiding questions for each section.

### 7.1.1 Earlier Forms of Knowledge Systems about Objects of Designing

Such a knowledge system (the Theory of Technical Systems -- TTS) on this high plane of abstraction is placed above the knowledge areas existing today, which treat predominantly the individual TS-families. In figure 7--1 some typical works of earlier periods from this area are brought together, to show quickly (on basis of mainly mechanical systems) the diversity of this literature. Similar arrangements are possible and accessible for other areas of technology, such as chemical engineering, civil engineering (buildings, mines, and water construction), electric machines, electronics, etc. Already the titles show the variety of content, which emerged as a result of the addressed readership and their needs. Based on a detailed analysis of the contents of more than 50 books (including the titles in figure 7--1), the following repeating elements of content were found:

1. in-depth treatment of the necessary basis from physics or other knowledge areas for the relevant mode of operation;
2. working processes (modes of operation) of the relevant machine species;
3. pertinent branch (domain) information for the treated area, e.g., specific problems about strength, raw materials, control, thermo-fluid dynamics, among others;
4. treatment of the special components of the relevant machine species;
5. mention of the total systems, their application, structure, description, evaluation.

The individual works treat these elements very differently in both extent and depth. For example, the books on thermal turbo-machines contain predominantly knowledge about thermodynamics and fluid mechanics (therefore content element 1), but in other areas, such

1) Billington, D.P., Thin Shell Concrete Structures, New York: McGraw–Hill, 1982
2) Severns, W.H., & Fellows, J.R., Heating, Ventilating and Air Conditioning Fundamentals (2 ed), New York: Wiley, 1949
3) Gaffert, G.A., Steam Power Stations, New York: McGraw–Hill, 1952
4) Horlock, J.H., Combined Power Plants: including combined cycle gas turbine (CCGT) plants, Oxford: Pergamon, 1992
5) Driggs, I.H., & Lancaster, O.E., Gas Turbines for Aircraft, New York: Ronald Press, 1955
6) Norbye, J.P., The Gas Turbine Engine, Radnor, PA: Chilton Books, 1975
7) Armstrong, L.V. & Hartman, J.B., The Diesel Engine, New York: Macmillan, 1959
8) Wankel, F., Rotary Piston Machines, London: Iliffe, 1965
9) Norbye, J.P., The Wankel Engine, Philadelphia, PA: Chilton Books, 1971
10) Shapiro, H.I., Cranes and Derricks, New York: McGraw–Hill, 1980
11) Horne, D.F., Dividing, Ruling and Mask Making, New York: Crane, Russak, 1974
12) Moore, J.H., Building Scientific Apparatus, London: Addison–Wesley, 1983
13) Jeffries, W.R., Tool Design, New York: Prentice–Hall, 1955
14) Munton, R., & Stott, J.R., Refrigeration at Sea (2 ed), London: Applied Science Publ., 1978
15) Phillips, L.N., Design with Advanced Composite Materials, London: Design Council, 1989
16) Shigley, J., & Mischke, C.R., Mechanical Engineering Design (5 ed.), New York: McGraw–Hill, 1989
17) Spotts, M.F., Design of Machine elements (6 ed), Englewood Cliffs, NJ: Prentice–Hall, 1985
18) Tuttle, S.B., Mechanisms for Engineering Design, New York: Wiley, 1967
19) Boothroyd, G., Assembly Automation and Product Design, New York: Marcel Dekker, 1991
20) ––, Timber Design Manual, Ottawa: Laminated Timber Institute of Canada, 1972 (Imperial Edition), 1980 (Metric Edition)
21) Faires, V.M., Design of Machine Elements (4 ed), New York: Macmillan, 1965
22) Christianson, L.L., & Rohrbach, R.P., Design in Agricultural Engineering, St. Joseph, MI: ASAE, 1986
23) Wakefield, L.P., Dimensioning for Interchangeability, Oxford: Pergamon, 1964
24) Liggett, J.V., Fundamentals of Position Tolerance, Dearborn, MI: Soc. Man. Eng., 1970

Figure 7--1 Treatment of Families of Technical Systems in the Literature

as cranes, the emphasis of the treatment concerns the partial and total systems (therefore content elements 4 and 5).

The analysis shows that many important questions remained unanswered or appeared only as disconnected remarks, for example statements about operational properties (e.g., reliability, service life, suitability for maintenance), appearance, ergonomic and economic characteristics, or suitability for transport or packaging of the products.

Still more unsatisfactory is the situation for some works which the authors declare as being about design, and contain no (or only very sparse) instructions for the procedures of designing. Further relevant statements, for example about designers, methods of representation, working means, organization, administration (management) and similar questions are often ignored (e.g. [109,148]).

The branch books about individual disciplines of engineering can be arranged fundamentally in three classes, which display characteristic features:

- *theory applied to machines* analyzes scientifically the laws of transformations (work process as causal chain) within the machine (the technical system) and also some means and conditions for their realization. The books 5, 15, and 21 from figure 7--1 belong here.
- *general machine texts* treat the available machines or machine families by describing them and trying to show their purpose (teleologically). They are shown as different kinds and classes of construction, ways of building and composition. As examples we show the titles 3, 7, 8, 10, 11, 20 and 22 from figure 7--1.

- *machine know-how* presents in simple, clear form the knowledge about a TS-species. The reader needs no deeper prior knowledge, or it is presented in an uncomplicated way. The treatment in individual works is not equally fundamental. Examples are titles 6, 9, 12 and 18 from figure 7--1.

The boundaries between the individual classes are porous and not well defined, the contents of the books are not homogeneous, and the definitions admit broader boundary areas. These books are partly discussed in the "map," figure 5--2, in the left (western) half.

For us is interesting, which of the classes are suitable as a knowledge system for designing. We can designate especially the theory applied to machines (useful for analysis) as a work directed at designers (which does not exclude entrance for other professional groups). Partially the general machine discussions can deliver valuable branch information for designing, especially because of the different kinds of construction and the often good surveys. The machine know-how can under circumstances facilitate entry to a new specialty.

In summary we can state that the existing knowledge systems (especially books) that are available about objects of designing (i.e. the process and technical system) only deal with for some specialties, and have remained on a plane which is accessible to the sense perceptions. The processing differs too much in content, form and depth, and in any case it does not (or only in individual partial areas) cover the information needs of the designer.

### 7.1.2 Earlier Attempts to Build more Abstract Knowledge Systems in the Area of Technical Systems

The branch knowledge systems (see also figure 7--1) as described in the last paragraph on the plane of the families of technical systems (TS-families) are not the only predecessors of the Theory of Technical Systems. Some more abstract works also emerged, however very sparsely, for example the theory of the machines and mechanisms (TMM).

This theory, formulated by Artobolevski [60,61], was based on works of P. Willes, P.L. Chebychev and F. Reuleaux in the second half of the 19th Century. Since this time three areas of the theory of the machines and mechanisms have been developed: Mechanism synthesis (linkage synthesis), machine dynamics, and theory of automata. These form the contents of this discipline to date. It is obvious from this enumeration of contents, that this theory delivers the necessary branch information only in fragments, because it reacted only to problems of the moment. Whilst Reuleaux was active, there is no doubt that especially kinematics and linkages presented the main problems of engineering, just as in later times the problems were in dynamics and automation. They did not, however, offer the complex answer to the design needs, and in addition these discipline do not belong directly to Design Science, but in the engineering sciences. Even if the limits between these two knowledge branches, Design Science and the engineering sciences, are not completely sharp, the range of problems refers to the completeness of the contents for designing.

The inventory of general machine theory would be incomplete, if we did not present some attempts in the direction of a general and holistic consideration of the "machine." Three facts are characteristic:

- first the contact to cybernetics and especially to system theory,
- second the attempts to create new terms for the "abstract machine," and
- third to create a combination with designing.

In England the expression "engineering system" was already used in 1962 by Gosling [174] for several different and complex technical formations (compare Section 3.1.3). An exact delimitation, however, is not indicated. The presented theory is limited to the terminology, the fundamental systematics of the main ideas, and representation and modeling (including algebraic and topological models).

### 7.1.3 Contents of the Partial Area Theory of Technical Systems

The goal in this partial area is the creation of a complete system of descriptive knowledge about technical systems. The structure of the knowledge system on this high plane of abstraction should serve also as model of contents for generating the knowledge systems on low planes, e.g., for TS-species.

#### 7.1.3.1 Fundamental Terminology

We must first establish clear definitions of the most important terms as exactly as possible, and so successively develop the scientific language of the area (as distinct from colloquialisms). A wrong conviction exists that several terms are "clear."

Using the example of the term "machine" we can show that the assurance with which these terms are used is not justified. This can be discovered by referring to some encyclopedias, lexicons, branch books or essays, to look for definitions. If we remember that the definition has to present the term by enumerating the features (properties), the first condition would be that at least the classes of properties agree among the definitions. Unfortunately we can postulate generally that this condition is not fulfilled. This emerges also in reference to "machine" by examination of approximately thirty documents from six countries: the lexicons belonged both to the general areas and to individual specialties or branches, e.g. [1,8,12,13,14,16,248]. From these definitions it can be seen that altogether six classes of features are mentioned.

For the "machine," most (but not all) definitions mention the purpose, sometimes however with relatively inaccurate description (for example completion of different operations), but other classes of properties are given only in some definitions. This must generally suffice for the terminology. We do not want to burden this book with many definitions, therefore we will restrict discussions to the fundamental ideas (*termini technici*).

Without immediately presenting an exact definition of the term "technical system" (which would not be appropriate at this point), we will just assert that *technical system* is the *inclusive term* for all technical things (such as house, machine, telephone), until a further and more precise definition can be presented.

#### 7.1.3.2 Purpose, Task of Technical Systems

Question:
For what do technical systems serve, and what are their tasks?

The purpose and tasks of technical systems can profitably be derived from the "model of the transformation system." Figure 7--2 defines the fundamental terms, starting from following premises:

1. the broad needs of people are satisfied by certain states (and conditions) of persons, things, energy and information (within a given environment);

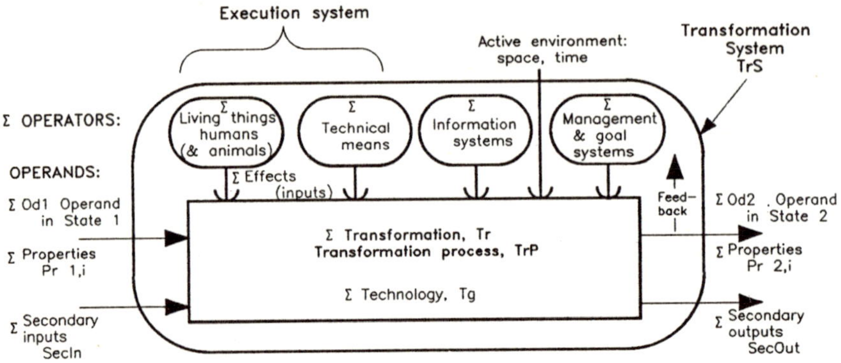


- Transformation (Tr) -- changes certain properties of the operand (as passive participant in the process). A transformation originates as a mutual interaction between the object being transformed, and means of causing the transfomation.

- Operand (Od) - WHAT is being transformed? -- object that is being changed in the transformation process (passive participant in the transformation) from an input state to a (preferably more desirable) output state.

- State -- sum (vector) of the values of all properties (Pr) of a system at a certain time. When observing a system, only the state of a selected group of properties is reported.

- Technology (Tg) - HOW is the operand being transformed? -- knowledge about the transformation, formulates the (input) effects needed to achieve the transformation.

- Effects (Ef) - WITH WHAT is it being transformed? Means of transformation -- effects acting (actions exerted) on the operand, including supply of the necessary energy, auxiliary materials, regulation and control.

- Secondary inputs (SecIn) -- (1) all necessary (desirable) additional inputs to the process, and (2) all undesired inputs (disturbances, contaminants, products of the environment, etc.).

- Secondary outputs (SecOut) -- mostly undesirable outputs of the process, their nature and composition depend on the chosen technology.

- Operators (Op) - WHO and WHAT delivers the necessary effects (as active participants in the process) to the operand:
  (output effects of operators = desired effects + secondary effects)
  Op (Hu) - living things, particularly humans, also (domestic) animals, bacteria, etc.;
  Op (TS) - technical (artificial) means, systems;
  Op (IS) - information systems;
  Op (M&GS) - management and goal systems (directing, setting and achieving goals);
  Op (AEnv) - active environment.

- Active environment (AEnv) - WHERE is the operand being transformed? -- takes part in the transformation (desired and undesirable effects).

- Space -- the main property of the environment (surroundings) of the transformation.

- Time - WHEN is it being transformed? -- time period during which the transformation occurs.

- Types of effects (acting on the operand), secondary inputs, secondary outputs, etc.:
  -- materials;
  -- energy;
  -- information (including signals).

| Figure 7--2 Part 1 of 2 | General Model of the Transformation System |
|---|---|

- Types of operand, objects being transformed:
  - -- biological objects (humans, plants, animals, etc.);
  - -- materials;
  - -- energy;
  - -- information.

- Transformation can change operand in:
  - -- structure (internal) = processing operations
  - -- form (external) = manufacturing operations
  - -- space coordinate (location) = transporting operations
  - -- time coordinate = storing operations

- Structure of the transformation process:
  - -- elements = operations (O), or groups of operations;
  - -- relationships = connections (and interactions) between the outputs of one operation (or group of operations) with the inputs of the immediately following operation (or group of operations);

  i.e. operations can take place sequentially (in series) or simultaneously (in parallel).

- Sub-classification (organization of elements, structure) in transformation processes:
  - -- preparing operations;
  - -- executing operations (the transformations proper);
  - -- finishing operations.

- Additional processes within transformation:
  - -- regulating and controlling processes = supplying signals
  - -- auxiliary processes = transforming auxiliary materials
  - -- propulsion processes = transforming energy
  - -- connecting and supporting processes = providing connections and support

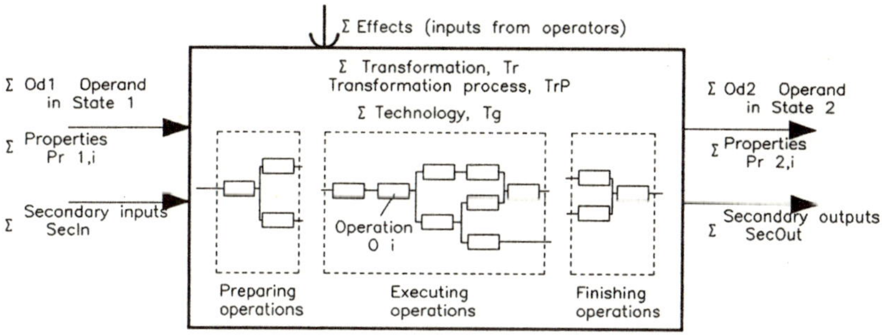


- Technical process (TP) -- transformation process (TrP) that is performed by using technical means (technical systems -- TS).

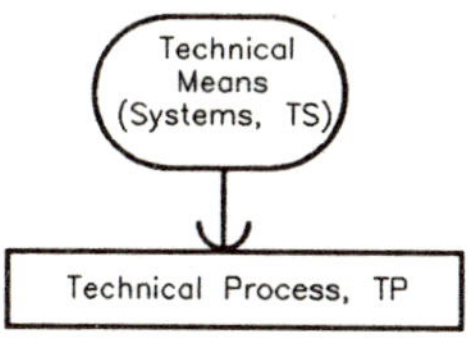


| Figure 7--2 Part 2 of 2 | General Model of the Transformation System |
|---|---|

2. the desired states (enumerated in 1) are usually not available, and are reached by transformation processes from existing states. Our life is marked by countless transformations.

The transformations take place in an (often complicated) transformation process. Such a transformation process can be subdivided down to *operations*, for which a further subdivision appears either not possible or not useful. Operations can progress in sequence (in series) or simultaneously (concurrently, in parallel). These relationships form one possible structuring of the processes. The outputs of an operation (or group of operations) are at the same time the inputs of the following operations. Matter and energy are subjected the laws of conservation.

From their arrangement, the transformations recognize operations (or groups of operations) of several types:

a) in an *operating sequence*:
- preparing (preparation phase),
- executing (execution phase),
- finishing (conclusion phase);

b) in *addition* to the working process (the desired main transformation):
- regulating and controlling operations,
- auxiliary operations,
- propulsion operations,
- connecting and supporting operations.

As shown in figure 7--2, technical systems feature among the operators of the transformation, together with humans, information, management and the environment. According to our way of modeling, the desired transformation of the operands proceeds outside the technical system. As an active member of the transformation system, the technical system acts on the operands by means of a certain *technology*, by delivering effects to achieve (drive) certain transformations.

Each technical system has therefore the task to deliver the desired effects (as outputs) at the desired time, and in a given environment, with a sufficient lack of sensitivity to secondary inputs (disturbances), and with an acceptable minimum of secondary outputs. This view of the technical system as *supplier of effects* brings new essential advantages for modeling and designing.

#### 7.1.3.3 Mode of Action and Structures of Technical Systems (TS - models)

Questions:

1. How can we achieve the effects (outputs from the TS) prescribed by the technology of the transformation?
2. How does a technical system work (what is its nature)?
3. Which kinds of structures does it display?

We expect therefore from the technical system a behavior, which consists of the sum of the expected effects: it moves, it protects, it warms. These effects demanded by the transformation are achieved by using certain established "natural" processes, which are executed within the technical system. Effects (e.g., warming) are outputs of natural processes (e.g., transformation of electric energy into heat). The processes give the object-systems (technical systems -- TS) a wholeness and unity, and determine the principles of their organization.

Effects can be achieved by different natural processes, e.g., heat not only through transformation from electrical energy (electro-processes), but also by means of burning of coal or oil (chemical processes), or through biological processes. The same effect (output function) can be realized in several ways, termed the mode of action of technical systems.

Each *mode of action* demands a certain structure which guarantees these internal transformations. According to the organization principle (construction principle) several kinds of construction emerge for the structure, which must all deliver the same output.

We see once again that the relationships between function and structure form the central problem of technical knowledge. And because designers decide about the mode of action and the organization principle, they decide also about the structure of the system. For that reason, Design Science must provide the necessary knowledge in suitable form for designing, particularly with respect to its structuring and content. In the design processes the effects are to be considered as *goals* and the structures as *means* to achieve these goals.

In connection with structure, another question needs to be clarified, namely what kinds of structures of technical systems can be distinguished. From the above explanations, the following kinds of structure can be derived:

- *process structure* -- what transformation processes take place *within* the TS; structure elements are the TS-internal processes;
- *function structure* -- what internal capabilities does the TS have (it only uses these capabilities when it it actually operating); structure elements are the functions;
- *organ structure* -- what active locations implement the capabilities; structure elements are the organisms and/or organs as function carriers;
- *component structure* (also morphological or anatomical structure) -- what physical (material) parts implement the organs; structure elements are the assembly groups, components (constructional elements).

These four structures, and an arbitrary number of further structures, can be represented for each technical system. Therefore a process, function, organ and component structure exists for each technical system, e.g., a heater for a building. Structures, from the point of view of origination of technical systems, are at the same time characteristic of individual origination stages, i.e. states (of existence) of technical systems, particularly during designing.

Relationships exist among these four types of structure. There is normally a coincidence between the TS-internal processes and the functions (capabilities) of the TS. Functions can be combined or divided as desired. A (more or less complex) function (as goal) can be realized by one or more organs, although several different organs can be found to fulfill that function. Each organ is usually realized by portions of several physical components.

All states of technical systems which accompany the path from the design task to the realization of the TS (therefore relationship goal -- means), are presented in figure 7--3 as corresponding TS-models. This way is often described as proceeding from the abstract to the concrete, which describes the nature of the process only in very crude outlines, and cannot suffice in any way as instruction for designing.

If we attempt to find relationships of Design Science to the other scientific areas, we must refer here to the method of morphology. Morphology (study of form after Goethe) is namely the study or knowledge of figures, forms, features, organization principles, particularly of living thing, but also of other systems (e.g., of historic and social phenomena, and of objects). It forms also a partial area of biology, as knowledge of the external

I)

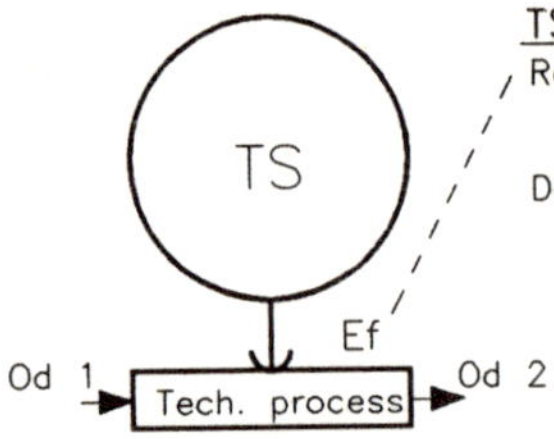


**TS Purpose** $\Sigma$ Ef $\rightarrow$ Od

Realizing the interaction between operand <---> operator, in order to achieve the desired transformation of the operand.
Delivered effects (Ef) conform to the technology used for the transformation.

Legend:
TS ... technical system
Ef ... effect
Od ... operand
Od 1, Od 2 ... states of the operand

II)

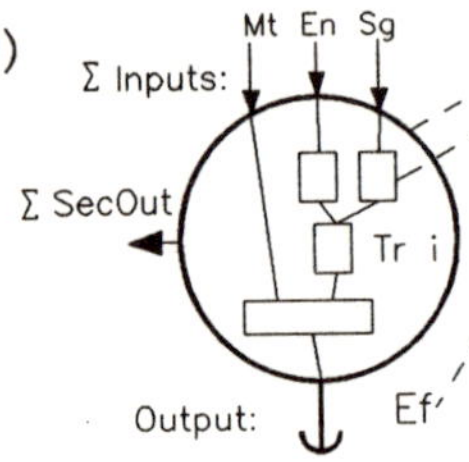


**TS Process Model -- Process Structure**

TS in working state -- operating.
Internal transformation of inputs to the TS (desired and undesirable) into desired outputs (Ef = effects) plus secondary outputs (SecOut = secondary effects).
Effect --> product (output) of the chain of internal transformations (Tr = mode of action based on laws of nature, e.g. physics, chemistry, biology, etc.).

Legend:
Mt ... material
En ... energy
Sg ... signal (Inf ... information)
Tr ... internal transformation
Tr i ... one of the n transformations of the process structure
SecOut ... secondary output, secondary effect

III)

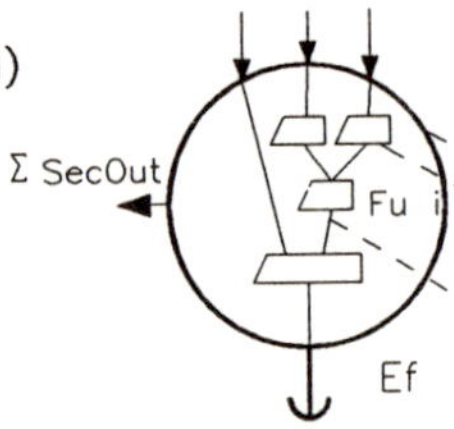


**TS Function Model -- Function Structure**

TS in state of being capable of working.
Function --> task (capability) of performing the TS-internal transformations, or permitting them.
Function corresponds directly to the internal process.
Relationship between functions.

Legend:
Fu ... function
Fu i ... one of the n functions in the function structure

IV)

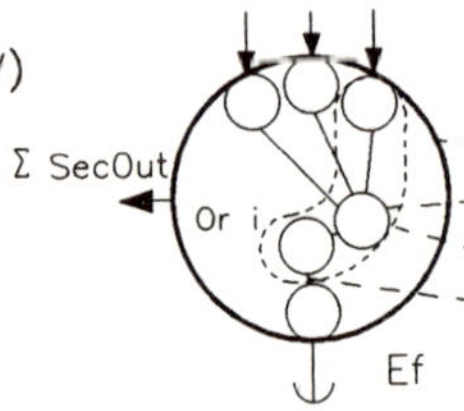


**TS Organ Model -- Organ Structure**

TS in state of being capable of working.
Organ --> means to realize functions. Main features of organs are the action spaces, surfaces, lines, etc.
Group of organs --> organism.
Relationship between organs.

Legend:
Or ... organ

V)

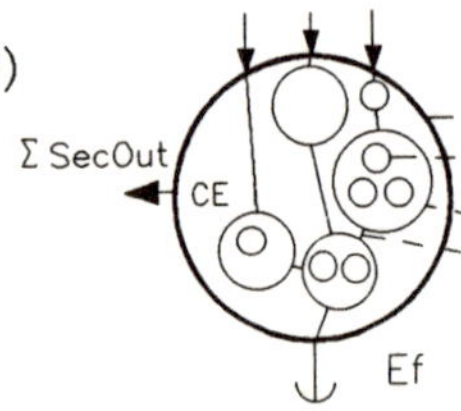


**TS Component Model -- Component structure**

TS in assembled state.
Constructional element (component) --> means of realizing organs.
Constructional group (sub-assembly, machine element).
Connection between constructional elements.

Legend:
CE ... constructional element

Literature:
Hubka, V., Theorie technischer Systeme, Berlin: Springer-Verlag, 1984, Abb. 5.4
Hubka, V. & Eder, W.E. Theory of Technical Systems, New York: Springer-Verlag, 1988, Fig. 5.4

| Figure 7--3 Part 1 of 2 | Models of Technical Systems (TS Models) |
|---|---|

Example: machine vice in operational state -- state of capability of working.

ABSTRACT MODEL — CONCRETE MODEL

I) PURPOSE

Design specification
Requirements specification
-- from customer, or potential user

Contract specification
-- company level

Designers' working specification
-- designers' approved interpretation

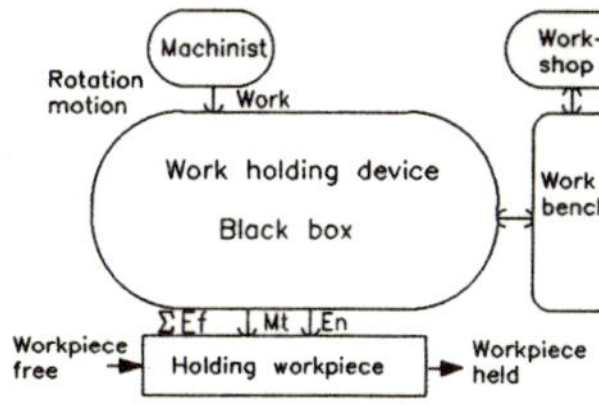


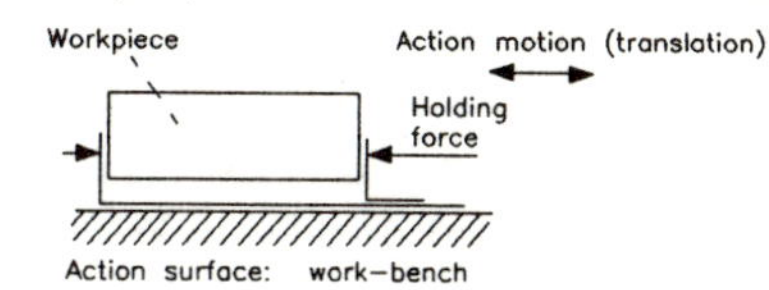


II) PROCESS STRUCTURE

not illustrated -- elements of the function structure and the process structure are in one-to-one correspondence, each function implements the capability to perform the corresponding process.

III) FUNCTION STRUCTURE

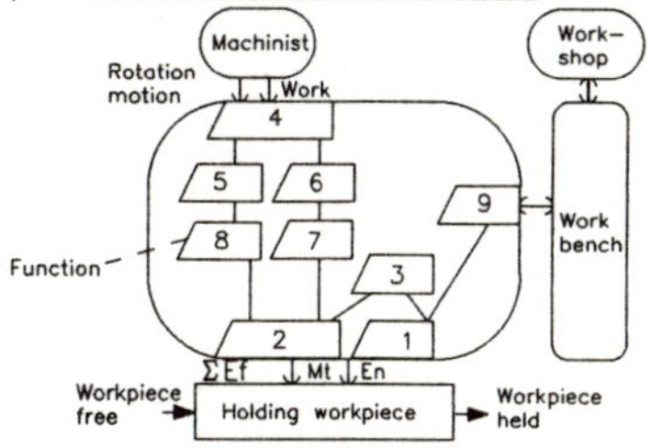


| | FUNCTION | ORGAN | |
|---|---|---|---|
| 1 | FORM fixed holding surface TRANSMIT holding action | Fixed jaw | FJ |
| 2 | FORM movable holding surface TRANSMIT holding action | Movable jaw | MJ |
| 3 | GUIDE movable holding surface rel. to fixed holding surface | Guidance for movable jaw | MJG |
| 4 | FORM the action surface ACCEPT work | Acceptance organ | AcOr |
| 5 | TRANSFORM rotation to translation | Motion screw thread organ | MST |
| 6 | AMPLIFY force | | |
| 7 | STORE energy for holding (strain energy) | Storage organ | StOr |
| 8 | RETAIN position against holding force | Motion screw guidance organ | MSG |
| 9 | FORM contact to fixed system PERMIT making connection | Connection organ | CcOr |
| 10 | Evoked addition: REACT radial force from work acceptance | Motion screw guidance organ | MSG |

IV) ORGAN STRUCTURE

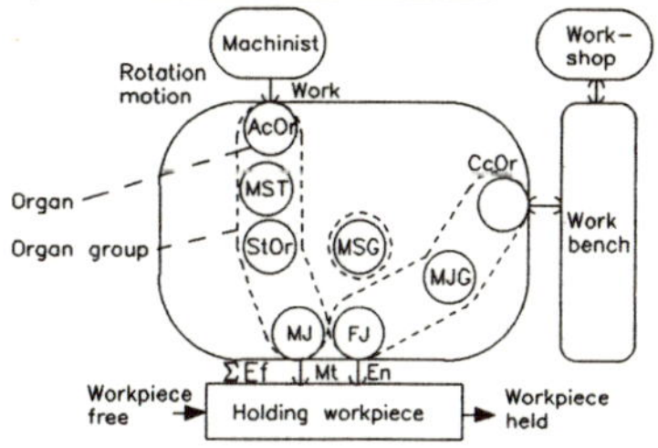


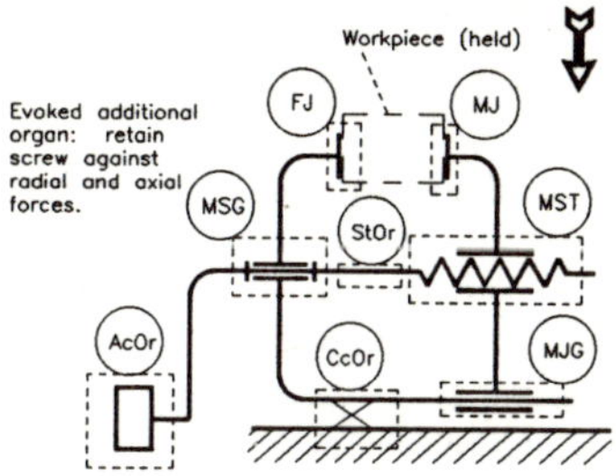


V) COMPONENT STRUCTURE (ANATOMIC STRUCTURE)

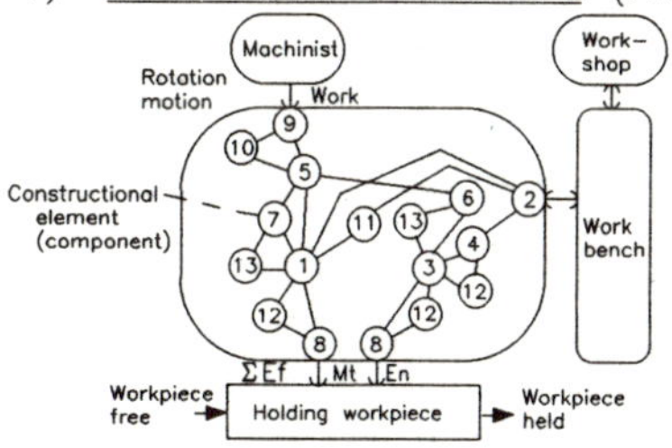


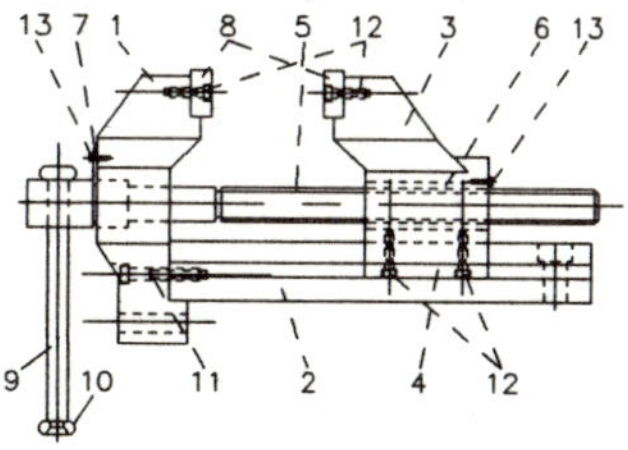

1 Fixed jaw
2 Base
3 Movable jaw
4 Guide plate
5 Motion screw spindle
6 Nut insert
7 Keeper plate
8 Jaw plate (hardened)
9 Handle
10 Handle knob
11 Screw
12 Screw
13 Screw

Detail drawings
Parts lists
Purchased components lists
etc.

Figure 7--3 Part 2 of 2 — Models of Technical Systems (TS Models)

construction (figure, shape, organization) of the organisms and their parts in the course of their development (ontogeny). The morphology of technical systems (following Zwicky [465]) must become an integral element of Design Science.

#### 7.1.3.4 Classification of Technical Systems (Systematics, Taxonomy)

Question:
Which meaningful systematic categorization of technical systems can be accomplished, such that all possible point of view (features, properties, characteristics) are considered?

Requirements for classification:

1. complete classification systems should be created;
2. the reason for each classification (ordering principle) should be captured (set -- sub-set);
3. the individual classes (taxa) should be mutually exclusive (from a particular point of view);
4. the division should be contiguous, breaks and gaps should be avoided as much as possible.

By classifying, an order emerges which facilitates orientation and overview in the area of technical systems. Exploiting the kinship of the elements mutually enhances the intelligibility and transfer of the knowledge about these elements. Order (classification) is an important principle of each science. A complete, hierarchical order in technical systems is essential, especially in design methodology or in expert systems, see for example the design catalogs [257,376].

The possibilities for classification of technical systems are numerous and diverse, especially because of the many available features, properties and characteristics of systems. For example, classes can be defined for function, complexity, mode of action, structure features, originality, etc. but also for manufacture quantity or recycling description.

The complexity of technical systems is the subject of Chapter 8 and shows the importance and possibilities as an instance for other ordering characteristics.

#### 7.1.3.5 Properties of Technical Systems

Questions:
1. Which properties or classes of properties does a technical system possess? (A complete list is required here)
2. Which relationships do individual properties or classes of properties have to one another?
3. When and how can the values of properties be established (parametrized) during the origination (designing) of a technical system?
4. If the mutual causal relationships among the characteristics are explored, at which "general causes" can we arrive?

– ***Fundamental terminology, fundamental terms***

*Properties* (attributes) of the technical systems are all those features which belong substantially to the object -- the object owns the property (it is proper to the object). The value of the property represents the measure of the property (size, condition, form of embodiment) in the concrete case. The value can be indicated quantitatively (*x* m/s -- number and unit of measurement), or only qualitatively (large, small).

The *value scale* forms a sequence of continuous or discreetly defined values. Continuous scales can be absolute, with a defined zero point, or relative, with an arbitrary zero point or none. Discrete scales can be selected values from a continuous scale, ranking (only defining larger or smaller), or merely "belonging" (to a set). The existence of scales with defined units presents the requirement for quantitative statements. The *total value* is composed of several values of properties to enable a total judgment (for example benefit or usage value).

### -- *Theory of Properties*

One of the most important parts of the Theory of Technical Systems is the theory of *properties*. Any technical system is built only because certain properties (including behaviors) are desired, used and valued. For example we do not in the general need "a system of steel profiles, which could form a bridge." The demands for the task are to create "a crossing over a river," one which has (possesses) a certain load capacity, functions (performs its expected duties) and is reliably safe and long-lasting. Additional demands are that it also has a pleasing appearance, is formed to comply with regulations and standards, and has a series of further properties. (Remark: a system of steel profiles with suitable connecting parts can be a *means* to fulfill this task).

The call to generate a complete and general list of all properties is old, but it is not easily satisfied. Some attempts to assemble such a complete list are known (see e.g., VDI R 2225 [26]), but they have not been successful. An estimate has suggested some hundreds of items in a such list. Such a large number of items would represent an impractical check list that is not useful for designing. The right way lies clearly in inquiring about a complete and general set of *classes of properties*. These classes can then be concretized for individual branches (domains), and sub-classes down to individual properties defined for a particular branch.

These classes of properties can be defined starting from the model of the transformation process (figure 7--2). Each process in the life cycle of a technical system (see Section 7.1.3.8) is a transformation that is driven by its operators. Figure 7--4 shows a matrix relating the individual processes and typical operations to their operators. Any technical system should possess properties that make it suitable for each operator in each life phase.

The theory explains that each technical system carries all the kinds of properties (see figure 7--5), especially those that make it suitable for a purpose. We distinguish between external and internal properties. The realized technical system owns (possesses) therefore all properties (features), whether they are deliberately planned, or have not been considered.

All properties (as fulfillment of needs, demands, requirements and constraints) must be established by designing. How is this possible? The fundamental and important knowledge is hidden in the mutual and causal relationships among the properties. Namely, the quality (value) of the internal design properties creates the quality of the external properties (compare figure 7--5, part 1 -- B). As a consequence of this relationship, designing can be seen as a *search for suitable design properties*. In the first place, the property of "structure" must be established (compare figures 7--2 and 7--3). The relationship (assignment) "function -- structure" (or in more general terms of "element -- structure") is presented in figure 7--5, part 1 -- C, and has been mentioned repeatedly.

Characteristic classes of properties can be defined to provide a complete coverage of the requirements which fill the matrix in figure 7--4. They can be merged into four collective classes, as shown at the top of figure 7--5, part 2:

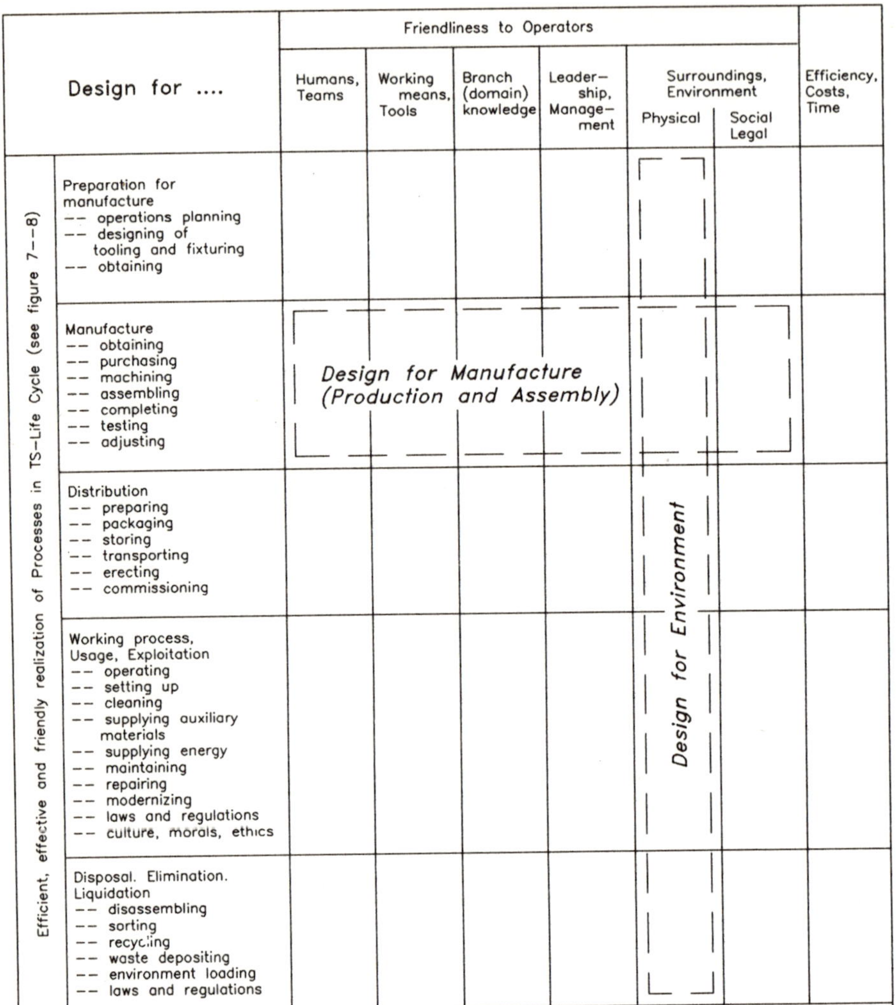


Figure 7--4 General Systematics of 'Design for ...' Classes

- classes related to the purpose of a TS, the life-cycle phase of using (exploiting) the TS (classes 1 and 2);
- classes that directly cover the other life phases (classes 3 to 7);
- classes that reference humans and society (classes 8 to 11); and
- design-related classes (class 12, the internal properties).

Figure 7--5, part 2, shows these twelve classes of properties, and examples of the properties contained in them. The boundaries between these classes are not well defined, each property can affect one or several classes. Typical relationships among the properties in these classes are shown in figure 7--5, part 3. Any class of properties, together with its

## A) External Properties (EPr)

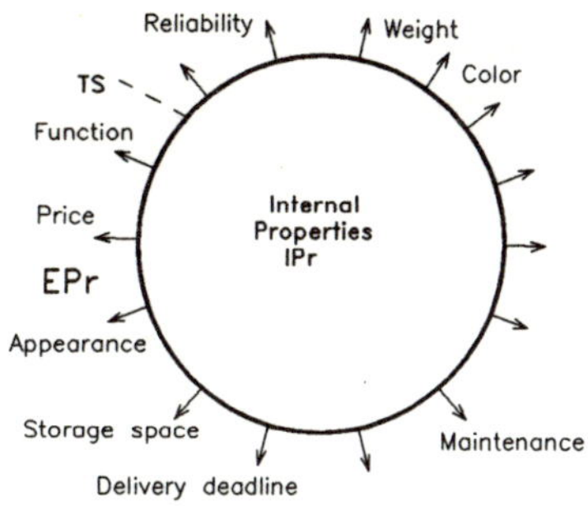


The technical system is carrier of all properties (Pr).

The external properties (EPr) which are observable and detectable by the user appear at the surface. The internal properties (IPr) detectable only by competent specialists are hidden in the interior.

THESIS: Each technical system is carrier of all kinds and classes of properties; the value and significance of the individual properties is different for individual kinds of technical systems.

Requirements (demands, needs, constraints) --> external (sometimes also internal) properties demanded or desired by customers/users/etc.

## B) Causal Relationship among Properties

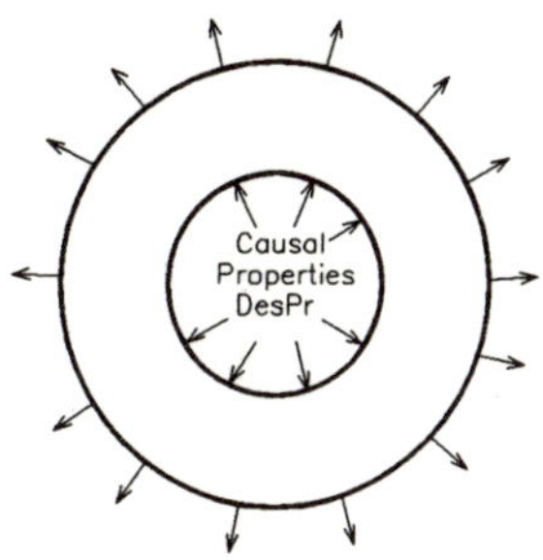


The 'causal properties' are responsible for all external properties of a technical system.

THESIS: All kinds and classes of properties are realized by means of the elementary design properties (DesPr -- structure, form, dimensions, tolerances, raw material, surface quality (finish), state of assembly, etc.).

In general: Pr(i) = f (some elementary design properties)
Example: Strength = f (raw material + manufacturing process, geometry, size, surface quality)

Designing --> Can also be interpreted as 'establishing the design properties for the purpose of realizing a technical system that meets the requirements.'

## C) Relationship (Mapping)

*Elements of a more abstract (higher) structure mapped to a more concrete structure (see figure 7--3).*

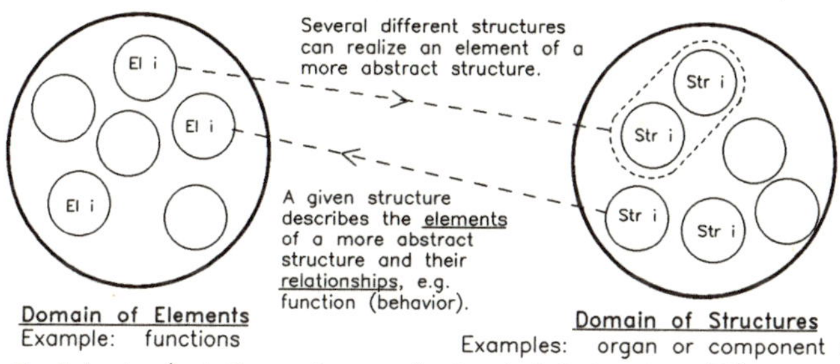


THESIS: The behavior (including performance) of a technical system (TS) is determined by the more concrete structure of the TS. All TS with identical structure possess the same kinds of elements of the more abstract structure, in particular the behavior is generalized by the elements.

THESIS: The behavior does not uniquely determine the structure. The same behavior (function and/or function structure) can be realized by several different organ or component structures.
Generally this is valid for each element of a more abstract structure in relationship to its more concrete structure, therefore also for variants in individual design properties within a component structure. This thesis is the basis for the law of formation of variants.

| Figure 7--5 Part 1 of 3 | Properties of Technical Systems -- Types |
|---|---|

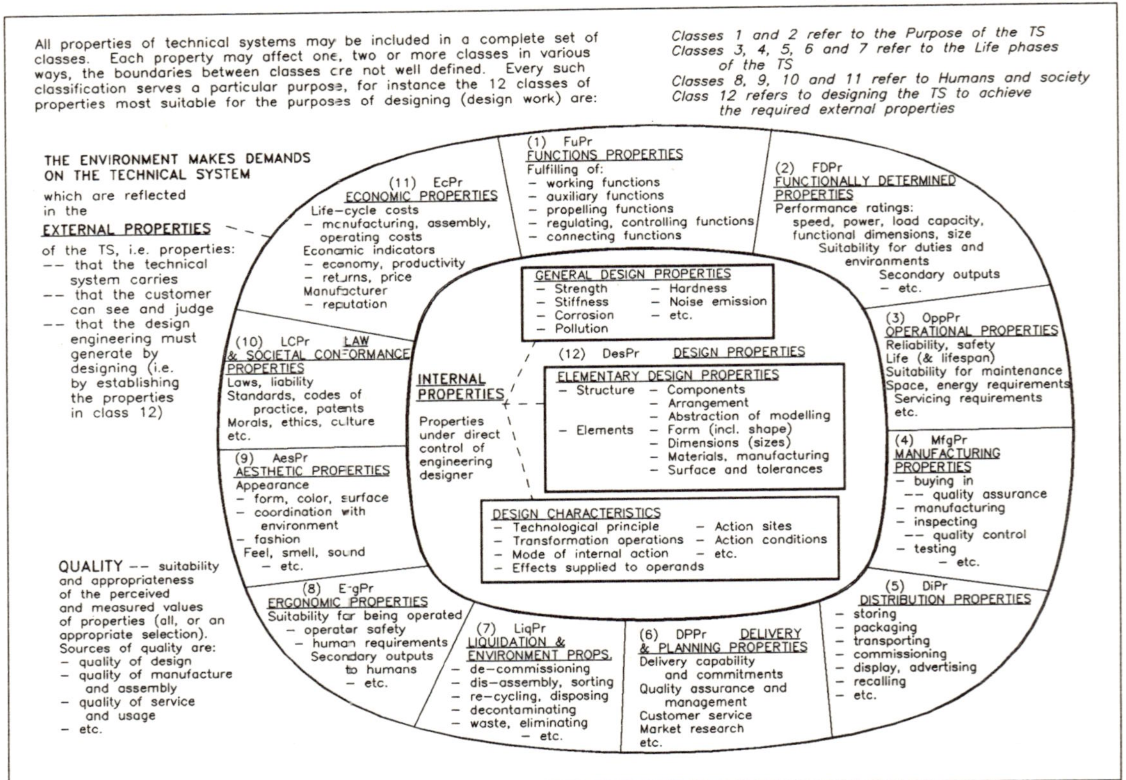


Figure 7--5 Part 2 of 3 Relationships among Properties of Technical Systems

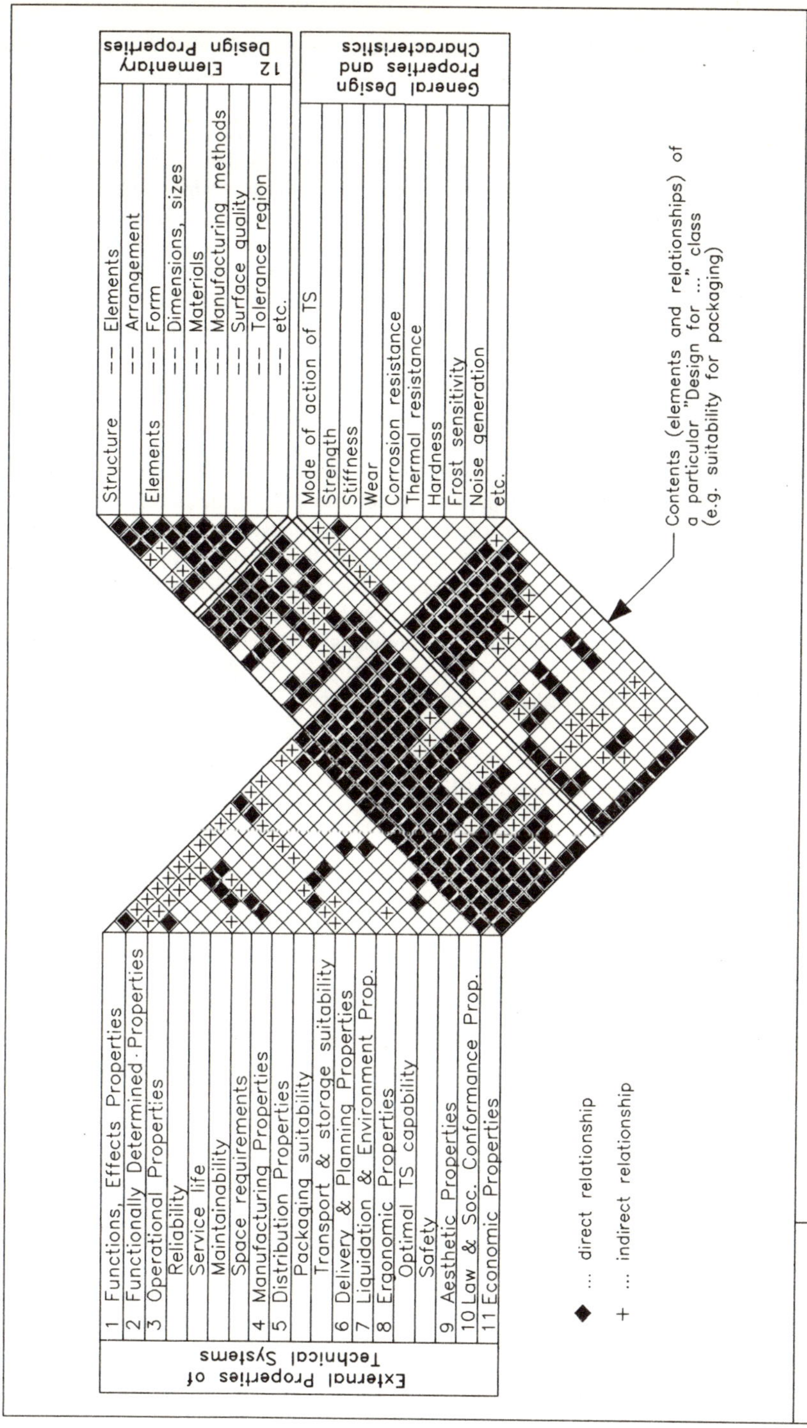


Figure 7–5 Part 3 of 3 Technical Systems: Survey of Classes of Properties and their Relationships

diagonal row of relationships to other classes, represents the sum of knowledge needed to be able to design for that property. This aspect is also indicated by the two typical areas shown in figure 7--4.

For instance, the property "beauty" of a system (a part of its esthetic properties) or safety (part of its operational properties) directly affects the operator "human." An initial influence acts on the makers of the product (using, exploiting). The influence comes also for a person acquiring the product (buyer), in the distribution processes. Then, an influence is felt by a user during the working process of the product. In addition, humans who are not directly involved with the product, bystanders as parts of the environment, may be affected.

However, the actual properties of the chosen structure manifest themselves as measurable quantities or assessable qualities only in the individual life phases of the technical system. Yet designers must establish these properties whilst designing, and as soon as possible, accurately and reliably, in advance (i.e. predetermine them), and compare the predicted values with the needs, demands, requirements and constraints (compare figure 7--5).

This necessitates good knowledge of the connections between those properties, sizes and values to be established by designing (the internal properties, class 12) and the properties desired for the product (its external properties, classes 1 to 11). For instance, "beauty" as an appropriate appearance of the product, or "safety" as a freedom from dangerous elements and behaviors must be "designed in" by establishing the form (shape), surface finish, accuracy and other design properties of the visible components. This technical knowledge rests in part directly on the laws of science, in part on the engineering sciences, on statistical inquiry, or only on the experience of the designers and their consultants (including knowledge-based or expert systems). This problem will be discussed further in the paragraphs about technical knowledge (Sections 7.2.2.6 and 7.3).

#### 7.1.3.6 Quality (Evaluation) of Technical Systems

Question:

1. How can a technical system can be evaluated, i.e. how can we make a statement about its quality?
2. How can the quality be influenced? (Quality assurance)

**– *Fundamental terminology, fundamental terms***

The German standard DIN 55350 understands quality as an entirety of properties (features) of a product or an activity which refer to their fitness to fulfill given demands. This definition implies a total judgment (total value, value vector) about a system of properties which make a thing to be that which it is (and whereby it differs from other things), for what it can serve and how suitable it is to fulfill all partial tasks and the total task. This total value is composed of the individual values of all relevant properties, as shown in figure 7--6. The total value may be constituted by a simple addition of values, but a synergistic influence of one property on another (one property enhancing or diminishing another property) is more likely.

The term *quality* is relative, as is indicated in figure 7--5, part 2. According to groups of properties used to determine quality, different kinds of quality can emerge, such as manufacturing quality, storage quality -- therefore quality referring to something.

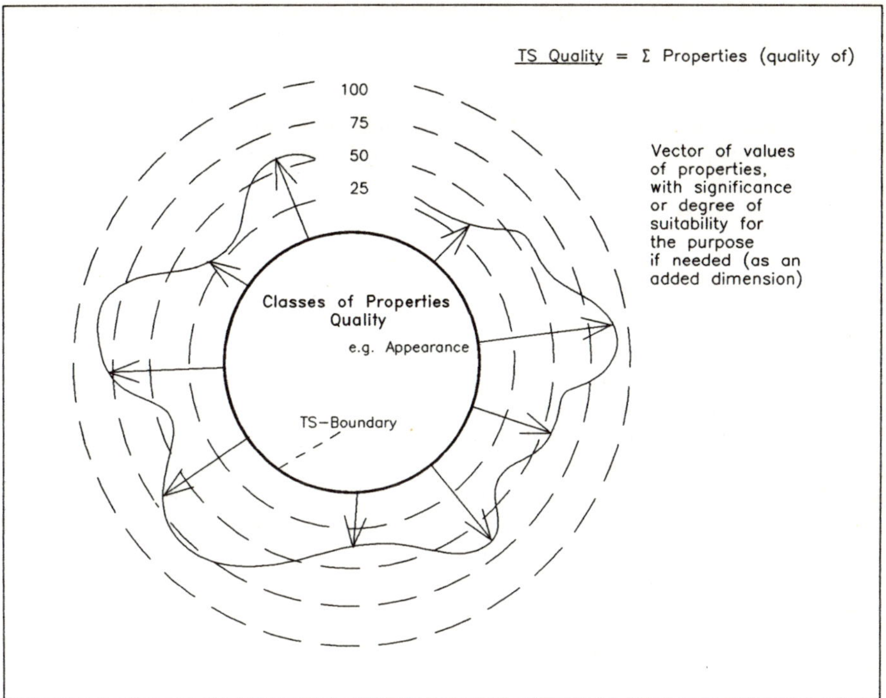


Figure 7--6 Quality Referred to Properties

### – *Quality assurance*

There are especially three partial areas within the development of a system which are the *causes* of the quality of a technical system (see also figure 7--5, part2):

1. *Quality of design* has the largest influence on the quality of technical systems. It can be assured through a methodical, systematic and transparent design process, which uses verification, tests and evaluation at suitable places during the design work and after its conclusion, based on the compiled documents ("design audit"). It is clearthat this quality absolutely demands an objective set of technical knowledge (Design Science).
2. *Quality of manufacture* (including assembly) is measured on the produced components (the parts and the whole) of the technical system. It is known as quality of conformity to the manufacturing specifications, i.e. detail drawings. The question is: Do these parts conform to the instructions given by the designers? After production of the components, measurement of this quality aims at rejecting the unfit parts (or systems), e.g., with the help of statistical processes ("quality control"). Before production or delivery, certain management processes and regulations can attempt to ensure the production of an adequate quality of the products, e.g., "statistical process control." This can be achieved even from the hands of outside suppliers, by "quality assurance" and "zero defects."
3. *Quality of application* (service, usage, exploitation) appears only when the technical system is employed. This includes also the necessary secondary processes, for example maintenance, repair, renovation, upgrading, liquidation (also recycling), etc.

Comments and knowledge systems have been developed for each of these three partial areas. The relevant standards for recognition and verification of quality assurance schemes are ISO 9000, 9001 and 9002 [11].

Various additional attempts to regulate these areas through management measures have been undertaken, e.g., methods QFD (Quality Function Deployment), TQM (Total Quality Management), simultaneous or concurrent engineering, integrated product development strategies, evaluation of statistical experiments (Taguchi), etc. (See Chapter 1 for more extensive references to literature).

– ***Value theory***

A suitable value theory forms the basis for evaluating (setting a value for, placing a value on) the realized or proposed technical systems. This theory sets up statements about the knowledge, justification, substantiation, validity and ranking of values.

The value philosophy (axiology) developed in philosophy has debated these questions in the subjects of logic and especially in ethics, but also in psychology (will and feeling). The results are not direct or only partially applicable for the purposes of technology.

The value theory of national economies (macro-economics) forms the starting point for our considerations. This theory is based on the question: To what laws do the relationships of product exchange conform?

Four propositions for a value theory are distinguished:

1. the conventional-tautological value theory, according to which supply and demand determine the exchange relationships;
2. the classical cost theory (A. Smith), according to which the products are exchanged for those values which are determined by the costs for the three production factors: capital, work and land;
3. the work-value doctrine (Ricardo), according to which the products are exchanged at the values which are determined by the work expended for their production;
4. the subjective value theory (benefit theory), according to which the products are exchanged at the value as assessed by the partners involved in the exchange, i.e. according to their respective real or apparent benefit, which can also include an esteem value (pride of ownership).

For us, the benefit theory, combined with the cost theory, can form a logical basis for the judgment of quality. This combination directly agrees with the goals of technology: "to satisfy the needs of society, especially the customer." The principle of efficiency (relationship of benefits to sacrifice) is usable, and the sacrifices are measurable through the costs. The manufacturers see the benefit, if their sacrifices are covered and the possibility of additional gains (profits) emerge. Profit only appears as calculated remainder at the end of an accounting period, in which all incurred costs are covered from the available income (earnings) and asset credits (including loans).

For different beneficiaries, the importance and meaning of individual properties is often not equal. Attempts have been made to resolve these differences through the weighting of individual properties. Assessors will set weightings from their own viewpoint. In this respect the originators of products (manufacturers) can also be considered as beneficiaries and undertake an evaluation from their own viewpoint.

*– Display of quality*

In addition to descriptions and tabular assemblies, quality can be represented graphically through vectors of the individual properties (see figure 7--6) and is visually quickly understandable.

### 7.1.3.7 Representing of Technical Systems and their Models

Question:
In which manner can technical system be modeled and represented in all their states of existence, so that the resulting models support different functions of designing?

Main demands:

1. models should serve all purposes necessary for designing, such as communication, information, experiment, calculation, thinking aid (What purpose does the model serve?);
2. models should support methodical and systematic procedure, and the application of computers;
3. the representation of models should guarantee uniqueness (lack of ambiguity) of interpretation and efficient "reading;"
4. the representation of models should consider the efficiency and economy of the modeling and representation processes (meaningful and simple models).

*– Fundamental terminology*

It has been established repeatedly that engineering practice uses an inaccurate technical language. For scientific purposes we need an accurate terminology. This necessity is also valid for our specialty, where the relevant situation is demonstrably unsatisfactory:

In engineering practice a drawing or sketch of the future technical system (component or assembly) is executed, which must contain all information about the technical system necessary for its manufacture. Thereby the practitioners are not conscious that they are really dealing with a model. They forget that the description of the model can be realized in different ways. One (certainly proved) way is the representation of the previously imagined artifact, e.g., a gear wheel, through a drawing. For the description, in this case of the form (shape), further types are available.

Therefore the term "model" must be introduced to specify the exact content of a representation. We define "model" as follows:

A *model of a technical system* is a complete or partial picture of an original (Urbild, prototype) of the reality, visualization or idea. The degree of similarity (analogy) between original and model can extend from similarity (analogy) in only one property, up to identity (complete similarity).

However, the verb "to model" brings complications. It is used mainly to designate the manufacture of the model (and therefore also of the representation). According to our view, "to model" must include all activities in the development of the model, therefore also the development of the visualization and idea.

In some areas other expressions have taken root. For instance in the building construction sector we speak of a plan or a blueprint instead of a model.

### – *Representation theory*

Models are not only used in designing, but also in many sciences, and usually with quite different interpretations (ways of expressing meaning). Although models are fashioned for certain purposes according to area of use, we can accept and take over certain proven knowledge and experiences for Design Science.

Especially the cybernetic view of modeling approaches our view. Cybernetics tries accurately to specify the term "model" and distinguishes two fundamental kinds of model:

- behavior models, which model the behavior;
- structure models, which model the structure.

For designing we need both kinds: *behavior models* (also dynamic models) are useful for simulation, examination of function(ing), calculation of dynamic behavior. The reaction of a system to inputs is explored in a behavior model (set of variables in a given time or time interval) -- the structure, properties of elements, and behavior rules are assumed as given.

*Structure models* find a broader application, they were frequently represented in the engineering drawings. Engineering drawing practice deals almost exclusively with the representation of the component structure, in detail and assembly drawings, but also in schematics such as "computer architecture." In comparison, the task emerges in Design Science to also expediently represent models of other structures (process structure, function structure, and organ structures, see figure 7--3) and to enable and support the mutual transformations among these structures. As always, attention should be dedicated to computer application where this is possible and expedient.

In relation to the modeled systems, structure models can be isomorphic (of the same form) or homomorphic (crudely similar).

Each model and therefore each representation should support the methodical and systematic mode of working (see main demand 2 above), i.e. the movement from the abstract to the concrete, as well as from the possible to the optimal, from the unfinished (incomplete) to the complete, from the provisional to the definitive state of the model (and of the modeled product).

An important position among models is the class of *ideal models*. The goal of this class is the creation of a typical pattern figure, a meaningful original compared to perfection. Idealizing frees a system of its imperfections and approaches it to an ideal, considered from a certain viewpoint. An ideal model serves as gauge for effort, and as scale for judging the reality. The idealization results in an ideal structure, in which all essential functional or morphological elements of the system's construction are included (but only these), as well as the fundamental connections which describe the system. In our book many ideal models are utilized, with their elements and ideal relationships, to illustrate the range of problems of the area. Another area of the application of ideal models exists in the so-called "Masters" (paradigm patterns, stencils, proforma, guidelines, prototypes, archetypes, exemplars, Urbilder), which represent an ideal model of a product structure or requirements structure, see Sections 8.1 and 8.4.

The idealization of technical object systems (TS) always means the idealization of the (physical) processes proceeding in these objects, e.g., thermodynamics explores only the ideal heat machines and their processes.

With reference to *clarity and vividness* as properties of the models, only those models are clear and vivid which consider the order of the elements and the space-time relationship to such an extent that a similarity between model and original is readily recognizable. Models such as geographical maps belong here, as well as engineering plans and drawings. In contrast, mathematical models as a rule do not possess similarity to the original phenomenon, except in an abstract symbolic sense.

The model represents either a single object as a system (e.g., model of the “pump type LH5”) or all single objects belonging to a species (e.g., model of a centrifugal pump). If the model of a certain system is acknowledged as representing other single cases of a certain species, it is called a “model-case.” Such a situation often emerges for the above-mentioned “master-models” (types; see also idealized models).

In general, models serve very different purposes. For designing, the following purposes are especially useful:

- demonstrate the properties (characteristics, behavior) of a technical system (induction);
- optimization of the organization (e.g., of a more abstract structure);
- testing (verification) of a hypothesis, i.e. also of a constructive solution of the structure;
- concrete planning, projecting (constructional plan), designing.

In the representation (i.e. in the model), individual properties of the technical systems are depicted. A complete representation of a technical system emerges through depicting (picturing) all design properties (compare figure 7--5--B).

The kind, code, technique, means and producer of the representation are chosen depending on the purpose, the represented object, the addressee of the information and further elements, so that modeling is justifiable also from an economic view (figure 7--7).

Design practice uses many kinds of representations and models. Certain kinds are more usually called “representation” or “projection,” others “model,” as for example orthographic, isometric or perspective projection, graphic, iconic representation, mathematical, constructional or function model, and so on.

***-- Normative statements to representation***

Because of demands 3 and 4, many statements about implementation of representations must be uniformly standardized in an organization (enterprise), a country and internationally. That affects above all the nature of engineering drawings, and (digital) transmission protocols for graphics-similar models of technical systems prepared on computers.

#### 7.1.3.8 Development of Individual Technical Systems (Origination Theory)

Questions:

1. Which is the “normal” development (origination) of technical systems, and which are its life phases (development structure)?
2. Which elements influence the overall development (origination), and the course of individual phases?

The whole originating development, beginning maybe from a vague, fuzzy idea at the start of planning, up to liquidation and disposal of the technical system, can be described through individual life phases. Meaningfully chosen life phases of the technical system form a

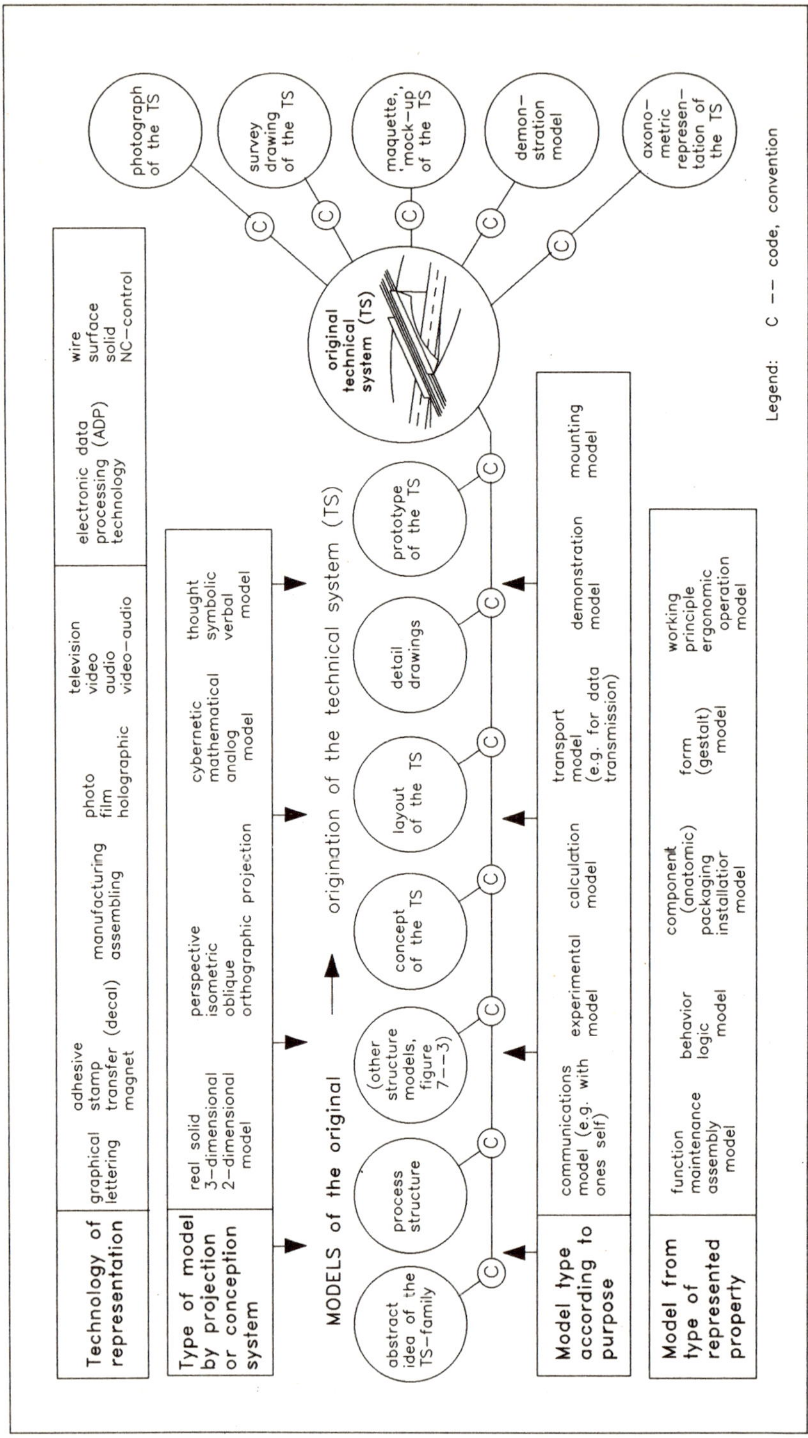


Figure 7--7 Modeling of Technical Systems

fundamental structure for the development (origination process) of TS. Designers are guided by these life phases to formulate the needs, demands, requirements, desires and constraints from each of these life phases and processes, and their operators (compare figure 7--4, and the related properties, figure 7--5). These lists of requirements also serve to control and evaluate the proposed (structure and behavior) models for their behavior in the relevant life phase of the technical system.

A technical system appears in different roles in the individual life phases. It is the operand of a process (compare figure 7--2) during designing, manufacturing, storing or disposing, but it is an operator of the desired transformation process during its usage (TS-application, exploitation).

The further elements, with which a technical system produces the transformation, can again be seen from the model of the transformation system, because the individual life phases are also transformations. All elements of the individual transformations are important for the quality of the product and must be considered in designing. For example, the process "transport to the place of application (usage location)" or "service and operation in application of the technical system" also influences the definitive form of the technical system to be designed and constructed.

Using the transformation model as pattern, the origination, development and life of a technical system can be represented as system of transformations as shown in figure 7--8.

#### 7.1.3.9 Development of Technical Systems in the Course of Time (Development Theory)

Question:
How does a technical system develop in the time, which are the laws of evolution and their elements, and how can the development be regulated and controlled?

##### – *Fundamental terminology*

The process of development (evolution) produces the successive new formations and differentiations of parts or whole technical system. We search for laws of development, so that the future total quality of systems can be increased.

##### – *Evolution theory*

The goal of general evolution theory of technical systems is the creation of a general evolution model, in which later forms evolve from existing plant to achieve growth and maturity of technical systems.

The evolution of technical systems rests on the maturation of their individual properties. This should lead to an increase of the total benefits with respect to the field of needs (including needs of the originators) and to minimizing the costs (i.e. sacrifices, see above) for the originators (manufacturers).

The historic development of technical systems (especially relevant for the TS-families) must form an important source of experience for the design and information systems. The evolution series of a TS-type reflects ideas, tendencies, but also reversals, and supports the formation of prognoses for future developments (see also Petroski [343,344,345,346]).

But also general development tendencies govern or influence the development of a certain technical system. Especially the stages of the technical-scientific revolutions present new demands and impulses for the existing technical systems and call new TS-types into

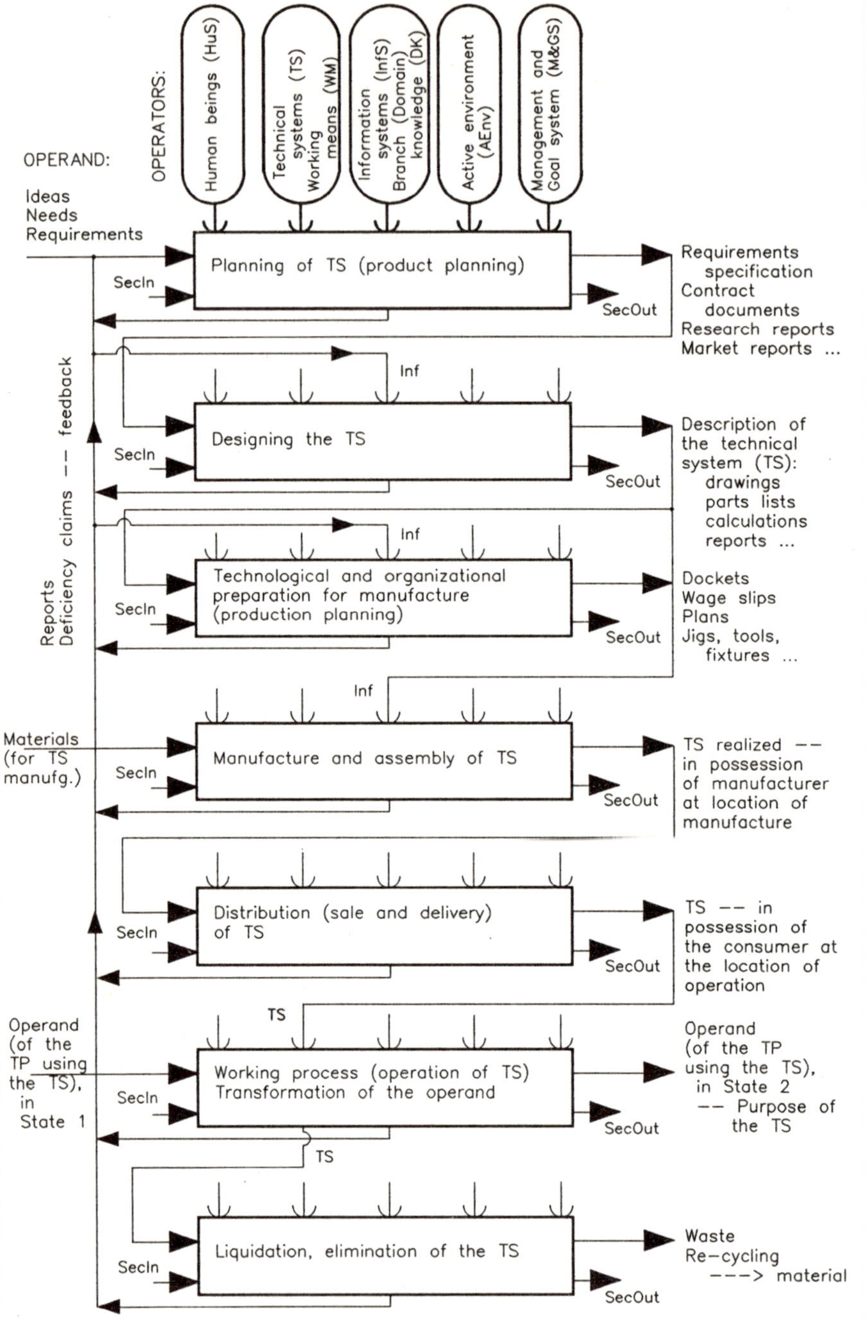


Figure 7--8 Life Cycle of Technical Systems

life. Examples of such (historically operating) renewing tendencies are:

- *mechanization* -- transfer of energy delivery from the human to the technical system;
- *instrumentation* -- increased application of instruments, measuring devices, gauges;
- *automation* -- transfer of the regulation and control functions from the human to the technical system
  - "hard" automation, using mechanical means of control,
  - "soft" automation, computerization -- transfer of some the anticipated decisions from the human to technical system, which can be extended up to flexible automation.

Also other influences, both regarding the relevant TS-family and in the design process itself, can affect the future line of development from the conventional (anticipated, expected) directions. These include new technologies (e.g., digital electronics), new raw materials (ceramics), new working means (miniaturized electronics and mechanisms -- mechatronics), new ways of constructing (fiber-reinforced composites), new means of regulation and control (digital computer), etc.

For guiding the development, the economic aspects must also find their place among the needs, requirements, demands, desires and constraints, together with the technical aspects. This means that the field of the influencing factors must be carefully explored from all aspects.

### 7.1.4 Summary

Figure 7--9 summarizes the partial areas of the theory of technical systems -- TTS with their thematic regions or questions. At the same time, some of those engineering sciences are mentioned which deal with detailed investigations of the relevant range of problems. The table contains in addition some examples of the knowledge areas in which the relevant range of problems is concretized.

Theory of Technical Systems (TTS) → *investigates the nature of technical products and processes (systems), whereby the general nature of technology is treated and can be explained.*

| *is organized into the following areas (structure of TTS):* | Technical process systems | Technical real systems (object systems) | Properties of technical systems | Representation of technical systems | Origination of technical systems | Evolution of technical systems |
|---|---|---|---|---|---|---|
| The area described by the column heading attempts to answer the following questions, e.g.: | Purpose, goal<br>– Model,<br>– Operand,<br>– Technology,<br>– Operators,<br>– Environment,<br>Systematics,<br>Classes,<br>Representation | Task, purpose,<br>Mode of operation,<br>Mode of action,<br>Structures,<br>– Concrete,<br>– Abstract,<br>Models,<br>Elements,<br>Systematics | Classes, categories,<br>Relationships among properties,<br>Design characteristics,<br>Determination (measurement) of properties,<br>Evaluation of TS, optimization | Representing, symbolizing of:<br>– Movement,<br>– Structures,<br>– Form,<br>– Function,<br>– Behavior | Origination phases,<br>Factors of origination,<br>Regularities, rules | Development process,<br>Factors,<br>Trends for TS development,<br>Regulation and control of TS development,<br>Motivation |
| Individual themes which form the objects of investigations in individual pure and engineering sciences, e.g.: | Physics,<br>Chemistry,<br>Biology,<br>Processing technologies,<br>Thermodynamics,<br>Technical optics,<br>Electrical engineering,<br>Control technologies,<br>Manufacturing technologies,<br>Food technologies,<br>Printing technologies,<br>Textile technologies,<br>Process knowledge | Cybernetics,<br>System technology<br><br>Object (branch, domain) knowledge about individual TS–families, e.g.:<br>Crane construction,<br>Electric motor building,<br>Instrumentation engineering,<br>Steam turbine construction,<br>Electric instrument construction,<br>Machine elements of all branches | Mechanics,<br>Strength (mechanics) of materials,<br>Manufacturing technologies,<br>Design (industrial design),<br>Ergonomics,<br>Appearance,<br>etc.<br>*(see figure 7--5)* | Technical drawing,<br>Modeling,<br>Representing<br><br>*(see figure 7--7)* | Planning,<br>Designing,<br>Work preparation (production planning),<br>Production, manufacture, assembly<br>Liquidation, disposal<br>Recycling<br>*(see figure 7--8)* | History of technology,<br>Predictions,<br>Politics |
| Concretization of the TTS results in a series of disciplines (and/or applications): | | | | Manufacturing technologies,<br>Company organization,<br>CAD, CAM, CAE,<br>Safety technologies,<br>Protection of the environment<br>etc. | | |

Figure 7--9 Region of the Theory of Technical Systems

## 7.2 Theory of Design Processes (TDesP)

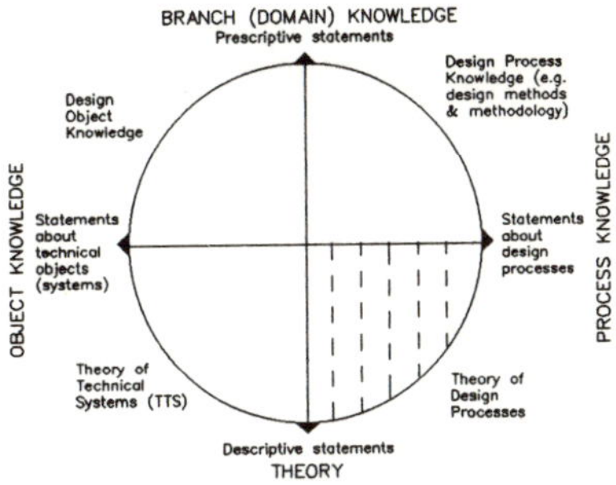


The Theory of Design Processes should contain the complete process knowledge about the transformation "designing" as a holistic unit. This means, in contrast to empiricism (the present condition), that each scientific knowledge unit should be processed with its knowledge (facts) or hypotheses into a whole, whereby the relationships among the individual knowledge elements (and those of the coordinated Theory of Technical Systems) are explicitly recognized and explained. The knowledge units should be systematically completed, until the overall knowledge system emerges.

The contents of the Theory of Design Processes can be outlined by following themes:

- task and importance of the design process;
- transformation technologies in design processes;
- structure of the design process;
- procedure in solving the design tasks by designers and by computers;
- tactical instruments (methods, resources) usable for designing;
- influencing factors for the results of the design process and its efficiency;
- evaluation, characterization of the design process;
- effects (actions of the operators) in the design process.

### 7.2.1 Earlier Views and Forms of the Knowledge System "Design Process"

In the treatment of designing in history (Chapter 3), only "designers" were considered as its operators. The design process was for a long time given almost no scientific regard, and at least that part of the design process starting from an idea and leading to the concept sketch was considered as a more or less irrational phase or "black box." The necessity of new knowledge was recognized, but not further explored, by Reuleaux. He wrote in 1856:

> *The knowledge of principles borrowed from mechanics does indeed not suffice, to execute the draft layout of a machine. Only through a combination of the theoretical results with the practical demands can rules be formulated, such that their application is an easy matter and which in the ordinary case lead to correct results. Generating such rules for designing of machine parts and of the complete machines, and the explanations for applying those rules, form the task and the object of design instruction for machine engineering.*

**A: Definition -- Designing:** (selected from among many available definitions)

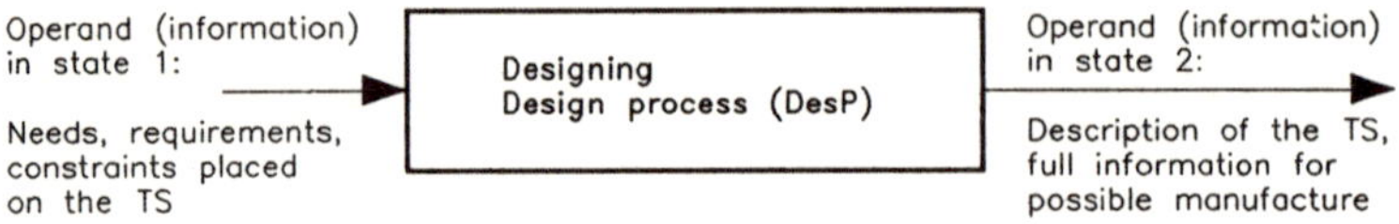


Design Process -- Definition: Designing means transforming the given problem statement into a full description of a technical system. The direct content of the design process consists of thinking out (conceptualizing) and describing the structures of a technical system (see figure 7--3).

Consequences:

1) The region of action of designing comprises the full range from the simplest component to the highest level of complexity of technical systems.
2) Designing also includes a set of various 'foreign' operations (e.g. management of design, see also figure 7--12).
3) The given problem statement and the description of the TS are items of information, the design process therefore has the general characteristics of an information-transforming process.

**B: Related Terms:**

Designing, according to the above definition, must be regarded as a collective term for other words in normal use, e.g.:

| Designing means: | When: |
|---|---|
| Developing | new and unknown TS are to be designed; |
| Adapting | prior examples of related TS are available; |
| Project engineering | complex TS (plant) are to be designed; |
| Planning | TS-constructional systems are to be designed; |
| (Organizing) | humans are present as operands (to be managed); |
| Conceptualizing | concepts (organ structures) are the object of action; |
| Laying out | component structures are being designed (as layouts); |
| Form-giving | form (e.g. shape) of components is the aim of designing; |
| Designing (artistic) | external form of a product (appearance) is being desiged. |

**C: Design Process -- Model** (Design System)

This model results when all elements of a general transformation (compare figure 7--2) are correlated into the system of designing.

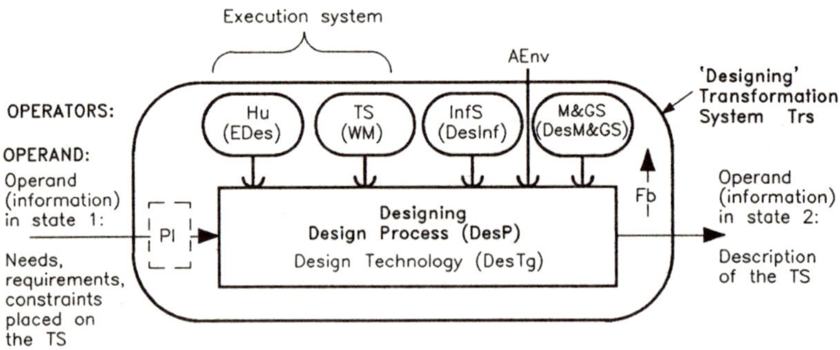


| Figure 7--10 Part 1 of 2 | Engineering Design Process -- Definition, Model, Terminology of Designing |
|---|---|

Element 1:
Operand (of designing) ---> the TS to be designed (and possibly manufactured)

Element 2:
Technology (of designing) ---> design methodology,
design methods, strategies, tactics, principles

Element 3:
Hu (EDes) ... engineering (and other) designers
TS (WM) ... working means
InfS (DesInf) ... design information
AEnv ... design office, time
M&GS ... design management

Additional elements:
Fb ... feedback
Pl ... planning (of product)

D: Scope of Design Process:

broader view (see also figure 1--2)
---> needs + product planning (Pl) + designing (in the narrower view)
narrower view (see also figure 1--1)
---> designing as main task

| Figure 7--10 Part 2 of 2 | Engineering Design Process -- Definition, Model, Terminology of Designing |
|---|---|

In the course of time some new works were written in which the authors wanted to transmit to the reader, under the title "design instruction" (*Konstruktionslehre*, design apprenticeship, study of designing), the knowledge about designing. From that view, design instruction (machine engineering basics) should teach how a machine system (TS) which carries all necessary properties can economically be generated, originated, designed. In contrast to most of the books in figure 7--1, the object of observation is not only the machine, but particularly also the design process. The knowledge of engineering sciences and the machine theory are appended in necessary breadth.

If we compare the contents of several works on design instruction, we find:

- With regard to contents a very broad spectrum of themes is treated, which offer good reference points for determining the common structure elements. With few exceptions, many themes in individual works are only touched, especially the design process and design activity, design methodology and working techniques in general, form-giving (form-determination, rather exhaustive and with emphasis on manufacture), the requirements for engineering designers, for the design office and its organization, as well as problems of strength and the available raw materials.
- The proportions of the individual structure elements in the total extent are very different, according to experience and branch direction of the author. In general, however, the sections about e.g. design methodology are very short. With the exception of *The Constructor* [365], the intent of the authors goes more in the direction of filling out knowledge gaps (predominantly after their own opinion), to create as homogeneous an information system as possible. The arrangement of the material is not uniform and hardly lets superior classes be recognized. Transferability is thereby complicated.

### 7.2.2 Contents of the Partial Area Design Processes

We will derive the structure for the contents from the model of the design system, figure 7--10--C.

#### 7.2.2.1 Fundamental Terminology

The definition (see figure 7--10--A) is formulated very generally, and in accordance with the "transformation process." The many available definitions (for example those quoted in Chapter 1) deal with almost all possible aspects of designing. However, they often operate with certain quality attributes (optimal, favorable solution), which is fundamentally wrong. The result of designing can also be an imperfect, unfavorable or even impossible technical system (however undesirable that may actually be).

In practice, several designations exist for such a transformation "design process." According to our definition, figure 7--10--B shows that all of these other designations can be subsumed under "designing" as an inclusive term. Even organizing can be interpreted as designing of human systems, whereby it is subjected to the general design theory.

It is important to stress that our explanations concern the general plane of designing, this is a consequence of the definition.

The model of the design process (or design system in figure 7--10--C) is derived from the transformation model, if all elements (according to figure 7--2) are concretized for "designing." The interpretation of the model is: The transformation of the needs, demands, requirements, wishes and constraints into the (manufacturable) description of a technical system is reached because of the inputs (influence, effects) of operators. The transformation proceeds in a certain space (environment) and time.

The last part of figure 7--10--D shows that the design process can be interpreted in a wider or a more narrow sense (compare Section 1.2.3 and figures 1--1 and 1--2). In the wider sense the interpretation starts from the (customers') needs, if the product planning and development process is integrated within the design process system, for transforming these needs to demands, requirements and constraints on the technical system. In the next sections we will treat the partial areas of the design process, derived from the "design system."

#### 7.2.2.2 Design Processes in General

Questions:

1. Which are the tasks (purposes) and the importance of the design process?
2. How is the design process described (properties, sizes, characteristics)?
3. How can the design process be evaluated?
4. Which classes (classification) of design processes exist?

Even if these questions may not have as high a value in comparison with the range of problems of designing, they must be completely answered and justified by the theory, and any variants of answers must be evaluated.

We have up to this point stated many concrete facts especially to these questions (compare Chapter 1). To sum up:

- The *task* of the design process is discussed in the definition (compare figure 7--10) in its fundamentals, and we can immediately start with detailed considerations here.

- Closer attention should be dedicated to the *importance* of the design process; the branch specialists (experts) in management still do not recognize the broad importance, range and scope of designing for the organization. The consequence is found in many aspects, including salaries offered to designers.
- As basis for the *characterization* and *evaluation* of the design process, the main question concerns the results of designing, i.e. the quality of the designed technical system. Further groups are formed by process quantities: characteristics are process duration, process costs, and others. A possible list of such characteristics is shown in the first column of figure 7--18 (objective: characterization of the design process).

An important description of each concrete process emerges through its assignment to a characteristic class. The classes can be arranged according to several features like degree of originality or complexity of the operand (the system to be designed), or of the state of embodiment of individual operators. Some examples are offered in Chapter 1, figure 7--10, or figure 7--16.

### 7.2.2.3 Transformations in Design Processes

Questions:
1. How can a transformation in the design process be realized, and which are the classes of the transformation technologies?
2. Which operation structures of the design process can be derived expediently and purposefully, i.e. into which design operations can the design process be divided?

Requirement, goal: As the main result of designing, the optimal quality of the proposed technical system should be reached with respect to the given requirements for the TS, but within a shortest possible design duration, lowest possible design cost, and considering the organizational constraints of the company.

In this phase the *technology* of the design process should be ascertained, which should make possible not only the transformation of the information (from the needs to a description of the TS ready for manufacture), but at the same time also its optimization. We must find the laws of this transformation, as well as the influences of individual elements. To guarantee that the process progresses in an effective way, all the psychological, technical and organizational elements and their mutual relationships must be standardized.

#### *7.2.2.3.1 Terminology*

The problem treated here is usually discussed in the literature about design under the term "method" (in different combinations) rather than that of technology. Two reasons may be given:

1. for the present, consideration is given only to the improvement of design work (executed by designers) through design methodology;
2. the problem remains in the area of the methodical, and one strives for a broader solution field.

However, more general goals have been formulated for Design Science, and the term technology is then more suitable as a general designation, an inclusive term for the ways of performing the transformation.

Two further remarks are useful. The first affects two possible interpretations of technology:

- Technology (used in the narrower sense, see terminology in the preface) was originally applied to the study of the production and processing of raw materials to create technical formations (artifacts). Currently technology strives not only for determining the progressions but also the solution of the entire range of problems, i.e. establishing the means of transformation -- work implements and tools, work organization, among others. Thereby the term "technology" has acquired a broader and more specific interpretation in the 1960's, namely as the science of technical processes. This breadth almost reaches the boundaries of technology in the broader sense (this in the European area), which is understood as the entirety of all technical objects as well as all means and process. According to this breadth, institutes for design at Technical Universities (particularly in Germany) are often designated as "Institutes for Design Technology." On the other hand the related term "technique" (in accordance with usage in the English language) is used to designate certain skills (measuring techniques, game techniques).

The second remark refers to the term method:

- The designation "method" for a certain defined way to accomplish a certain goal is not uniform, neither in science nor especially in practical life. On the one hand complex instruction systems which should be designated methodology are called "method," for example value analysis, modeling method. On the other hand a simple rule of behavior is also called a method, as for instance "cover the field," or "doubt." For our purposes we want to tie the name "method" especially to an operation (activity) in which only one change of the state of operands (e.g. information about the technical system to be designed in the design process) takes place in one property.

The designation of the contents of technology is also implied by other expressions, such as algorithm, but they generally cover only some aspects of the term.

##### *7.2.2.3.2 Transformation Technologies*

An analysis of present and future design processes allows us to recognize at least three typical classes of transformation technologies (compare figure 7--16 and Section 7.2.2.8):

- traditional, as is predominantly used in engineering design practice (mostly intuitively, compare Section 1.4.1 and figure 7--16 rows A and B);
- methodical, if the process is (discursively) guided by methods and method systems (figure 7--16 row C);
- mixed, if both intuitive and methodical classes of procedures are used together (figure 7--16 row D).

Supporting the transformation of information in the design process by technical means (e.g. computers) is a different aspect of the problem (see figure 7--16 rows E, F and G).

The tasks of the individual operators (see figure 7--10) are obviously different within these types of design process and technologies.

Other parameters of the design process are also influenced by the design technology, particularly quality (of design of the product), time to completion of the design work, and especially the transparency of the design process -- this latter factor can permit or hinder such procedures as team cooperation, design audits, product liability litigation (and preparing in anticipation to defend against it), passing a project on to another designer, etc.

We need some added considerations in this respect about the position of Design Science:

- Design Science aims to further above all the methodical, systematic, structured procedures, and of course also the machine-based (computer) procedures -- which also corresponds to the principles. These two classes of design procedures have a certain causal connection, because a methodical process (i.e. a logically structured process) creates the necessary conditions for using the computer effectively.

In addition, the traditional (intuitive) way should be explained in and by the theory, its apparent mysteries discussed and clarified, and its mythology destroyed. Some of this task was performed in Chapter 1.

### *7.2.2.3.3 Design Method as Technology*

Without doubt many successful technical system ("designs" and constructions) were realized even when design methodology as such did not yet exist. Observations and analyses of good design works could have led to certain generalizations and to introducing a tendency towards improvements. That was the first consideration for rationalizing.

In the attempts to describe the individual activities (operations) within the design processes, as well as the outputs, different individual "leaps of thought" were discovered which did not allow generalization, and were not able to be performed by a method. This shows that the technology of "intuitive" processes (compare Section 1.4.1) as a promising basis is not transferable (i.e. not teachable) because the process remains obscure and closed.

Transparent and repeatable are only discursive operations. The formulation of the system of discursive operations and their methods, used as technology of design processes, has become the goal of design methodology. It is also known as methodical, systematic, planned or structured designing.

The method should guide the process purposefully and as planned. In the available procedural models, three classes can be found according to the type of guidance (regulation and control):

- strictly algorithmic instruction, i.e. strictly regulated procedure, for example, the generalized procedural model in figure 7--13;
- heuristic instruction, i.e. relatively flexible procedure (compare Section 1.4.6);
- relatively vague instruction with no clear references, i.e. fairly free procedure, guided only by some relevant principles.

A planned process of designing cannot be built only on the general, logical methods, but it must develop particular, empirical methods, which are determined by the individualistic characteristics of the object of designing (compare also figure 5--4).

Methods, as instructions, have strictly speaking a prescriptive rather than descriptive character. Therefore we will discuss methods again in the prescriptive parts of this Chapter, Sections 7.3 and 7.4. In contrast, the theory must treat the study of methods as basis for the whole area. In figure 7--11 we try to present the most important facts of methods, especially the terminology.

To make the area of methodology clearer, we distinguish between *design strategy*, which

A: Purpose

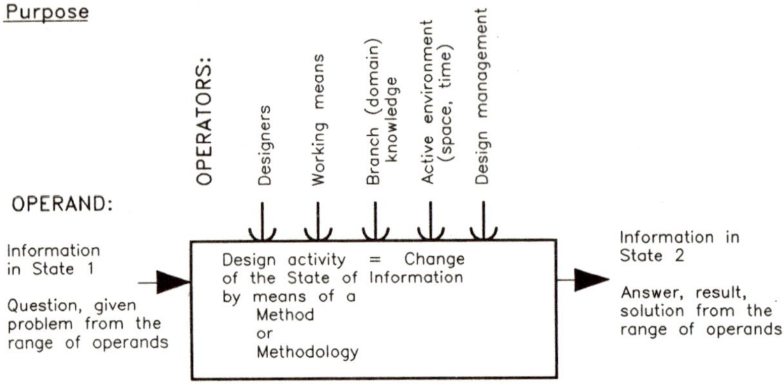


## B: Method, Methodology, Procedural Plan, Procedural Manner and Working Principle

We define the term design method to mean a system of methodical rules and instructions. These are intended to guide and/or determine the way of proceeding for executing a certain design activity and to regulate the interaction with the available technical means. They are valid under the assumptions that a particular 'normal' designer, with 'normal' technical knowledge, is acting within certain 'normal' environment conditions. A model which reflects all these elements is represented in this figure as process model of a design activity.

A coordinated set of methods is termed as a methodology. This term can cover different sets of methods, for example for a region of activities or branch specialization (product family).

Each element of a method can be regarded as a working principle. It usually delivers a universally valid instruction for appropriate action (conduct) in certain situations. Examples of such a working principles are: the guideline of providing constant wall thickness when establishing the form of a casting, or the endeavor to achieve minimal manufacturing costs in all design activities. As different levels of a hierarchy, a system of working principles plus further references can represent a working method, and a set of methods forms a methodology.

The existence of a method permits the development of a plan of action (procedural plan), which establishes the rules for behavior in a design activity for a particular (concrete) case. A method can deliver a starting point for a series of plans of action which will change according to an altered task definition (referring to branch and kind of design activity) as well as deviations from 'normal' branch knowledge, and the technical and organizational conditions of the task.

The individual way (of a certain designer) in which a design task is performed is called a procedural manner or designer's mode of action. It can be derived from a method or an action plan, and depends on how far a certain designer deviates from the 'normal'. The extent of the knowledge and ability of a designer, tiredness and similar factors influence the utilized procedural manner.

Each design process can fundamentally be structured with the help of a general procedural model into more or less complex partial processes, phases, down to design operations and steps. The elements of procedure that emerge are also processes, within which the state of information is changed. Each of these elements is therefore directed towards a (more or less) precisely formulated goal which is evident from the procedural model. In order that these processes can proceed methodically and according to plan, directed towards the goal and under the given boundary conditions, corresponding rules of behavior and methodical advice must be available. These are contained in the methods or in the working principles, which can regulate the work as reference points.

| Figure 7--11 Part 1 of 2 | General Study of Methods -- Design Tactics |
|---|---|

C: Characteristics of Methods

Each method carries a number of features which characterize that nethod from different points of view. Some of the features can also be used as organizing criteria, which help in generating classifications. These features of methods can be arranged crudely into two classes:
- Features which characterize the method as a tool, i.e. with respect to its application;
- Features which characterize the method as information.

The features with respect to the application can be obtained with the help of the following key-words or questions:
- which goal (purpose) should be served by the method? For which areas of the designer's activity is it applicable?
- what is its breadth of application for a product family? Is it usable only for one field (branch, domain) or for several?
- is it only usable for certain conditions, or must certain conditions be fulfilled? What resources are necessary?
- for whom (type of operator, designer) is the method intended? For an individual person or for a group? What preconditions must the person fulfill?
- from what source did the method come? In which branch, area of knowledge, discipline, branch or domain did it originate?
- how does the method work (behave)? On which phenomena it is based? What is the 'mechanism' of the method?
- which time demands does application of the method make?

A method can be characterized as information by means of questions about general properties of information:
- are the contents correct?
- is its effect verifiable? (Sources, experiences)
- are the contents complete, not only excerpts? Are further details available?
- is the format of the description unequivocal, clearly understandable?
- how old is the information about this method? Is its status still current?
- is the method of unrestricted application?

We term the system of design methods and working principles as design tactics; in contrast to design strategy (e.g. a procedural model) which has the task of providing the basic structure. Design tactics and strategy are the fundamental areas of design methodology.

| Figure 7--11 Part 2 of 2 | General Study of Methods -- Design Tactics |
|---|---|

should determine the general direction of the procedure, and *design tactics*, which treats the methods and working principles of the individual design steps.

In the context of the systematic discursive instructions for procedure, the question must emerge whether the whole design process is algorithmically solvable. Franke [157] denies this on the grounds that "algorithm" should only be interpreted in its strictest sense of machine instructions. Even so, procedural models of designing have emerged and continue to emerge, as "flexible algorithms," not only in design strategy but also in design tactics.

On which considerations, bases and knowledge do these methods rest? Because we are dealing with thinking processes, knowledge from the psychology of thought is utilized, and especially association is explored. In addition, we look for means to eliminate thought errors and fixations. Contacts are also made to general works from the study of knowledge, and logic, as was reported in Chapter 1 (Descartes, Polya, see also Chapter 5).

Scientific organization of work is also an important source of references. Especially the principle of decomposing a complex task into a series of part-tasks is consistently followed.

The essential knowledge applicable for methods must, however, originate from the technical system, the object being designed. As figure 5--4 shows, the method is influenced especially by the object. Because designing tries to establish structures, the essential task consists of finding suitable structures of technical systems (compare figure 7--3) and empirical laws of transformation.

#### *7.2.2.3.4 Structures of the Design Processes*

Seen from the structural side, the design process can be recognized generally as system of operations. The type, quantity and arrangement of the operations depends on the technology of designing -- the technology creates one possible way of structuring the design process (see figure 7--13) by describing a recommended procedure (*how should or can the transformation proceed*, what should a designer do?)

Another possibility aims at determining all operations with respect to their complexity. This hierarchy (see figure 7--12) offers a favorable starting point for many considerations.

Each operation at a particular level in figure 7--12 contains the operations in the next lower level, and forms simultaneously an element of the operations in the next higher level. Level 1 presents the overview of the design process, with a crude subdivision into the three main stages (the general viewpoint suitable for management):

*Conceptualizing, laying out (embodying), and elaborating (detailing).*

The design operations on level 2 are repeatedly utilized parts of the main stages, which fulfill certain purposes of changing information in the design process, including the work of designing. Level 3 contains the group of the basic operations which is generally known as the problem solving cycle (compare figure 6--1), with additional operations referring to providing information, verification and representing. The elementary activities on level 4, although frequently used during design work, and operations on level 5 are no longer specific for designing and problem solving.

Apart from the hierarchical relationships, other relationships can be seen within each level, namely the ones among blocks on each level, as is clear on level 3 for the basic operations. The secondary and leadership (management) operations must not be forgotten as necessary elements of the design process.

#### *7.2.2.3.5 Forms of Presentation for the Design Knowledge*

The design methodological knowledge can with advantage be presented in form of *flow charts* (with explanations). These can clearly show the complete process and define the design steps and their inputs and outputs. Such representations are especially suitable for the strategic design knowledge, and partially also for the tactical knowledge. In figure 7--13 this type of representation is illustrated using the example of a general procedural model for designing -- this figure presents a finer subdivision of level 1 in figure 7--12, and contains the operations at level 2.

The top block in figure 7--13, part 1, shows that the task should first be defined, and indicates some steps suitable for generating a design specification (refer also to figures 7--4 and 7--5). Within the larger block labeled "conceptualizing," the upper part describes steps to establish the transformation process according to figure 7--2, and the applicable or available technologies. This continues by establishing the needed effects, figure 7--3, level (I), and the TS-internal and trans-boundary functions and their relationships, the function structure, figure 7--3, level (III). The second part of "conceptualizing" intends to establish inputs to the TS, modes of (internal) action, and classes of organs (as function carriers). This part continues by progressing to suitable arrangements of organs into an organ structure (concept), level (IV). Laying out (embodying) to establish the component structure, level (V), takes place in three distinct stages. In the first of these, progress is made by preliminary investigation of arrangements, sizes, forms, etc., usually in the form of sketches -- mainly

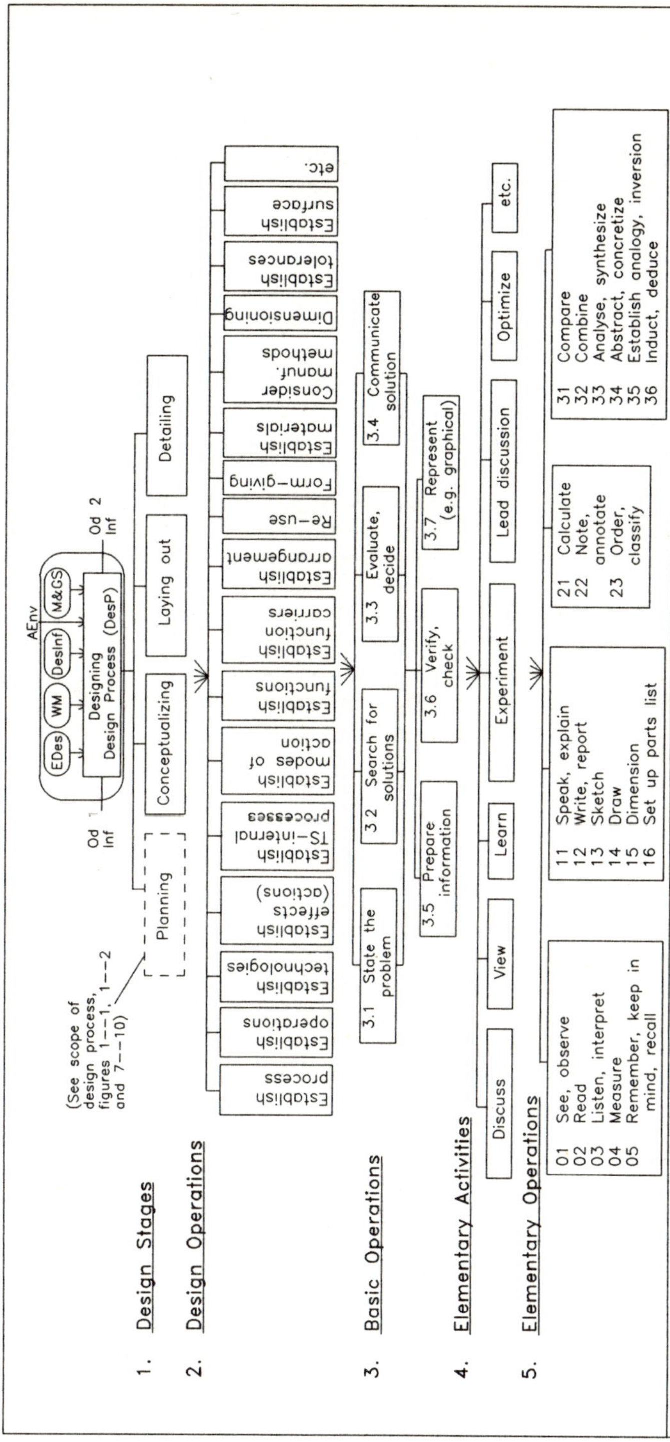


Relationships: A. hierarchical -- every activity at one level (e.g. '3.6 Verify, check') contains the activities at the next lower level (e.g. 'Learn'), and is simultaneously contained in every activity at the next higher level (e.g. 'Form-giving');

B. block dependency -- some activities are repeatedly applied in a certain (flexible) sequence, particularly the cycle of problem-solving operations at level 3 Basic Operations.

Note: 'thinking' is not included in this hierarchy, this process takes place inside the human and is not directly observable or verifiable by documentation (e.g. written or graphical records). Thinking could be incudec at level 4 Elementary Activities.

Figure 7--12 Structure of Possible Activities in the Design Process

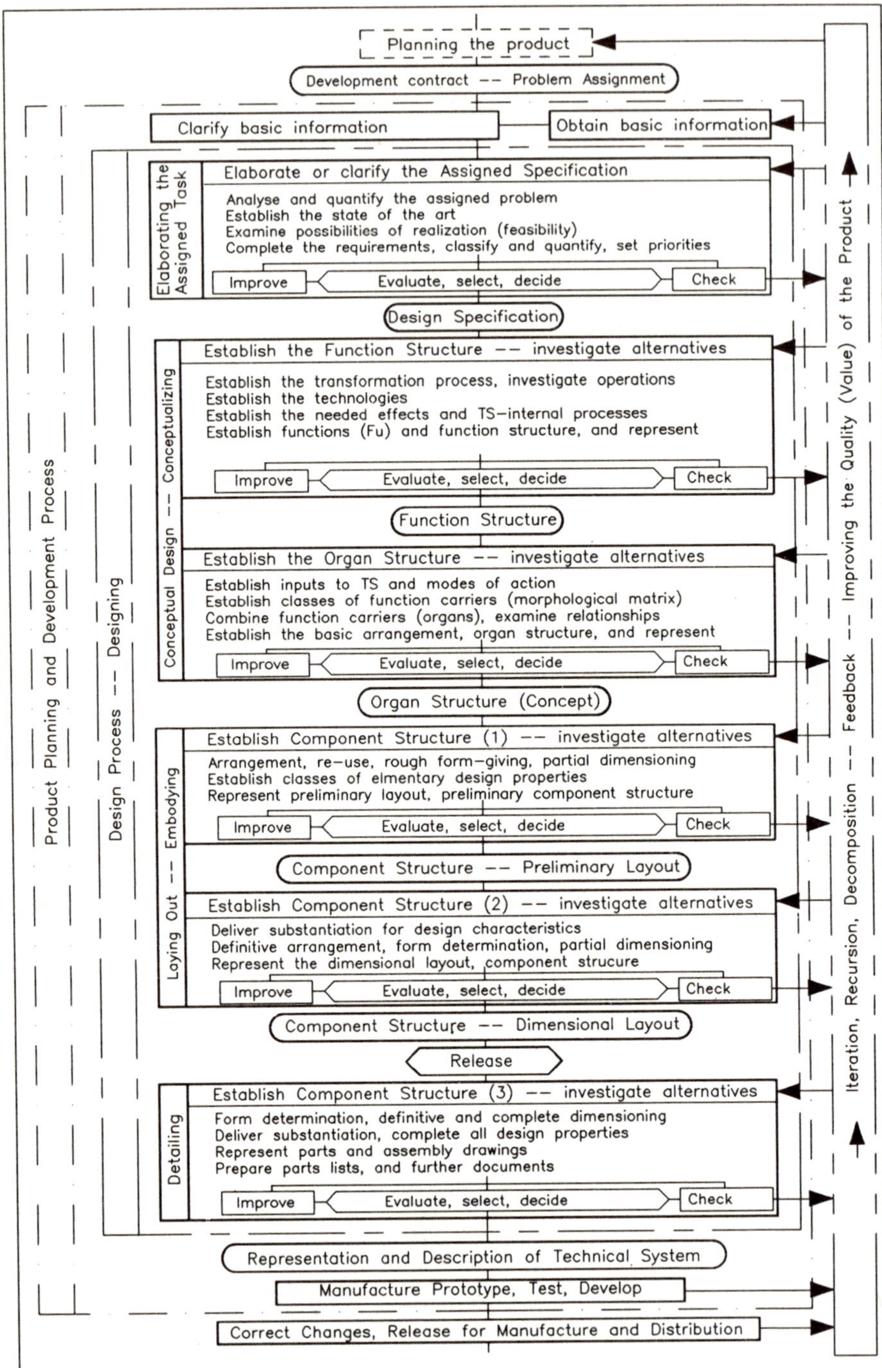


Figure 7--13 Part 1 of 2 — General Procedural Model of the Design Process

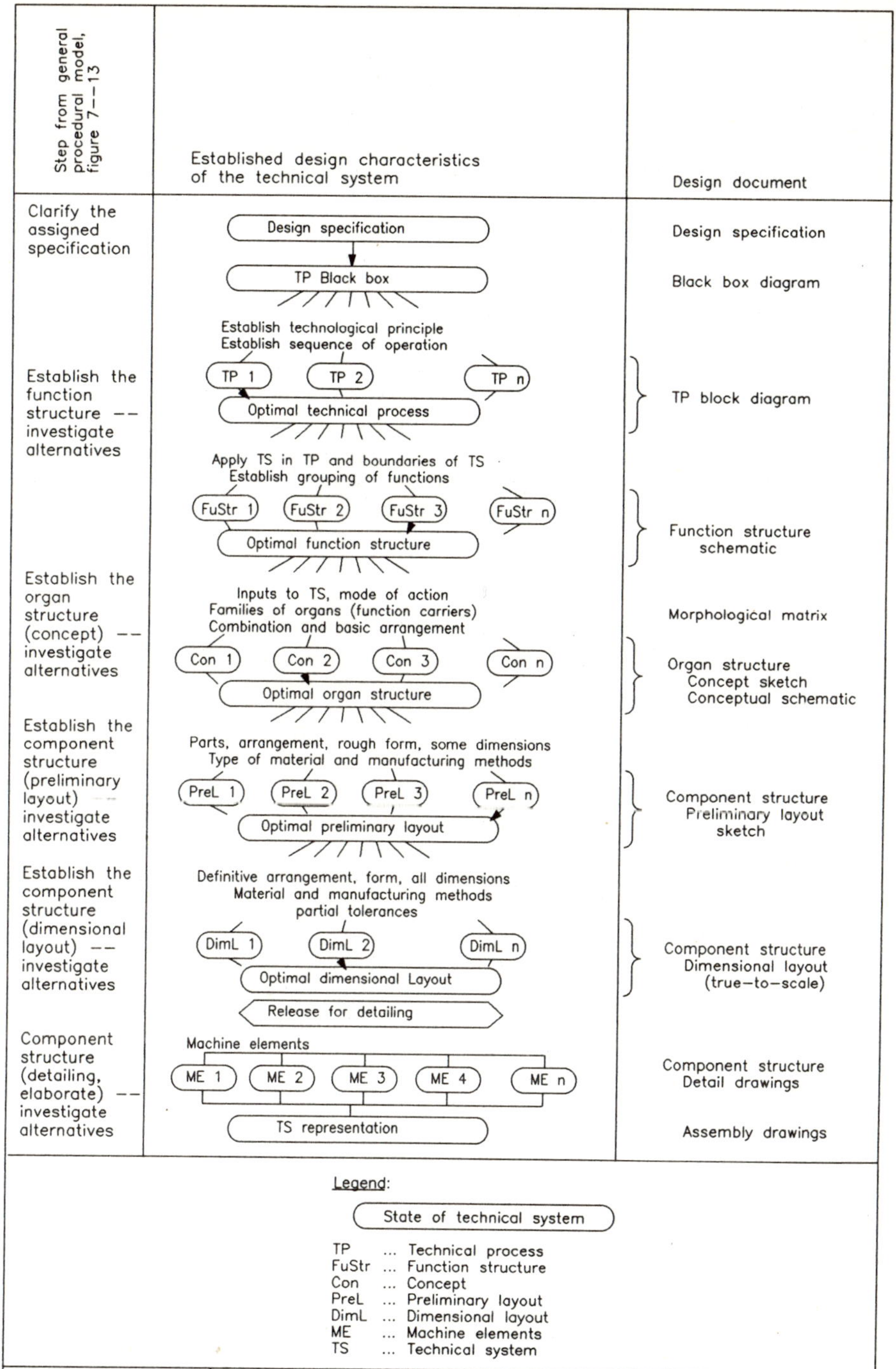


Figure 7--13 Part 2 of 2 The Design Process: Graphical Representation of the States of Technical Systems during Design Work

executed by hand. The second stage transfers to a largely dimensional representation, the common layout, supported by calculations, simulations, etc. A management procedure is usually inserted here to audit the design results and approve further expenditure on this project -- a connection to the second "house of quality" in QFD is suggested. The third stage allows production of all detail and assembly drawings and parts list, and completion of the instructions for manufacturing. For further explanations to these operations see [219,228,229].

Each such design step should permit and encourage an investigation of alternative solutions. It should also support and encourage validation and verification (checking), iteration and recursion (decomposition), evaluation and selection (decision making). The results of these processes are indicated in figure 7--13, part 2. Selection of the most promising alternative(s) at each modeling level helps to control potential "combinatorial complexity," which could yield an inordinate number of combinations of alternatives -- the numbers of alternatives at each successive level of abstraction are, in effect, multiplied.

All conditions for which such an idealized model is valid must be indicated, e.g. that it covers the general direction for progress in *novel* design situations. Further the possibilities for entry into the procedure should be shown, indications should be given about what iterations and recursions can be executed with the help of feedback loops, and what information is needed.

A researcher would normally consider such an idealized general model with broader validity as a satisfactory result. It can, however, serve a practitioner only as preliminary model for working out concrete procedural plans. These must reflect the situation and design problem to be processed. Figure 7--14, part 1, shows some directions for moving from the general situation (procedural model) to specific cases, which the designer must traverse. We can speak of different *depths of concretization* in this context.

The ability to perform the described concretization is apparently not as widespread in engineering practice as would be necessary. The concretization steps demand much knowledge about the individual design operations. The lack of knowledge leads to not using these possibilities, and thereby also to unsatisfactory application of design methodology in engineering practice. Most likely, working designers can be effectively helped if they are given appropriate advice and references for the respective situation, which should answer the current question "What should I do now to make it more likely that I can find a suitable solution to my design problem?"

Such a form of methodical knowledge is not available at present. It demands, beside the classification of such "design situations," a costly revision of the existing knowledge. An important question usually remains open, i.e. whether this form of knowledge is meaningful at such a level of abstraction, or whether the necessary concreteness and quality is achievable only on the level of product families.

### *-- Design Situation*

A further possibility for structuring the design process starts from considering the design process as a system of design situations.

To describe a such situation in the design process (when designing), we introduce a new term, "design situation," and determine suitable parameters. Again the elements of the design system (figure 7--10) help us to find the features. They are, as examples:

| *in the area*: | *characteristics, features*: |
|---|---|
| Operand (TS): | TS-Family, degree of originality, degree of complexity |
| Technology: | Methods and methodologies, intuition, creativity |
| Designer: | Education, experience; |
| | Lone designer, group, team, cross-disciplinary team |
| State of knowledge: | Which state of knowledge is available for the particular problem? |
| Working means: | Types of available means |
| Management: | Organization; location and time when designing |

Another guideline for establishing an effective information system uses the hierarchical classification of design situations. Some classes of design situations occur perhaps once in ten years, other classes occur much more frequently. Accordingly, the information systems which rest on design situations should be expanded only for such meaningful, concrete situation levels which appear sufficiently often in the design process.

Very pragmatically, according to their use in designing, design situations are established either from the design operations and basic operations (figure 7--12), or by certain features of the operand. We can describe the structural element (figure 7--20 A) as follows:

- in a certain branch (domain, specialty) and at a certain level of originality, a certain element of knowledge is associated with a particular design situation -- e.g. when searching for solutions to a certain organ. In this case the position of the design situations is clearly determined within the procedures.

Another class of design situations emerges when considering the design operations dimensioning (size calculations), form-giving, determining of raw materials, and others. These are repeated several times during designing (also as an iteration process), and different methods, but also different elements of knowledge are utilized in each situation. That means that different elements of the branch knowledge must be available to the designer, according to state of the relevant design situation. Thus, completely different stratagems are used for example in the conceptualizing stage for the first assumptions about form, dimensions or raw materials, than in the solution of these problems in the preliminary layout or in detailing. We underline the importance of determination of property values as exactly as possible in the early design phases (shorter or longer control circuit and feedback loop, compare figure 7--15).

Every design situation can be described by a model such as figure 7--14, parts 2 and 3: a contract which contains particularly the characteristics of the ordered TS and its delivery, is met by a certain *design potential*, consisting of the company's design system with its capacities and capabilities for performing design work. It should be clear from this model that the contract contains information about the operational and environmental factors. If some of these factors are unknown, a more or less abstract situation results that must be concretized in successive steps.

#### *7.2.2.3.6 A Particular Task for Methods in Designing*

Designers establish the individual structures of technical systems. These structures become carriers of the necessary and desired properties of future real products. The difficulty in designing lies in the uncertainty of predicting the future properties, which are detectable (measurable) predominantly only when the technical systems is put to use. Other properties come to light in production, storage, transport, maintenance, etc. All reveal themselves (as

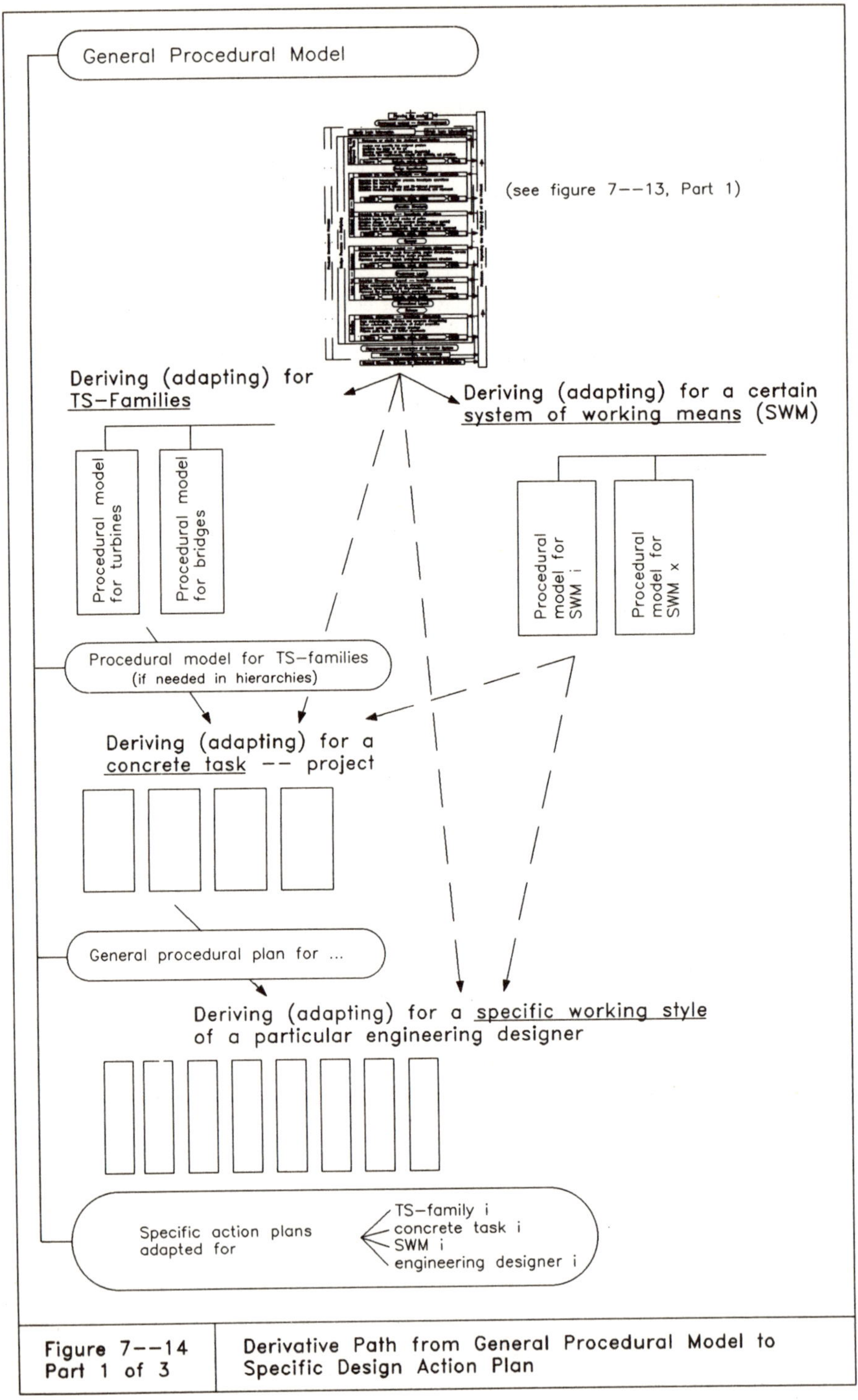


Figure 7--14 Part 1 of 3 | Derivative Path from General Procedural Model to Specific Design Action Plan

| Class | | Designation | Example of Form of Embodiment |
|---|---|---|---|
| FU | V/W | ENVIRONMENT FACTORS<br>V ... Local Economy<br>W ... World Economy | |
| | 1 | Politics | Export–, Import–, Taxation Policies |
| | 2 | Society (Social System) | State of Education, Incomes, 'Fashion' trends |
| | 3 | Science, Categorized Knowledge | State of Research, Scientific Knowledge |
| | 4 | Technology (wide interpretation) | State of the Art (Engineering) |
| | 5 | Market for Goods | Competition, Prices |
| | | for Personnel | Employment Rate, Expertise Availability |
| | | for Acquisition | Materials, Devices, Semi–finished Goods |
| | | for Services | Transport, Services, Repair |
| | | for Financing | Financing Availability, Interest Rates |
| FB | | COMPANY FACTORS | |
| | 1 | Production Programme | Breadth of Assortment, Relationships |
| | 2 | Development and Market Strategy | Long–term Market Plans |
| | 3 | Size of Company | Small, Medium, Large |
| | 4 | Types of Manufacturing Processes | Single, Series, Batch, Mass, Continuous |
| | 5 | Potential of Company | Human, Working Means, Financing |
| FA | | TASK FACTORS | |
| | 1 | Type of Product | Breadth, Variety, Consumer Groups |
| | 2 | Novelty of Product | New Products, Repeats, Variants |
| | 3 | Degree of Complexity | Plant, Machines, Groups, Details |
| | 4 | Degree of Difficulty | Requirements for Functions |
| | 5 | Particular Requirements placed on the Product | High Safety, Reliability |
| | 6 | Deadlines, Time to Design | Pressure of Deadlines, Systematic Work |
| | 7 | Particular Requirements regarding Procedures | Task, Documents, Prior Permissions |
| FK | | DESIGN POTENTIAL FACTORS | |
| | 1 | Designers (engineers and others) | Education, Experience, Attitudes |
| | 2 | Working Means (tools, equipment) | Toole, Equipment, Computers |
| | 3 | Branch (Domain) Knowledge | Type, Form, Carrier (Medium), Reliability |
| | 4 | Management | Procedural, Hierarchical Organization |
| | 5 | Working Conditions, Environment | Working Climate, Motivation |

Figure 7--14 Part 2 of 3 — Survey of Factors that Influence the Design Situation

measurable or assessable) only a considerable time after conclusion of the real design work, when a substantial amount of money has already been invested.

In order for designers to be able to plan all properties in advance, by establishing the needed design properties, they must know the relationships among all kinds of external properties and the design properties (see Theory of Technical Systems [219]) and the right "master" of procedural technology. A good available method should then become clear. A method for strength calculation can serve here as example, it allows determination of which loadings are associated with the corresponding values of material, geometry, surface quality, etc. The existing methods affect predominantly the component structure, i.e. they are only usable from the late parts of the layout phase onwards.

The effectiveness of each method is measurable from the magnitude (time span) of the feedback loop which must be traversed to be able to change unsuitable values of the design properties. This is therefore the length of path between establishing a design property (in the design process) and determining its real value (in TS-use, etc.) from the achieved properties. This determination and measurement can be done either by a reliable predictive method or by factual measuring, experiments, trial operation, or other ways. This statement can be demonstrated on the example of *manufacturing cost calculation*, as shown in figure 7--15.

The manufacturing costs for a component depend generally on material, geometry, anticipated manufacturing process, surface quality and treatment, among others. The values

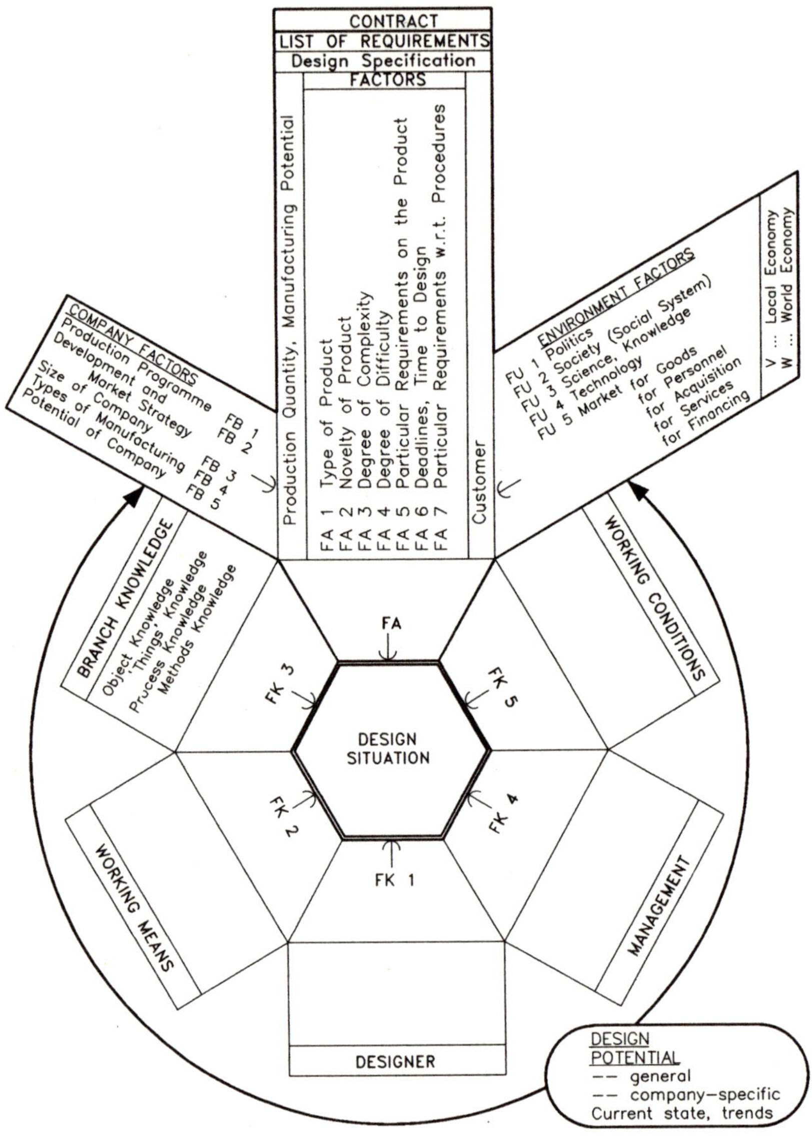


Figure 7--14 Part 3 of 3

Design Situation -- General Model

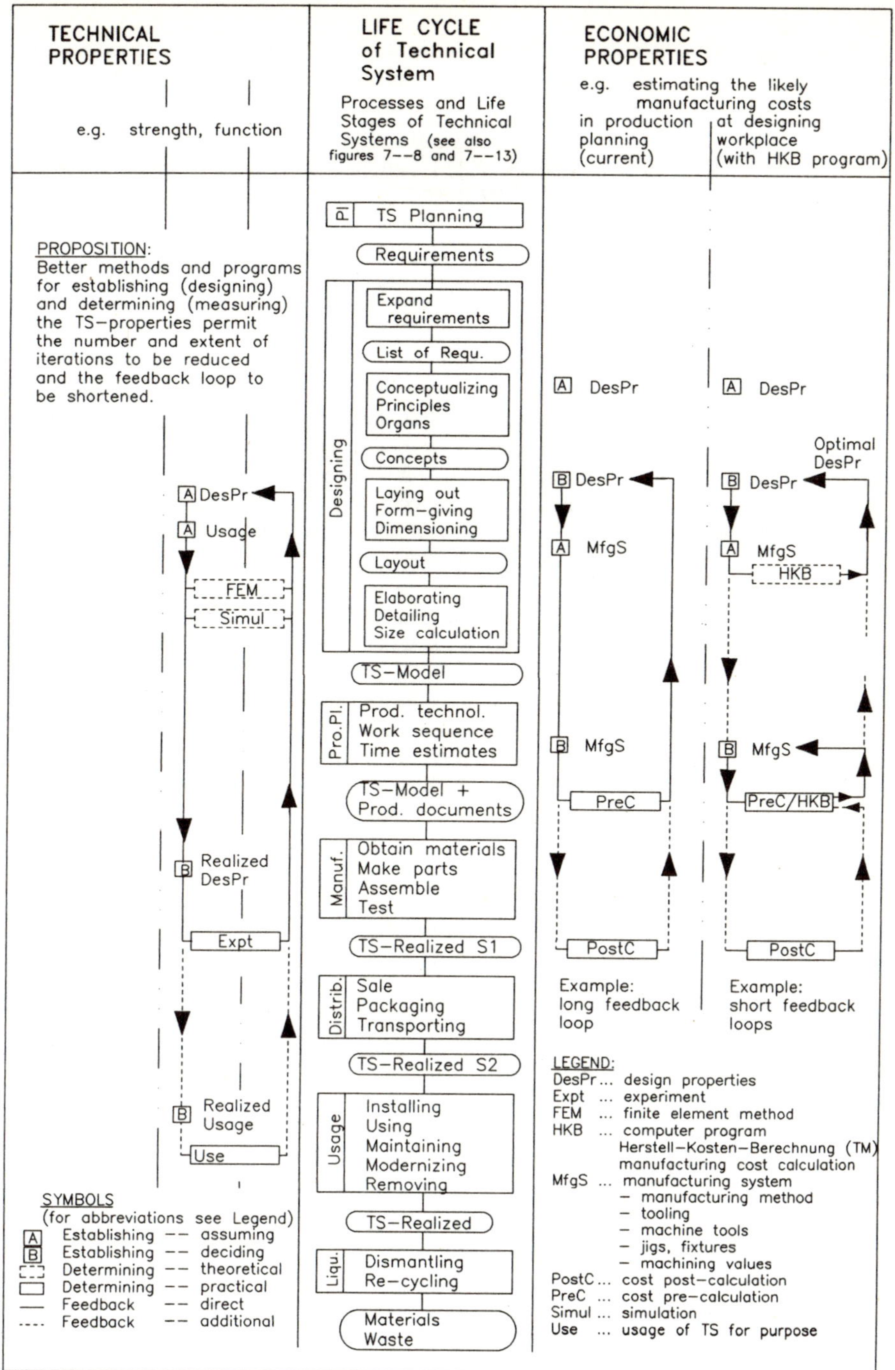


Figure 7--15 Interval between Establishing and Measuring the Properties of Technical Systems

of these design properties are established during layout, detail design and drafting. The question of whether the manufacturing costs correspond to the cost goal or whether they are acceptable, can in fact only be answered after manufacture, by means of post-calculation of cost. If a negative result should occur (i.e. the costs are too high), the designers would have to go back to the layout phase, redesign, write off all investments, and invest new money in design, production planning, material and production. The feedback loop is too long and costly, and negatively influences the delivery time.

The situation is much more favorable if the anticipated manufacturing costs can be determined in production planning (work preparation) by means of a reliable pre-calculation method, and if necessary to send the proposed components back for revision. But even here we must reckon with loss of money and time, because the feedback loop is still too long.

The goal of determining the properties must be to bring the inquiry as near as possible to the point of establishing these properties -- designing. With this objective in mind, the computer method of obtaining manufacturing costs during designing was developed (HKB [150,151,152,153]). It allows the designers to determine directly the manufacturing costs of proposed components and assemblies, and to optimize the design properties with reference to the costs. This computer program can augment and partially replace the advice and cooperation needed in *concurrent engineering*, to "design the product and the manufacturing process at the same time." Manufacturing enters and maintains its expertise in this knowledge-based system, designers call on the expertise when needed. Face-to-face negotiations are still important, the system can only maintain a limited range of expertise (in this case about costs), but especially because personal interactions can improve the quality of considerations and ideas.

This effort can be further continued in this direction, to eliminate the possible causes for the unfavorable design properties. In our case the function structure would have to be examined with respect to potential costs (function costs [150,151]).

The same guideline is valid for all other properties listed in figures 7--4 and 7--5.

#### *7.2.2.3.7 Recapitulation and Expansion*

As recapitulation of this section we will specify the kinds of the design processes which possess characteristic properties. The kinds of process in figure 7--16 are described from several viewpoints, especially from applications of design methodology and computers.

The machine process, when the design process is (completely or partially) performed by a machine (e.g. a computer), in other words "automation of the design process," is discussed in more detail in Section 7.5 about CAD.

### 7.2.2.4 Designer -- Operator of the Design Process

Questions:

1. Which tasks do designers have in the transformation of the design process?
2. Which kinds of designers exist?
3. Which qualifications can we expect from the individual species of the designers?
4. According to which criteria are designers evaluated (rated)?
5. How does one attain the qualifications for designing?

| | Design Processes (DesP), Types | General Characteristics of Procedure | Results |
|---|---|---|---|
| A | Trial and Error | A series of trials (attempts) should find the solution for individual steps. Appraisal (evaluation) of the solution for fitness is then necessary. | TS: accidental quality, possibly no result<br>DesP: time intensive<br>Prep: none |
| B | Intuitive DesP (traditional) | The experienced designer spontaneously finds a fairly useful (but not optimal) solution. This procedure is mostly connected with creativity, inspiration, feeling. | TS: usable, not optimal<br>DesP: very quick<br>Prep: long experience |
| C | Methodical, systematic DesP | Procedure of a designer working systematically, using methods and a plan, considering all aspects of the object to be designed and boundary conditions for the problem. | TS: good quality possible<br>DesP: could be slow<br>Prep: schooling, training |
| D | Methodical–intuitive DesP | Combination of types 'B' and 'C', in which some steps are accomplished intuitively and by leaps, others methodically and according to a plan. | TS: usable to optimal<br>DesP: quick<br>Prep: experience, schooling |
| E | Partially computer supported DesP | A computer is used to support some operations of the design process, for example calculation, drawing, storage. Better suited for methodical design work. | TS: fairly good quality<br>DesP: possibly quicker<br>Prep: schooling, esp. in branch |
| F | Integrally computer supported DesP | A computer is intensively applied, and the majority of steps are supported by computer. The designer remains the key operator of the DesP. | TS: fairly optimal quality<br>DesP: quicker<br>Prep: long, in branch and organizational |
| G | Automated DesP | Automated total or partial process of designing, in which only the data for the case are entered, and the solution is output with no further human intervention. | TS: optimal quality<br>DesP: quick<br>Prep: long, in branch and organizational |

Legends:
TS ... Technical System, DesP ... Design Process, Prep ... Preparation

Figure 7--16 Characterization of Design Processes -- Typology

### – *Terminology*

Starting from the transformation system (figure 7--2), the operator "designers" can include all staff members which cooperate in and contribute to the transformation in the design process. In our model (figure 7--10--C) the inclusive term "designer" covers therefore not only development engineers and layout engineers but also draftsmen. This view differs from the normal professional designation, which defines only a particular employee class with rather exact duty statement ("terms of reference" or "job description") as designers.

Besides the people directly active in the transformation (i.e. designers) the complete design system (figure 7--10--C) contains also several further staff members who fulfill the general functions necessary for the progress of the design process: providing information (including standardization), managing (leadership), administrative processing, archiving, etc. This group of employees is understandably not included in the collective term "designers."

#### *7.2.2.4.1 Tasks of Designers*

The design process (designing), viewed as execution of the transformation, demands from the operators a large quantity of different effects, which are described in figure 7--12 through the activities of the designers. The individuality of designers meant that the predominant share of the effects were accomplished by designers themselves. The fulfillment of all qualitative characteristics of the design process has measurably influenced the designer-operator. Only in the later phases of automation of the design process (as part of the CAD efforts, see Section 7.5) can we expect that a part of these tasks is taken over by the computer. For this reason the great importance of designing for the company rests with the designers. (This circumstance is, however, in blatant contrast to the reality, where the designers are generally held in low respect, as previously mentioned. Even so, a positive world view on the part of the designers, both socially and technically, is indispensable.)

The duties of the designers are, however, not only limited to the transformation "designing" itself. Many other activities exist to support designing, but cannot be allocated directly to the real design process (see also Chapter 1 and figure 7--11, two lowest levels). Further ranges of tasks are formed in this sense by activities such as planning, preparing and supplying information, consultation with production planning or assembly, experimentation section (e.g. research laboratories or development sections) or company operation [180].

The aspiration to minimalize the proportions of such tasks in design work should be supported by the theory through corresponding knowledge. It is also expected from the theory that it thoroughly explores all task ranges of the designers and places at their disposal a basis (or a pattern) for concrete cases.

#### *7.2.2.4.2 Kinds of Designers as Operators*

The answer to this question can be given especially under consideration of two points of view. On one hand the designer-operator is embodied from the point of view of structure as:

- individual designers who should each solve a task or partial task;
- a group of designers who have different specializations;
- a team of designers of equal status and capability should solve the task as a community. This can include people (experts) from other specialties taking part, consulting, advising.

On the other hand it is possible to form certain species of designers according to their employment and qualifications. The result will not be equal in different places or countries. In the German language area, mostly the following species are named:

- design engineers;
- design technologists;
- technical draftspersons;
- detail draftspersons.

The definitions also contain complete conditions for each species.

Another point of view can be obtained from the hierarchy in management and/or in the designation of salary groups.

The species and classes of designers have a close relationship to the problem of specialization and organization in the design office, and these are discussed in the operator "leadership, management" (Section 7.2.2.7).

#### *7.2.2.4.3 Demands on Designers and their Evaluation*

The demanding activity of designers puts correspondingly high requirements on them, their abilities, skills, knowledge and attitudes. In this part of the theory we deal especially with the general specification of all knowledge (about the general and about the branch/domain/expert knowledge), abilities and skills, and the personal characteristics and outlooks, which are necessary for the profession. This "model of the ideal designer" should serve to derive pictures of the profession and terms of reference. A possibly useful model is contained in figure 7--17, in which the necessary knowledge, abilities and personal characteristics are listed.

| Ideal (engineering) designers possess: | | |
|---|---|---|
| Knowledge, understanding | Abilities, skills, faculties | Personal characteristics, attitudes |
| General knowledge<br>languages<br>literature<br>history<br>geography<br>etc.<br><br>Mathematics<br>Geometry<br>Physics<br>Chemistry<br>etc.<br><br>Technical branch knowledge<br>basic knowledge<br>specialized branch knowledge<br>design knowledge (design tuition)<br>manufacturing knowledge (manuf. technology)<br>materials knowledge<br>etc.<br><br>National economics<br>Legal knowledge<br>Psychology<br>Technical esthetics<br>Ergonomics<br>etc. | Memory<br>Logical thinking<br>Synthesis ability<br>Cost awareness<br>Visualization ability<br>Combination gift<br>Creativity<br>Mental flexibility<br>Methodical working mode<br>Information procuring<br>Decision ability<br>Representation ability<br>Observation ability<br>Concentration ability<br>Punctuality<br>Leadership<br>Organization<br>Orderliness<br>Personal bearing<br>Precise expression<br>Persuasive power<br>etc. | Productivity<br>Perseverance<br>Willpower<br>Honesty<br>Responsibility<br>Duty awareness<br>Openness<br>Thoroughness<br>Conscientiousness<br>Care<br>Contact readiness<br>Broad horizons<br>Objectivity<br>Critical attitude (including self-criticism)<br>Self-confidence<br>Enthusiasm, delight in designing<br>Readiness for cooperation<br>Constant study<br>Fairness<br>Psychological typology<br>Psychic stability<br>etc. |

Figure 7--17 Model of Ideal Designers

#### *7.2.2.4.4 The Education of Engineering Designers*

The development of a student to a successful designer should be guided, and should not be left to chance (compare Section 1.5). Therefore it must be one of the goals of this theory to propose suitable models and didactic instruments, which consider all elements (school, practice, etc.) and integrate them into a total system. The main task remains, however, to process the Design Science generally and particularly for students (further discussions appear in Chapter 9), and to develop the necessary pedagogic and didactic principles, tools, procedures and teaching materials.

The assignment of this task designation to Design Science could be disputed (does it not belong more into the sphere of education?). The task is so special that it can be solved only by design specialists. Of course educationalists must also be included as cooperators. This cooperation shows an important role for specialized organizations combining engineering and education elements, for instance the American Society for Engineering Education (ASEE), the Internationale Gesellschaft für Ingenieurpädagogik (IGIP -- International Society for Engineering Pedagogics), and Société Européenne pour la Formation des Ingénieurs (SEFI -- European Society for Formation of Engineers).

### 7.2.2.5 Working Means -- Operator of the Design Process

Questions:
1. How can designers be supported in the completion of their work?
2. Under which suitable circumstances can individual working means be applied?
3. Which effects in the design process can be taken over or fulfilled by machine?
4. Which requirements and conditions must be met for machine application of such technical means?

The application of working means is in general very different for individual kinds of design transformation. It is particularly dependent on the kind of transformation, the quality of the available technical means, and the knowledge of the technology. From all these aspects, the application of working means in the design process, before the origin of Design Science and before the invention of computers, was limited to resources for the designers" own activities (compare figure 7--11).

The importance of the operator "working means" increases with the new resources, especially with the computer as independent supplier of effects.

The theory should especially consider questions 2 and 4. The creation of conditions and requirements belongs to the most important tasks of Design Science. The regulation of the conventional means (for example through good classification) brings also positive consequences in the rationalizing process. Further in the section about CAD (Section 7.5).

### 7.2.2.6 Branch (Domain) Information -- Operator of the Design Process

Questions:
1. Which knowledge is required by designers for designing?
2. Which forms of knowledge are important for designers?
3. Where can the necessary knowledge be found?
4. How can the quality of the information (the knowledge) be evaluated?
5. How can the information activity be rationally organized?

Branch (domain) information is (apart from the designers) the most important operator of the design process. Designing in engineering under conventional circumstances can be done without methodology, almost without working means and without management (except possibly self-management), but by no means without knowledge of the laws of nature, materials and their characteristics and processing, without standards, patents, etc. Also, rapid provision of important information is decisive for the effectiveness of the design process.

As already indicated, from the point of view of terminology the words *information* (in the arts and colloquially), knowledge and awareness are used practically as synonyms, even if deviations of content are often implied.

As *branch information* (technical or domain knowledge) we consider above all the contents of the Specialized Design Science in that certain area (see Chapter 8). The general knowledge and ability is regarded as a general requirement.

Technical knowledge is divided into branch object knowledge and design process knowledge. Branch object knowledge concerns the object of designing (the technical system to be designed), process knowledge concerns designing itself (as activity and technology).

These definitions answer the questions 1, 2 and 3, because we have now arrived back at the discussed content of Design Science (see Chapter 5). Essential ideas to the range of problems of branch information are explored also in the next sections, in which the object and design process knowledge are treated, and in particular the form and sources of knowledge (see figure 6--2).

In this connection we must refer to a further area, *information and documentation science* (library science). Their object is also information, they look for optimal methods and means of presenting, registering (referencing, cataloging), processing, storing, retrieving (searching), evaluating and interpreting of information, manually and/or with the help of computers (almost without regard to the content and its meaning for a special branch of application). The technology of these activities is treated in the disciplines of documentation (e.g. library studies). Especially the operations of recording (cataloging), organizing and retrieving of documents and providing them to interested people are explored. Because this area deals also with operations which are at times executed by designers, this knowledge should not be foreign to them. This is especially true because they must maintain contacts with libraries, for which information and documentation presents a fundamental discipline. The questions about information sources and evaluation are answered in these areas, and further important information like cataloging (ordering) systems, databases and data carriers (media) are delivered (i.e. in answer to questions 4 and 5).

All rationally working designers must unavoidably find out that they must set up their own *information system*. Whether they use computers, files, or other working means is not significant. They should, however, get instructions from Design Science about how such a data base can be set up, because the cataloging (ordering) system should be harmonized with the knowledge from Design Science.

### 7.2.2.7 Leadership (Management) of the Design Process -- Operator of the Design Process

Questions:

1. Which are the special tasks of leadership (management) in the design process?
2. Which are the tasks in the enterprise in reference to the design processes?
3. Which kinds of organization for the design office, and which leadership strategies are possible? What are the Conditions?
4. Which are the specialization problems and possibilities?
5. How is the work of designing planned?

The leadership and management systems and techniques are organized very differently, according to kind of the design process (see figure 7--16) and the state of the individual elements of the design system (figure 7--10). It is especially difficult to lead and to organize the individual persons.

Design work is surely a particular kind of human activity, which can also be described with attributes such as creative, artistic, etc. (compare Chapter 1). The leadership strategy in the design process must consequently be different from that for other person systems and will hardly be found in a manual for management. Leading designers up to the supreme position must always be branch experts, who understand the work and are ready to take part in discussion "at the drawing board" itself, for otherwise they can hardly find the necessary recognition from their staff members.

All activities of the design leaders, such as determining tasks and assignments, planning, giving orders, determining working methods, demanding effective and complete documentation (e.g. of decisions), work coordinating, organizing and others, must be conducted and influenced by conscious leadership tactics. Continuing education of the staff members should not be forgotten, in branch (object) knowledge, in design process knowledge and in personality development.

It is a further task of Design Science to support the design manager with corresponding knowledge and to make available for design processes the particular theories of specialization, organization and planning. The following often opposing characteristic feature of the design process must be considered:

- creative-innovative tasks which can be very complex;
- long transition times of projects with intermediate stages which are often difficult to control;
- staff members who are often individualists with distinctive character;
- work often transacted under time pressure;
- responsible work, coupled with risks (liability);
- lack of capable staff members (the job "at the drawing board" is not preferred);
- psychological barriers when new methods and techniques (technologies) are introduced;
- space for personality development, career prospects and chances for promotion (for example ascent without transfer into enterprise management);
- simultaneous processing of several tasks which are respectively in different states of origination (design maturity).

#### 7.2.2.8 Active Environment of the Design Process -- Operator of the Design Process

Question:
How and with what does the active environment influence performance and progress of the design process?

The design process proceeds in a material and social environment. The influences which these exercise on the design process, especially on the person, are included in terms such as working conditions, working milieu or working climate, and are generally explored in work science. Included in this field are psychology, physiology, technology of work, as well as work medicine and study of culture -- this implies also the breadth of this range of problems. This range of problems, which is coupled to the job, must be recognized consciously as a micro-situation within the broad framework of the macro-situation (i.e. the company and general society). The influence of the macro-situation on the design process can gain great importance in certain phases (e.g., war, economic crisis). Our investigations will be limited to the micro-situation. Three main cases can be distinguished, which create completely different boundary condition:

1. the case of traditional designing -- this includes A to D of the design processes in figure 7--16 (see also Section 7.2.2.3.5);
2. the case of computer application -- includes E and F of these design processes;
3. the case of automation -- G of the design processes.

The currently important situation is the one of computer application. In a few TS-families (e.g., VLSI -- very-large-scale-integration of computer chips) this is already close to full automation of some parts of the design process. This stage includes the physical and organizational working conditions (position, size, equipment and arrangement, lighting, climatic conditions, noise in the design office) and the psychological working conditions, but also the range of problems of work with the computer (relationship human-computer, interfacing).

The importance of the discussed working conditions is larger than one could suppose. This is proved by questioning designers, who often rate these elements relatively very highly. The influences of the active environment appear also indirectly, working conditions have important relationships to other elements (e.g. to leadership, to working means, to information), whose influences they either strengthen or weaken (brake).

Design Science should put at the disposal of the design process the appropriate results of work science and special research, and emphasize their coupled interactions.

#### 7.2.2.9 Summary -- Design Processes

The treatment of the design system, with the design process at its center (figure 7--10) shows the range of problems in breadth, with all their elements and relationships. An orientation is offered by the matrix which presents the relationships between the goals and features of the design process and its influential factors. In figure 7--18, these relationships are rated on a four-point scale (decisive, medium, small, or no influence). From this matrix we can see that they determine more or less the place values of individual factors for the chosen characteristics.

| Aim concerning Characterization of the DesP ↓ / Operators -- factors influencing the Design Process (DesP) → | Engineering Designers | Design Information (Branch, Domain) | Design Methodology and Methods | Working Means (Equipment, Tools) | Management of the Design Process | Working Conditions, Environment | Techniques of Representation |
|---|---|---|---|---|---|---|---|
| 1 Quality of the TS to be designed | decisive | decisive | medium | small | small | small | small |
| 2 Design time (duration) | decisive | decisive | decisive | decisive | medium | small | small |
| 3 Efficiency and effectiveness | decisive | decisive | medium | decisive | medium | small | small |
| 4 Reduction of risk for designers | decisive | decisive | decisive | small | medium | none | small |
| 5 Ratio of skilled to routine work | small | medium | none | decisive | decisive | none | none |
| 6 Maturation time for designers | medium | decisive | decisive | small | medium | none | none |
| 7 Design costs | decisive | decisive | decisive | decisive | medium | small | small |
| 8 Team work | decisive | medium | decisive | small | decisive | small | small |

Measure of Influence: decisive, medium, small, none

Figure 7--18 Influence of Factors (Operators) on the Aims of the Design Process

## 7.3 Branch Knowledge

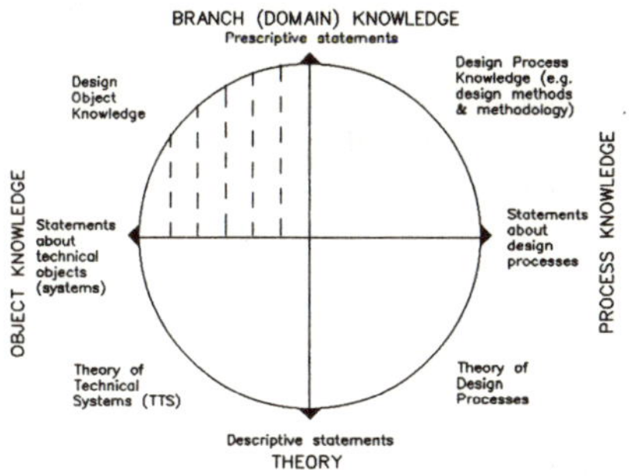


With the two following partial areas of Design Science (Sections 7.3 and 7.4) we transfer our considerations away from the theory (descriptive statements, see Sections 7.1 and 7.2) to the area of the *prescriptive statements* which are closely related to engineering practice (compare figure 5--2 upper "northern" half). The prescriptive areas (together called prescriptive design-technical knowledge) should contain the practical knowledge which should give answers directly and quickly to those immediate questions asked by designers in all situations when designing. In analogy to the theory, the prescriptive technical knowledge is treated in his two partial areas:

1. in the branch knowledge (knowledge about more specific technical systems), and
2. in the design process knowledge (knowledge about designing).

Designers, whilst they are creating, repeatedly ask questions such as:

- what is the problem, how is it formulated, what does it mean?
- "how" (what process and procedure), and
- "with what" (means) can I solve a problem?

The multiplicity and variety of the problems encountered corresponds to the breadth of the possible design activities (compare figure 7--12). Therefore the necessary technical knowledge should not only be available, but also quickly findable, and should answer the questions directly (both in object and in process knowledge) as they emerge in designing. Especially in realizing the individual TS-requirements (which later appear as the properties of the realized technical systems) all influencing factors must be known, and their relationships must be available, if possible in quantified form.

### 7.3.1 The Development of Forms of Branch Knowledge and its Sources

Questions:
1. Which forms of branch (domain) knowledge are favorable for designers or for computers?
2. From where does the prescriptive branch knowledge originate?

Such a knowledge system as an independent area exists only in attempts and fragments. The knowledge which designers need today and will also need in future, once found, remained in earlier times scattered in some relevant sciences. In additional it was in most cases only vaguely formulated and, as merely personal "experiences" of individual designers, not integrated into a system. The retrieval or transfer of knowledge into other specialties was therefore extremely complicated. Experience was the main source of knowledge. Only later, about the 1920's, engineering sciences became the dominating source of this knowledge (figure 7--19, path 1).

This meant for designers that they (i.e. each designer, mainly for themselves and partially from their colleagues) had to derive their concrete specific knowledge for their own specialty from the general knowledge. Where no engineering sciences had emerged, the search for technical knowledge was possibly extended to the natural and "pure" sciences (physics, chemistry, biology) and in further areas. The "field of search" of designers consisted therefore -- and still consists -- of the two hierarchical knowledge levels shown in figure 7--19 (path 2) in columns 1 and 2, and of course also experience knowledge (column 4) and standards (column 3). It should also be mentioned that some TS-features or TS-properties (figures 7--4 and 7--5) could not be associated with an existing knowledge area. This is explained in the following on some examples.

Approximately in the 1940's this disadvantage came clearly to light. It was discovered that too large a transfer of knowledge for designers was to be overcome, so that information from the engineering sciences could be utilized in engineering design. The area of realization of technical systems which showed the necessary future direction, when the discipline "design for manufacture" was formulated (compare [48,49,80,81]). The basis for this partial area of prescriptive branch knowledge actually emerged in this way, and is situated in column 6, figure 7--19 (compare figure 7--4). For design practice, this knowledge system should serve as a substantial reservoir of necessary knowledge (path 3).

### 7.3.2 The Structure of Branch Knowledge

Question:
What is the structure of branch knowledge?

The structural elements of branch knowledge should be adapted within the objectives (goals), on the one hand to the situations in the design process (see design situation, Section 7.2.2.3.5), on the other hand to the kind of operand (type of technical system to be designed, the different aspects of the technical systems), as they have already been explored in the Theory of Technical Systems (Section 7.1 and [214,219]).

The knowledge about the origin, development in time, and taxonomy of technical systems remains fairly constant, regardless in which design phase we ask about them. Therefore they can form independent areas, which fully agree with the corresponding partial areas of the Theory of Technical Systems. The essential difference exists merely in the kind of knowledge, whether descriptive or prescriptive.

The knowledge about the nature, effects, modes of action and structuring of the technical systems are in contrast specific to and dependent on the branch (domain) or specialty, working stage and the kind of design activity. Substantially different information is required for the synthesis of the total system when conceptualizing, than in the layout stage, and again different sets of information for evaluating the proposed or realized systems.

Following these insights, this branch (object, domain) knowledge should be categorized primarily from the kind of design activity (design process knowledge), and then from the procedural stage. The assignment of the relevant branch knowledge to the individual design activities can be accomplished in very differentiated ways, according to the importance of individual characteristics for the concrete situation. We must first find out which

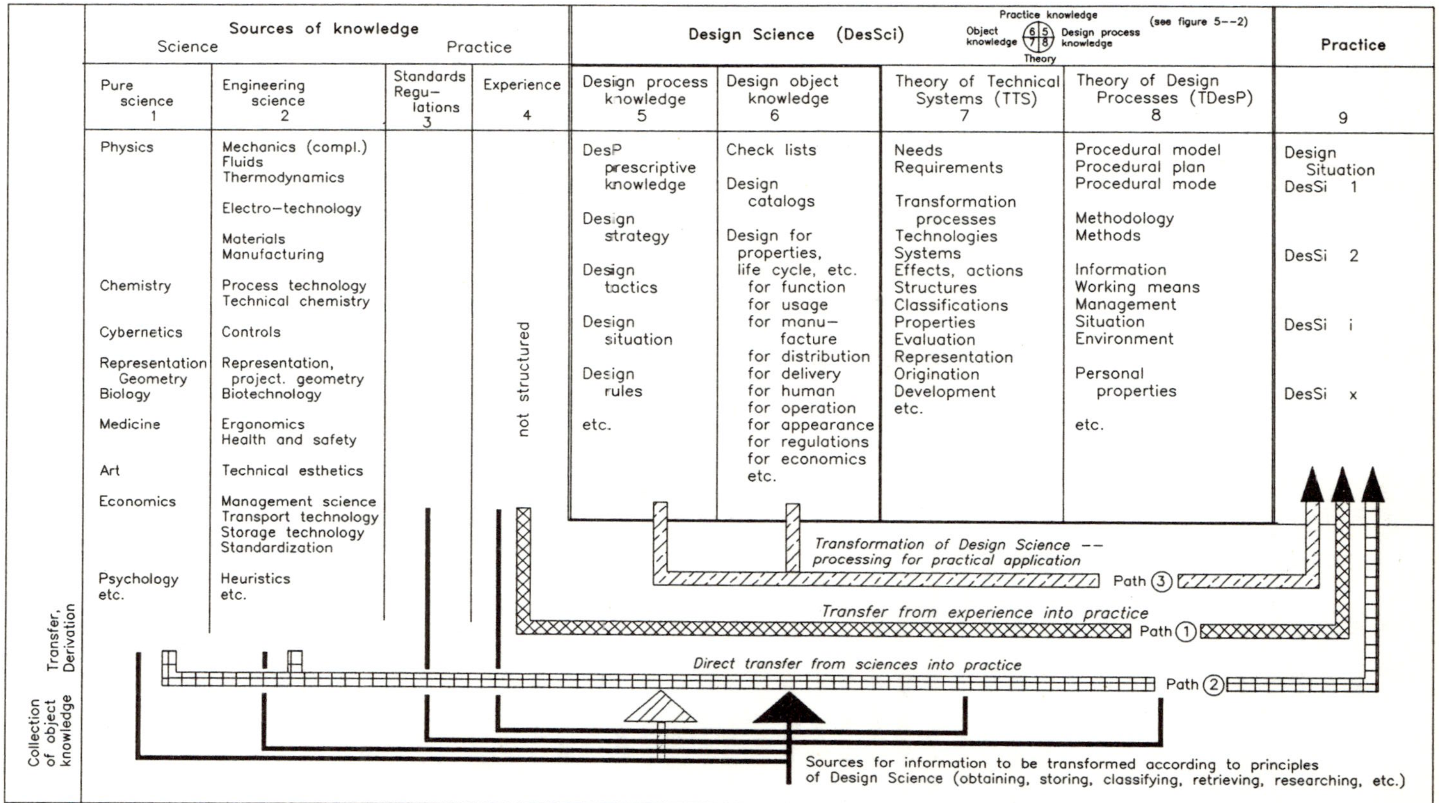


Figure 7--19 General and Specialized Knowledge Systems and their Relationships for Practical Design Work

characteristics exert the decisive influence on the categorization (order) of the knowledge; only afterwards can the primary ordering characteristics be used. If for example the form (shape) of components is to be established, then it is obvious that (beside the function) the production processes are the prime influence on the form. Therefore the production processes can be chosen as key indicator for the categorization (ordering) of the knowledge about form.

But also the choice of the design activities as key indicator for the categorization is not free of problems. This has already been demonstrated in the treatment the design situation (Section 7.2.2.3.5). The design activity may neither be set too broadly (compare figure 7--12, plane 1, design stages), nor too narrowly (figure 7--12, planes 4, elementary activities, and 5, operations). Thus we come selectively to the predominant planes 2 (design operations) and 3 (basic operations, problem solving thinking and actions) as main categorizing feature for the design activities.

In the area of the operand (the system to be designed), the first key characteristic is the branch of industry (the functions of the relevant TS-group, therefore for example turbines, machine tools) in which several more hierarchical planes can be distinguished. Following this, the second key indicator is the level of originality of the system. We could add the complexity level between these two, and with it form a special group of "elements" (this coincides for practical purposes with the conventional machine elements).

The design situation offers a further possibility for defining categories. The possible arrangements of design situations and the associated structural elements of the branch knowledge are obvious from figure 7--20.

If the branch knowledge is independent of design operations, then an aspect of the operand can become the key indicator for categorization. The area (B) from figure 7--20 contains such cases. So the certain elements of branch knowledge can be derived with the help of the general theory of origination (life cycle, figure 7--8), for example in the framework of a certain specialty (industrial branch), further for different levels of originality, and still further for different kinds of production (mass, large or small batch-series, and single-piece "one-of-a-kind" production).

An important categorization of technical systems develops also from the degree of complexity. Plant, machines, assembly groups, and individual components cover the four levels for machine systems (compare Section 8.1), whereby each level displays several particular characteristics. Especially the latter two levels contain constructional units which are generally applicable across many specialties (branches). For this case it would not be favorable to categorize the branch knowledge about constructional units according to design activities. A better categorizing feature is the order derived from TS-organs. For that reason, elements of the branch knowledge are arranged according to elementary organs or organisms (such as connections, transmissions) and can contain the complete knowledge about these specific constructional units. The design situation which is applicable for such elements of the branch knowledge can be described as follows:

> an organ structure (solution concept) has been established for the system to be designed, and the component structure should now be established.

The knowledge in this area can also be processed in a rather general fashion, for instance in the way in which it is used in books about machine elements, or published for a certain branch and a particular manufacturer, in the concrete form of an industrial guideline or even a standard.

The current considerations about the structure of branch knowledge prove clearly that no generally valid solution of this problem can be recommended. Some fairly clear classes of the elements have emerged, but they do not cover the whole content of the area. We can

| | Design Situation | | | | | Design Object Knowledge | e.g. Figure |
|---|---|---|---|---|---|---|---|
| | Design Activity | | Operand -- TS to be designed | | | | |
| | Design Operation | Basic Operation | Branch, Domain | Level of Originality | Further Aspects, Characteristics | | |
| A | Establish mode of action | Either all (when optimizing) or individual | | | | Catalogs of modes of action | 7--21 |
| | Establish function | | | | | | |
| | Establish function carrier, organ | Elaborate the assigned specification<br>Search for solutions<br>Establish arrangement<br>Evaluate etc. | for every branch or domain | for every level of originality | | List of requirements (master check-list)<br>Catalogs of solutions, arrangements | |
| | Establish material + Dimension + Form-giving + .... | | fairly independent of branch or domain, only its 'specialities' | independent of level of originality, only specialized methods | | Design for materials and manufacture (ordered according to manufacturing methods) e.g. welded parts<br>Design rules | 7--22<br>7--23 |
| B | for every operation or complex | | for every branch or domain | for every level | Origination plan, action plan | | 7--8<br>7--14<br>7--11 |
| | | | | | Development plan – usually extrapolated | | |
| | | | | | Taxonomy | | 8--4 |
| C | every | | as desired | fairly independent | Components, taxonomy | | 8--11 |
| | | | as desired | | Facility planning | | 8--2 |

Figure 7--20 Design Situation and the Forms of (Branch, Domain) Object Knowledge

therefore expect a mixed solution for engineering practice. And the main reason is that guidance does not come from a total conception. It is directed by a successive process (origination of individual areas one after another), which is intimately connected with other areas, especially with that of the construction of knowledge-based systems (expert systems).

### 7.3.3 The Form of Branch Knowledge

Question:
Which forms of the branch knowledge are favorable for application in the design process (when designing)?

Establishing the most favorable form of branch knowledge is the main requirement for its effective application. The knowledge should always bring a direct answer to the immediate question, whether it is asked by designers or by the computer. The series of design questions is formulated not only by the procedure, but also by the design activities (see figure 7--12).

Fortunately the diversity of the possible questions does not lead on the equal diversity of forms. These appear by chance only in some classes. We can therefore cover and position the total area from some few classes of questions. The most important and frequent forms are discussed here.

*A) Class "Which are .... ?"*

Many design steps ask about solutions of a certain problem, e.g.: which demands or requirements, which action principles, which organs, etc., fulfill this function?

For given boundary conditions, the branch knowledge should therefore place all solution possibilities at the disposal of the designer or computer. We can differentiate several levels of such knowledge material for delivery to designers, for example:

- enumeration (a simple enumeration without further statements);
- enumeration with statements of the conditions;
- enumeration along with further documentation, e.g., formulae;
- enumeration with the properties of a solution for evaluation and comparison.

Such knowledge is stored, for example, in design catalogs [257,376] and in guidance sheets, which use mostly tabular form. They are usually compiled in three or four parts: categorizing (classification), presentation, extension (reference), and possibly an appendix (according to Roth [376]). Further details are shown in figure 7--21, where some examples illustrate the possibilities of this knowledge system. The successful application of design catalogs is predicated on the systematic order of the categorizing arrangement. Often it is not the concrete solutions, but the classes of knowledge, which bring further important inspiration. A complete enumeration of the classes is more easily possible than a complete enumeration of the solutions (examples: classes of properties, or mechanical, hydraulic, electric amplifiers).

It would be conceivable to store such knowledge in knowledge-based systems, so that the necessary information (including advice) is accessible from the computer. Advice could also contain heuristic instructions for the selection of values for properties, e.g., as rules. A part of these instructions are already available within the design catalogs, but the extent could be expanded much further.

Design Catalogs: A collection of useful knowledge for engineering designers, external to human memory. They are usually presented in the form of tables, which are arranged according to methodical viewpoints, with information as complete as possible (within a given framework) and systematically categorized. They permit targeted search for particular content. Most catalogs contain parts that categorize, present, and extend the information, and appendices if needed. Operation catalogs do not usually contain extension parts.

Categorization Part: That part of a design catalog that determines the systematic arrangement and completeness of its contents. It should be finalized during development of the catalog, such that it also delivers a framework for theoretically possible contents, even if such contents are unknown at the time (they yield empty fields in the presentation parts). The categorization part contains exclusively the characteristics for classifying potential contents.

Presentation Part: That part of a design catalog that presents the content, i.e. the objects, solutions, or operations. The contents are represented by terminology, laws, symbols, formulae, sketches, drawings, etc. Those contents that are known do not necessarily fill all available fields in the presentation part, they can be completed at a later date.

Extension Part: That part of a design catalog which contains searchable information that is not classifed by the categorization part. The extension part (a one-dimensional catalog) is adapted to the needs of the user, can be extended as desired without changes to the categorization or presentation parts.

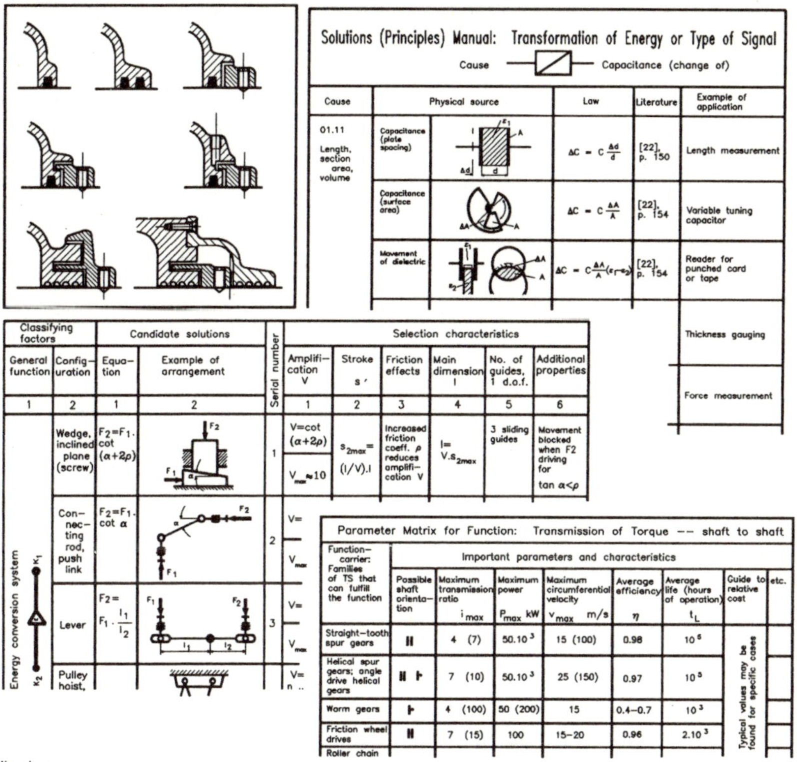


Literature:
Roth, K., Konstruieren mit Konstruktionskatalogen (Designing with Design Catalogs) 2 ed. 2 vols., Berlin/Heidelberg: Springer-Verlag, 1994
Koller, R., Konstruktionslehre für den Maschinenbau (Study of Designing for Mechanical Engineering) 2 ed., Berlin/Heidelberg: Springer-Verlag, 1985

Figure 7--21 Forms of Design Catalogs

*B) Class "how should form-giving be performed?" (or any other branch activity)*

Especially during form-giving (of components and assembly groups), the most frequent questions refer to design properties. How should form-giving be performed, or dimensioning (establishing sizes), choosing raw materials and determining surface quality, so that the given demands (function, strength, friendliness for production, transport, storage or operation, etc.) are fulfilled? The instructions and information for this kind of question are recorded as branch knowledge in the disciplines usually known as "Design for ..." classes (see figure 7--5, part 3), for the individual properties of technical systems (e.g., design for manufacture and assembly [48,49,80,81], ergonomics [27], maintenance, and others, compare figures 7--4 and 7--5). We could use the expression "property-friendly designing," since operability or transport friendliness can be included in the TS-properties, and can be classified into the normal set of properties.

Using the example of the metal casting process (figure 7--22), the large difference in content and form of the information is made clear between manufacturing technology (in section A of this enumeration) and the information for casting-friendly design (section C). Designers (or computers) get a direct indication about how the forms (shapes) and dimensions (sizes) should be established, so that a casting (metal cast component) suitable for modeling (casting patternmaking), forming (in sand or other form material), casting (pouring), fettling and machining can exist. The prevailing knowledge form in this area consists of "rules" as text and pictures (either examples of good practice, or good and bad for comparison), as well as diagrams or equations.

*C) Class "One should do it this way"*

Often no large difference exists between the prescriptive and normative statements, unless the obligatory nature and depth of treatment for the instruction are different. As a further pattern of the form of branch knowledge, the instructions for establishing the design properties of a certain technical system can be considered (figure 7--23). These are usually related to a set of scientific theories of the relevant phenomena, and suitable heuristics and experience values. The individual properties -- form, dimension, raw material or surface processing -- can be recommended (guidelines -- prescriptive statements) or prescribed (standard -- normative statements).

These three classes of branch knowledge and the examples of the appropriate forms do not exhaust the possibilities. In connection with the contents, we will always find new forms which present the branch knowledge more completely and make retrieval and use still more effective. The other element, the computer, will make further demands, and enrich the offering through its abilities, independently of whether the knowledge serves designers, or is installed in a knowledge-based system.

### 7.3.4 The Generation of Branch Knowledge

Question:
From where does the branch knowledge originate?

One of the largest problems in planning the branch knowledge lies in the acquisition of the sources. As shown in figure 7--19 (the lowest comments), all kinds of knowledge sources must be used so that a complete system of design knowledge is realized. The hope to be

A -- Knowledge of manufacturing technology and faults:

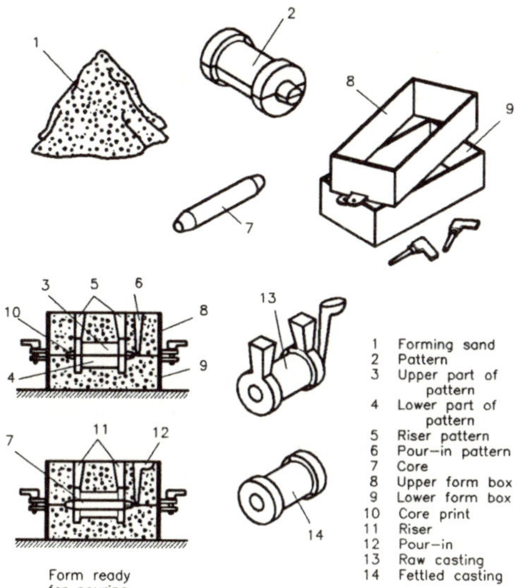


Form ready for pouring

Defects in castings, fault occurrences on finished cast item. Casting faults are so varied that a complete listing with characteristics and advice for avoidance is not possible in this space. Casting faults can consist of deficiencies in form and dimensions, but can also reduce the strength properties of the cast item, cause leakage, make machining and surface treatments more difficult, influence corrosion resistance and appearance, and lead to scrap. Frequently an interaction between different causes makes recognition and avoidance of casting faults more difficult. Even with careful procedures, the levels of scrap in casting production can vary over wide limits, scrap levels depend on materials to be cast, and forms of castings.

Casting faults include:

faults in shape due to forming errors;
shifted cores;
fettling errors.

B -- lead to requirements for castings:

i.e. regarding structure + form (shape) + dimensions + materials + tolerances + surface

Suitability for casting

- Suitable for pattern-making
- Suitable for forming
- Suitable for pouring
- Suitable for stresses from solidification
- Suitable for fettling
- Suitable for machining

C -- lead to rules for form-design and recommendations for sizing (dimensioning):

*e.g.*
- maintain constant wall thicknesses,
- avoid sharp curves,
- avoid collections of material,
- use smooth transitions,
- split up large surfaces for machining,
- provide draft angles for lifting pattern parts.

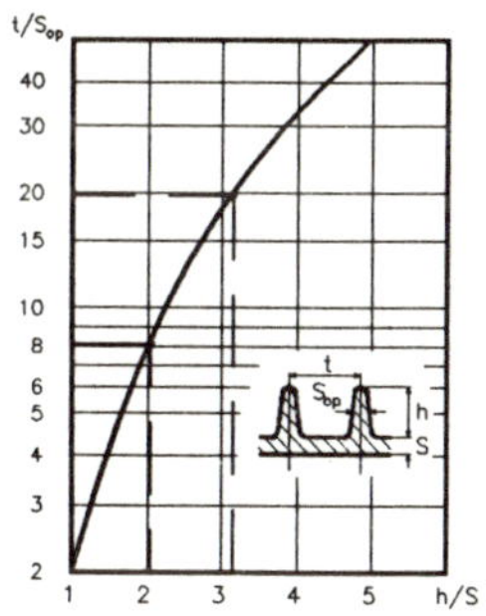


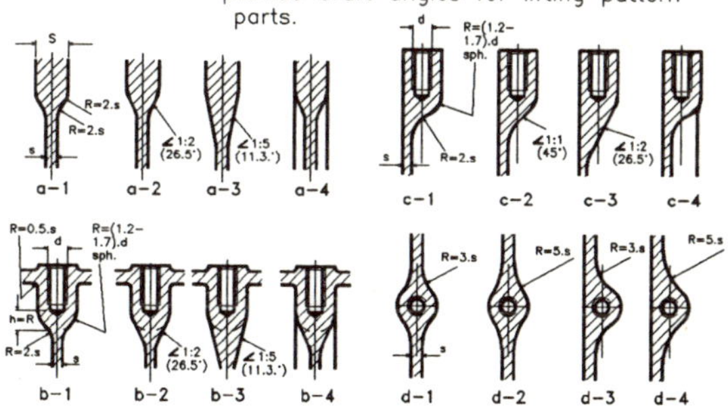


Figure 7--22 Knowledge -- Designing for Casting

A -- Requirements and typical application

e.g. crankshaft and connecting rod bearings

B -- Dimensions - calculation of bearings see Guideline Sht. SB ...

Select e.g.

A =
B =

C -- Form-giving

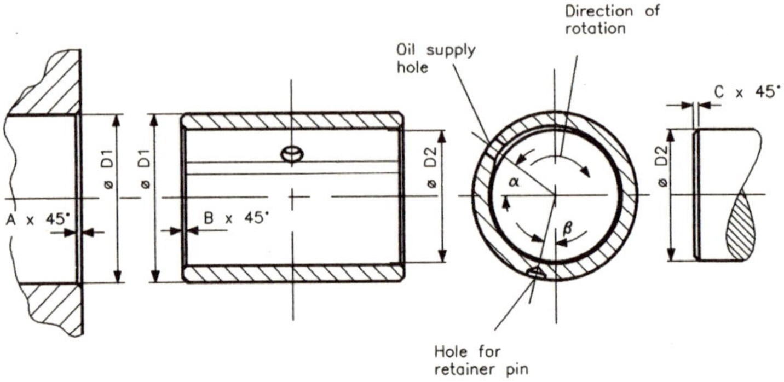


D -- Material

e.g.

phosphor bronze 2B 8 for ...
lead bronze LB9 for ...

E -- Machining

e.g. surface

a) ..... grind, Rs = 0.4
b) .....

F -- Tolerances, fits

Diameter D1

bore H6
bushing r5

G -- Type of manufacture

| Figure 7--23 | Design Guideline -- Sliding Bearing (scheme) |
|---|---|

able to solve the problem of generating a branch knowledge system only from the available literature, is a pure illusion. A very extensive source of knowledge that does not appear in the published literature consists of experiences -- a very individualized system -- and some of it is proprietary, subject to security precautions by an individual enterprise.

Especially two problems arise. One relates to the difficulty of formulating personal experiences (subjective knowledge) and to find important dependencies and laws. The second problem is in some ways the consequence of the first, that often vague (fuzzy) knowledge must be processed. These are in particular transient entry problems.

It is further necessary to transfer the available branch knowledge between specialties. Designers should direct their particular attention to published research, and fault events (e.g. [346,448]), in own and other specialties, and to draw consequences and derive guidelines for their own work from such sources. All data and statements should be subjected to tests for correctness, with regard for continuity, reliability of the source, verifiability of the statements, agreement of the data in transitions between sources, etc. That these tasks are too demanding for individual designers is evident. More reliable information is accessible in part from different organizations, e.g., from ESDU [6,7].

In generating the branch knowledge, the situation has changed in essence. Through the new expertise and laws in the areas of theory, especially the Theory of Technical Systems, a positive influence has been found; the right direction of development from the theory to the guidelines is made possible.

### 7.3.5 Branch Knowledge -- Areas of Validity

The branch knowledge is by nature a concrete form of knowledge. Its dissemination is therefore to be expected in the branches of industry (fields of use) and even more in the individual enterprises. The "know-how" can assume clear and unique forms from the completely defined elements of the design process (compare figure 7--23).

There are also areas which can contain concrete references, and can serve a broad consumer circle with rather limited "specialties," although they were developed on a general plane. As example we can consider the areas of "design for properties," as described in Section 7.3.3, especially figure 7--5, parts 2 and 3. Figure 7--22 illustrates how the indications obtained from knowledge of manufacturing technology, as rules or form recommendations (part C), have an extensive validity, and represent a very concrete form of knowledge when the special additions for the enterprise are made.

### 7.3.6 State and Prospects for Development of Branch Knowledge

If they think about their demands for technical knowledge, branch (domain, object) knowledge appears to all designers as something self-evident. Yet only a fraction of this knowledge was available before the introduction of methodical (systematic) designing. The general difficulty in constructing and maintaining these knowledge systems hinders their broader application; and it is a fact that experienced designers are generally not convinced of the benefits likely from constructing these systems. The application of a system of branch knowledge could, beside other consequences (discussed in this section), also substantially shorten the expected "maturity time" of designers.

Practically, an essential change is to be expected, in our opinion, only when the knowledge-based computer systems demand a storage of knowledge, and "discover" that the branch knowledge is an essential element of the necessary designers" knowledge.

## 7.4 Knowledge about Design Processes

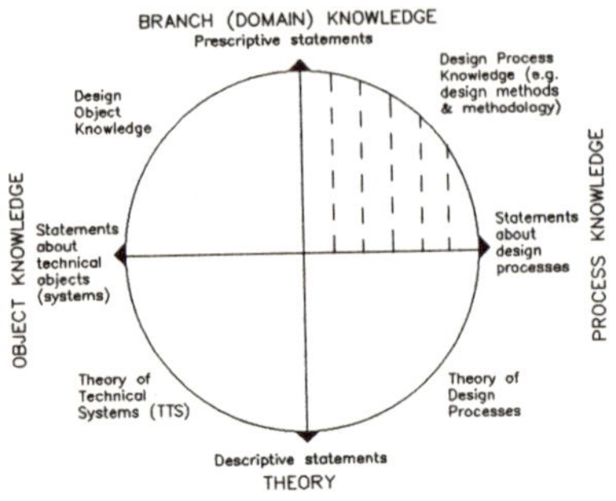


The second partial area of technical (prescriptive) knowledge -- knowledge about design processes -- deals with the transformation according to figure 7--10. The Theory of Design Processes was treated in Section 7.2, and questions relevant for the theory were asked. In contrast, these questions for a designer in engineering practice assume a different, more concrete form. According to the functions and rank of the individual designers, the emphasis of the work and the formulation of the questions can be found in different places. On average, the following question formulations are frequently repeated:

1. How should designers proceed for a concrete design task?
2. Which method should designers choose (reference Section 7.2) for solving a certain problem (under the given conditions), or how should they behave?
3. Which working means can they use in a certain design situation?
4. Which branch (domain) information is needed for particular design situations, and where can it be found? (Coordinating with branch knowledge)
5. How should certain design tasks be led (managed) and organized?
6. How can evaluations (of proposed TS) be better performed?
7. How can the work of designers be improved?

Compared to the questions in the theory, the difference is clear. We do not ask about a systematic treatment of a knowledge area, but for a recommendation about how a problem can be solved in a certain situation (or certain class of situations), or how one can behave and which measures should be used. Such a designated task for a knowledge system is difficult in comparison to the theory, especially because of the very large number of possible situations (if all variations are to be considered).

In order to keep the knowledge system within concrete boundaries, we must not only deal with the knowledge about classes of situations. We must be more specific, and deal only with selected, characteristic classes. In this case the restriction is not too difficult, because analysis shows that the individual questions are not asked with equal frequency. This insight serves now as starting position for considerations on the structure of the knowledge about design processes.

The required technical system is only rarely treated as a system to be newly designed, in which case the problem then consists of designing without an available model (precedent). The product is (in the ideal case of a design process) continuously more concretely defined. It experiences therefore a development in a crudely linear progression [310], from the given task to the concrete description of the system ready for the manufacture, see figure 7--24, part 1. Deviations from linearity are necessary because of iteration, recursiveness, and the need to work on one section (or a few) of the problem at a time. Further problems can occur which can be solved through a similar, crudely linear procedure. This is valid even if a design problem is decomposed into smaller partial problems, and during all the necessary iterations, recursions and revisions.

This scheme indicates that it should be natural for designers to incorporate information about manufacturing into the documents for designed systems. Concurrent (or simultaneous) engineering is supported by systematic work, especially because of the need for iteration and recursiveness to optimize a proposed system. This assumes that sufficient will and encouragement to cooperate is shown by personnel in other specialties and management.

According to the presented scheme (see figure 7--13) this solution path should traverse a series of different modeling forms which bring a flowing, continuously more concrete description of the technical system (TS). The crudely linear procedures are supported by numerous shorter cycles of detail problem solving (see figure 7--12, level 3: basic operations): The partial task is defined, information is procured, different solution possibilities are established, explored (analyzed), evaluated, selected, verified, improved, represented, then re-introduced into the main current of the solution procedure and documentation, communicated, etc.

The predominant number of cases of designing deals with alterations, changes, adjustments, variations, adaptations, etc.: A solution already exists, it should be adjusted or improved in one or another aspect of the designated task. Here too we need to initially designate and clarify a task. Then a more concrete structure can be taken over, either from an existing TS-model (see figure 7--3) as a break or jump in the crudely linear procedures, or obtained by abstracting from an already designed model, see figure 7--24, part 2. Following such a step, we can again follow the procedural model (see figure 7--13) to establish the revised manufacturing documentation. Through abstracting and subsequent concretizing a reversing loop through the structure models is traversed.

For alterations in the constructional details (configuration and/or parametrization) and their assembly, i.e. changing only the component structure, the reversing loop through abstraction from a model is short (or a break or jump from the task designation very long). If larger regions of the component structure are to be changed, especially when newly developed components or organisms are available, the reversing loop can be extended in time, and reach higher into abstractions of the structure. Still longer time and higher abstraction in the loop is needed if changes are demanded in the organ structure, e.g., if the functions of the technical system should be solved electronically instead of mechanically (e.g., for regulation and control, up to automation). Changes can even occur in the function structure or in the chosen technology, whereby the reversing loop for some regions becomes still longer and higher. Under this last condition, the design path can probably be made more rational by adopting or newly developing a higher, more abstract structure (small jump or break in a "novel" process), rather than by abstracting. The case for concurrent engineering is again shown.

### 7.4.1 The Structure of Knowledge about Design Processes

If our question catalog to Section 7.4 is used as basis for the inquiry about a structure, the possibility to organize the questions in three classes arises, namely:

(1) questions of methodical character, asking about available and suitable methods and behaviors (e.g., questions 1, 2);

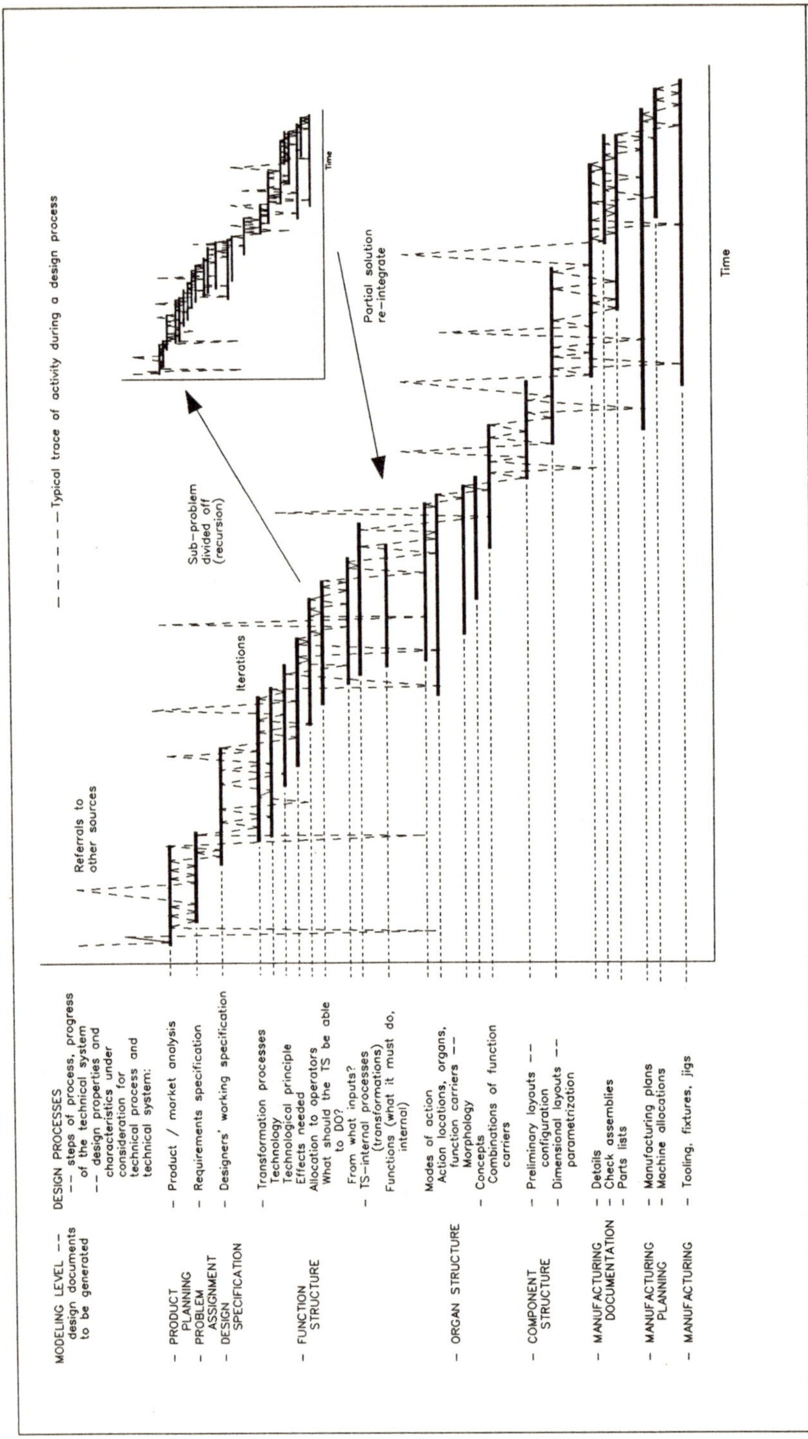


Figure 7--24 Part 1 of 2 Design Situation -- Procedural Aspects and Concurrency -- NEW DESIGN

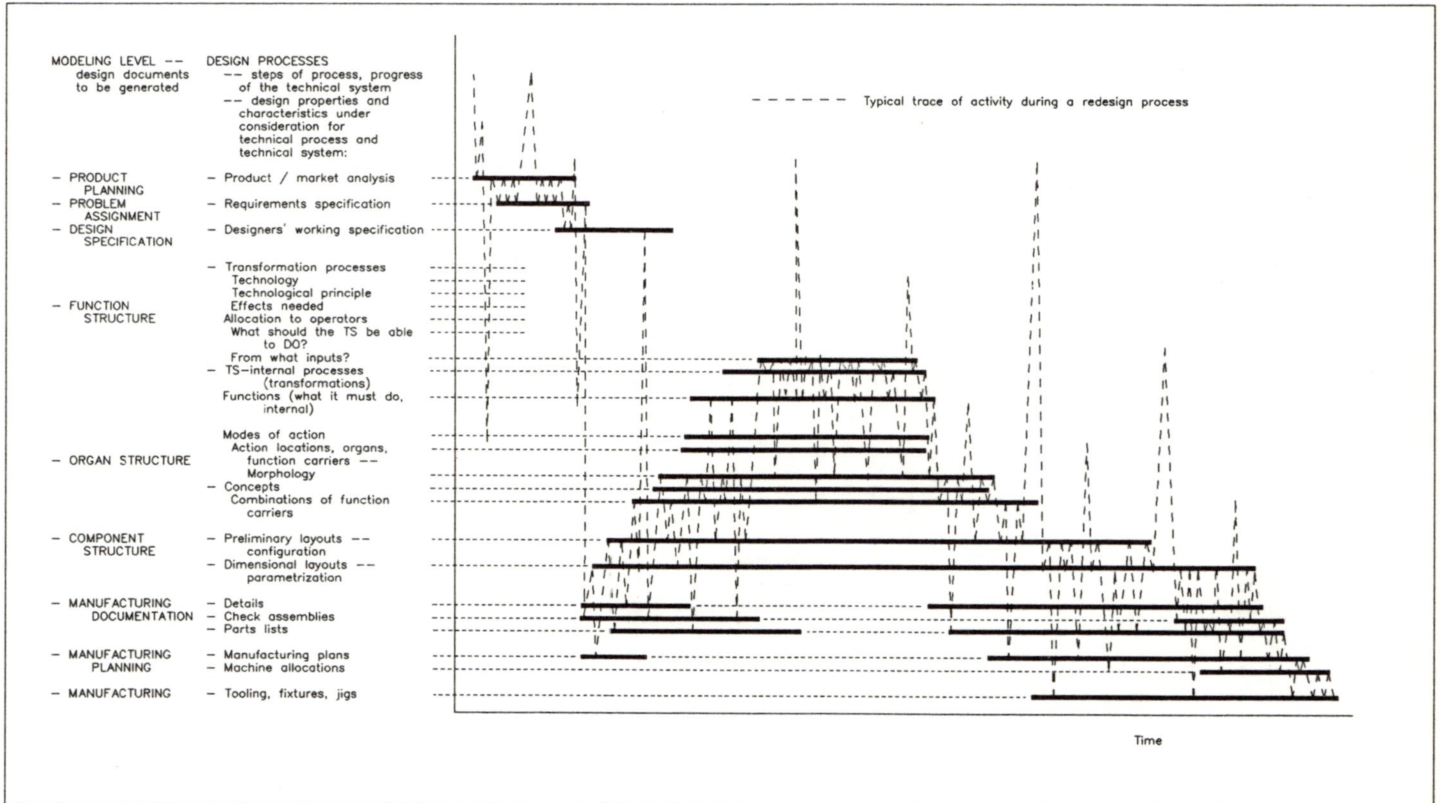


Figure 7--24 Part 2 of 2 Design Situation -- Procedural Aspects and Concurrency -- REDESIGN

(2) questions about relationships in the design process, about its characterization, evaluation, and especially about operators of the design process (e.g., questions 3, 5);
(3) special questions, which are tied to a certain situation (e.g., questions 6, 7).

The class (1) can be described as a range of problems that is interesting for many designer (and computer programmers). Questions are frequently asked regarding this range of problems. Answers are available in two basic forms, see figure 7--25. The first form contains listings of available methods, usually with descriptions of the procedures for operating them [98,228,239], typified by the upper section of figure 7--25. The second form relates possible and preferred usage of a method to the stages of designing, as indicated in the lower section of figure 7--25. As may be seen, the answers, i.e. the desired knowledge, can be organized fairly reliably using the design activities as classes. (This knowledge approaches that of the branch knowledge -- compare figure 7--20).

The knowledge in class (2) is predominantly of interest for a smaller group of personnel -- the leading designers (e.g., project leaders as part of the management hierarchy). This knowledge is seldom called for, it is for certain special situations, i.e. not for day-to-day use.

Apart from these, now and then particular singular situations emerge, which display little similarity with other situations. General knowledge for such cases does not suffice, it is necessary to obtain special knowledge for these situations.

From this analysis, we adopt the three classes (1), (2) and (3) as fundamental structural elements, and the characteristics of design activity are used as ordering features of class (1).

The analysis has also drawn attention to the frequency of application of the knowledge. From the viewpoint of effectiveness it is not rewarding to plan the construction of the information systems for rarely asked questions and seldom encountered situations.

### 7.4.2 The Form of Knowledge about Design Processes

The importance of the form in which the knowledge is presented has been emphasized repeatedly. For the knowledge about design processes, the following forms are beneficial:

1. Catalogs -- the design catalogs that were described in the area of branch knowledge (compare figure 7--21), can also present the knowledge about processes, in the form of operation catalogs (e.g., methods catalogs, or descriptions [98,228,239]). They contain predominantly the procedure, or its steps and rules, as well as the conditions of application and criteria for employment. In contrast to object catalogs they contain no extension part.
2. Models -- it is recommended, and confirmed by experience, that the procedure and other methodical indications should be presented in the form of models, not only by text. The engineer understands and “reads” diagrams considerably more easily than text (or than mathematical relationships) and retains graphical elements and relationships better in memory than words (compare figure 7--13).
3. Tables, decision table -- the representation of complicated relationships or combination possibilities can be clearly recorded in form of a table or matrix. Therefore this form is often utilized (compare morphological matrix, or “house of quality” in the QFD-method). A particular form of table is the decision table (logic states table), which is used for pre-determination of decision procedures, but also for the preparation of the logic flow for computer programs. The development of decision tables can be different, however they must contain rules and measures following the pattern “if -- then.”

Goal–directed searching for optimal solutions to a design problem, at any suitable level of abstraction of a TS–model, can be assisted by the use of appropriate problem solving methods. According to the type of procedure, we distinguish two groups of methods:

-- discursive problem solving methods;
-- intuitive problem solving methods.

Discursive means that the method recommends a stepwise procedure that builds systematically towards a solution or set of solutions. Using discursive methods, a previously defined goal can be approached according to a plan. By different combinations of known elementary solutions, new relationships can be discovered, and therefore new solutions can be found.

Solutions may also be found with the help of intuitive problem solving methods. These are methods that lead to finding new solutions by employing intuition and spontaneity. For the problem solver, approaches to solutions come suddenly into consciousness

Both kinds of methods can and should be used together.

Methods for finding solutions

Discursive methods
-- morphological matrix
-- use of laws of similarity
-- considerations of analogies
-- methods of structuring
-- method of reversal (Zwicky)

Intuitive methods
-- brainstorming
-- brainwriting (method 6–3–5)
-- synectics
-- induced creativity
-- contradiction–oriented procedure
-- heuristic principles (Polya)

Example:
Heuristic principles characterize a problem solving procedure as a battle with a problem that is fought under strict rules of play (tactics). These tactics cannot be of general validity, they depend on the subject. In order to learn them, each person must work through examples individually (and in groups, teams of individuals), then abstract the successful heuristic principles, and re–use them in adapted and modified form for later problems. Typical principles include, according to Polya:
-- introduce and solve an auxiliary problem -- auxiliary problems are problems that we do not approach for themselves, but only because we hope that considering them can help towards solving the original and important problem. The solution of the original problem is the goal, the auxiliary problem is a means by which we attempt to reach the goal.
-- search for analogies -- analogies are present in all out thoughts, daily conversations, and even trivial conclusions. They also appear in all artistic expression and high scientific achievement, whereby all kinds of analogies may play a role in discovering a solution.
-- generalize the problem -- generalizing leads from considering an object or process (or a smaller set of such objects) to considering a larger set (collective) of objects or processes that contain the goal object(s).
-- specializing, particularizing the problem -- specializing leads from considering a given set of objects or processes to considering a smaller sub–set.
-- induction -- induction is the method whereby general laws can be discovered by observing and combining particular cases.

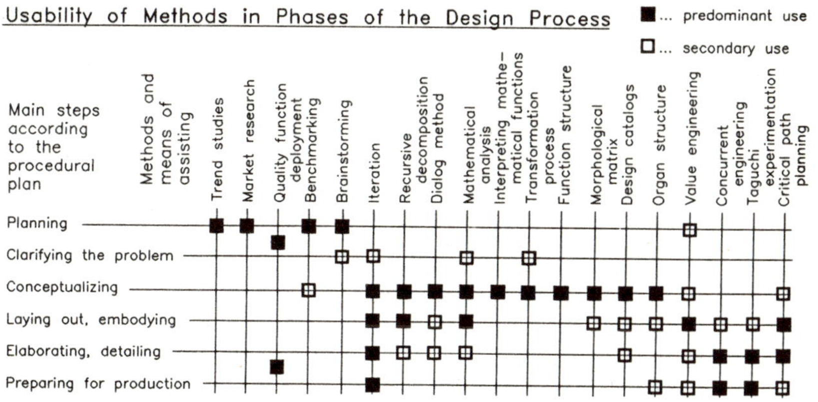


Figure 7--25 Design Tactics: Types and Usage of Methods

# 7.5 Quasi Main Areas

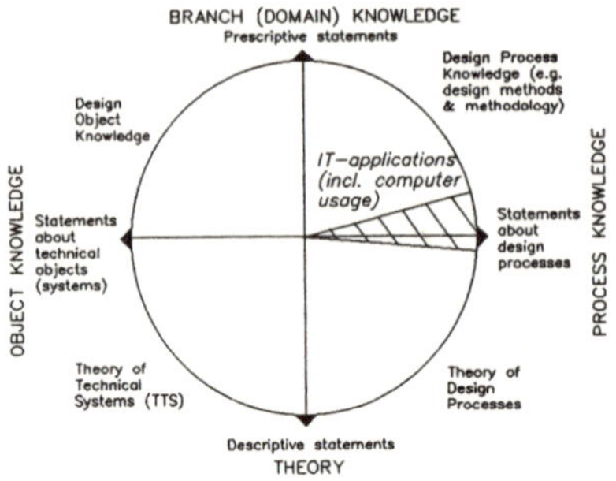


The four main areas of Design Science (as outlined in the previous sections of this chapter) with their knowledge in each level of concretization of technical systems cover the needs (and applications) of knowledge for designing. In certain conditions it can be useful to raise one or several partial area of these four main areas to a further quasi main area. In this fashion a partial area can be pragmatically categorized into the highest hierarchical level because of its immediate importance, timely relevance or because of the demands for knowledge from several main areas. As an example, "CAD-knowledge" is presented in the following paragraphs.

## 7.5.1 Knowledge about Computer-supported Designing -- CAD-knowledge

This partial area of Design Science is discussed in more depth for pragmatic reasons, because of the importance of computer application and of the novelty of the range of problems. To present the range of problems as clearly as possible, the descriptive (theory) and prescriptive statements are combined into a unit. But the emphasis of the useful knowledge for designers lies clearly in the prescriptive area. The knowledge in this area must always be presented to the users, and no fundamental theory is expected. The theory is treated mainly in computer (information) science and is adopted as necessary from there.

To be absolutely clear, we repeat here that this range of problems forms a part of the Theory of Design Processes. In this part, the working means (and therefore the computer) represent an operator of the design process (see figure 7--10--C).

Terminologically, despite all definitions, confusion rules in this area. The English abbreviations were formulated from completely different viewpoints than ones related to Design Science, and also their interpretation is not uniform. So for example, according to the perceptions of some branch experts the abbreviation CAD may cover applications in the whole design process ("computer aided design"), for others it only covers drawing ("computer aided drafting"). There are also other branch experts for whom designing does not need calculations; thus CAE ("computer aided engineering") would not belong to CAD ("design" or "drafting"), but this is not our conviction. Historically and pragmatically CAD/CAE-systems were written, prepared and developed particularly by computer programmers using techniques that at the time were "feasible" -- with only little regard to the needs and modes of operation of the engineering designers who were expected to use these program systems.

With reference to the *structure of CAD-knowledge* this sphere can be subdivided from the view of designers -- the computer *users* -- into the following partial areas:

### 7.5.1.1 Knowledge about Data Processing

#### *Basics of Information Science*

To understand the fundamentals of information processing and to be able to communicate with the information scientists, designers must master (at least at the level of awareness) the fundamental knowledge of information science.

If we think in this respect about digital computers, we imply for example an understanding of the binary number system and its realization through two conditions of a physical quantity (electric potential, voltage); understanding of information display with bits, codes for symbol display, and the computer-internal display of objects, their kinds of model and so on. Further we must distinguish different kinds of data (e.g., alphanumeric, graphic), files, data administration, data base systems, etc. That the data is entered, stored, processed and delivered to certain data carriers (media), is a further useful element of knowledge.

#### *Basic Knowledge about Hardware and its Functions*

Devices and material, all material and physical parts of the computer system which can be touched, are labeled as *hardware*.

The different devices can be assembled into several different systems (we speak of "architecture" and "configuration"), in which the peripheral devices for input and output, storage and central processing unit serve to process, control and store data.

#### *Basic Knowledge about Software*

Under *software* we understand all programs and data of a data processing system. The program prescribes to the computer a sequence of operations and commands which are to be executed to solve a given problem. This is also valid in general, e.g., independent of whether these commands run in one processor, or two or several parallel processors. Programs can also be stored in and accessed from "firmware."

For writing programs, the solution path must first be described with requirements and boundary conditions (an algorithm). The desired transformation must therefore be algorithmizable, the description can be mathematically formalized.

The program must control and instruct the computer in the real machine language -- binary. This computer-specific language, which transfers the commands to the central processing unit, is not suited for programming, except in particular and restrictive circumstances. For that reason, the higher level languages are used for programming. These are available in a wide diversity, and are suitable for certain specialized areas, e.g., FORTRAN for more mathematical and Pascal for more logical operations, LISP and Prolog for knowledge-based systems, assembler (mnemonic symbol) language or C++ for operating systems and programmable process controllers.

The ordinary user does not normally deal with programming. The current developments point in the direction that program administrators can enter their instructions by word processing, that means without or only indirectly through a programming language. These higher level languages are then translated by the machine into the machine language, with an interpreter directly from the higher level language, or with a compiler which generates a suitable running program in the machine language.

From another point of view we can divide programs into different kinds, for example system programs, user interfaces (shell programs, graphic -- GUI), applications (usable programs), or operating systems. A tendency exists that individual programs (calculation, simulation, search or optimization programs) are combined into program systems which have also stored the necessary knowledge (knowledge-based systems, expert systems) and which address the same database (integration).

With all programs, attention must be paid to transparency of the programming, verifiability, and references to progress and calculation paths. A feature of knowledge-based systems is the separation of procedures (the inference engine) and data (knowledge, rules and so on). Such systems can have additional abilities installed, as for example ability to learn (e.g., neural networks), control of the procedural direction from probabilities, and the possibility of inquiry about how the result was reached.

#### 7.5.1.2 Programs

The real tools for designing are resident in individual programs or program systems, which can be applied according to need. Application is not without problems, operation of or becoming familiar with these programs is not always obvious. The program documentation (which usually includes help for "on-line" inquiry) is generally not sufficiently informative or user-friendly with respect to language and content. The attribute "IBM'ese" points to unclear usage of language, and jargon-ridden form of expression, as was allegedly invented by the computer company IBM. Installing, maintaining, starting and using programs on a particular computer may also cause problems. Experiences show that the complete abilities of a program are used only in exceptional cases. Dependability (including accuracy) of the results is to be considered, and is not guaranteed by the tests performed on the program.

A further question is the effectiveness of use of the programs. Programs for single-user operations can be very powerful and offer technically favorable solutions. Working with them can turn out to be difficult and time-consuming. Therefore we find the tendency to use program systems (see above), in which the command structures are common among the units, and individual units use the same data base, demanding only a single entry of data.

Most program systems are open systems, which use extensive branch and company specific data for concrete situations (e.g., in an organization). Full exploitation then depends on delivering the data to the computer (shell-program).

### 7.5.2 Acquisition of Knowledge for Programs

The branch and company specific data used for the programs refer on one hand to the range of products and all life phases of these products. In contrast to the considerations in the technical knowledge area the questions concern the branch knowledge. On the other hand these data affect procedure knowledge (algorithms) as process knowledge. The program follows the work of the designers and in similar design situations the program asks the same questions as the designers. Consequently the knowledge remains the same in contents, only the form could differ for the computer application. Therefore the statements in the Sections 7.3 and 7.4 are also valid for this area. Once these close relationships are discovered, the view of Design Science as a system is confirmed as correct and profitable.

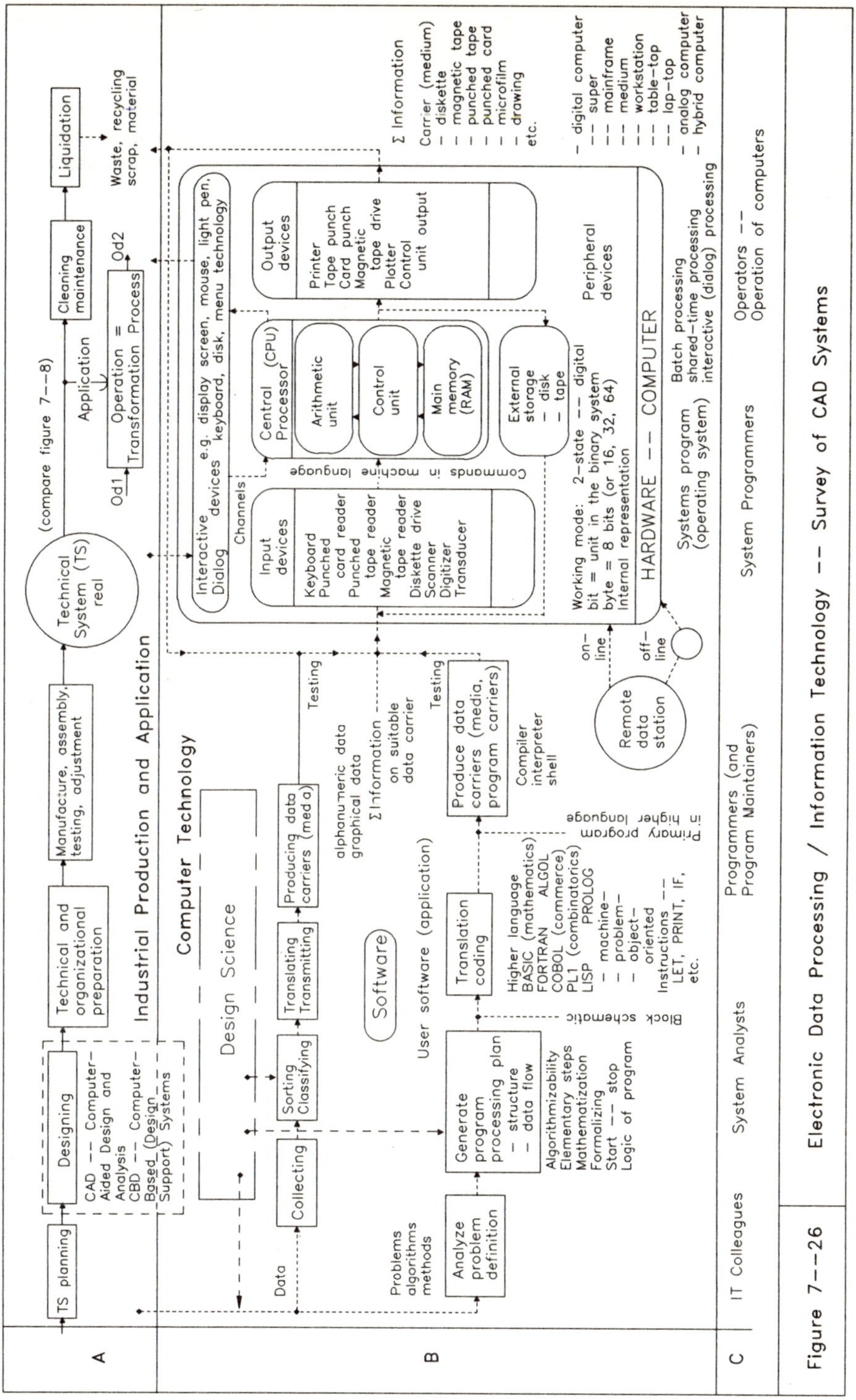


Figure 7--26 Electronic Data Processing / Information Technology -- Survey of CAD Systems

### 7.5.3 Summary

The CAD-knowledge is in its character a relatively new and rather foreign element in the whole knowledge system of designing. The depth of knowledge of the individual designers and design groups needs to be very different. It is profitable to select computer-specialists among designers.

In the previous sections, computer technology was described in crude terms with the goal of mentioning important ideas and processes, and to stimulate further study (acquisition of knowledge).

In figure 7--26, the areas of individual operations and hardware-elements of computer technology (part B) have been brought into a visual relationship with the sphere of the origin and the life of technical systems (part A), as well as with Design Science. We do not intend to describe all the presented processes, but only to present the system (with hardware, software, user etc.) as a whole -- interested readers can find the descriptions for themselves, and can research the relatively complicated system in this way.

*Part III.*

# Derivations from Design Science in Other Disciplines -- Specialized Design Science

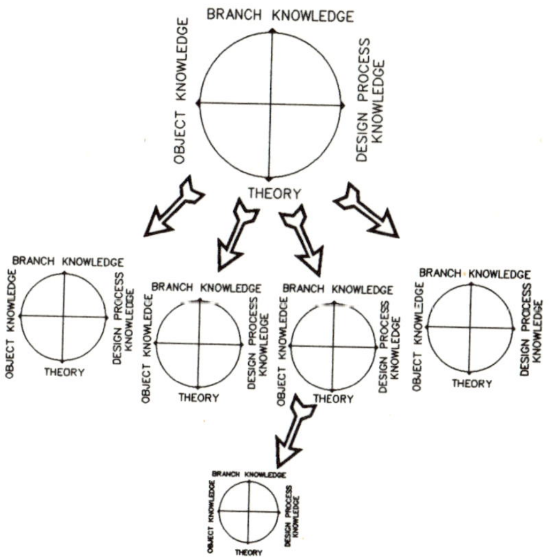

*Cleverness is not a substitute for knowledge.*
***Mary McCarthy***

General Design Science is not only consumer of the knowledge of other sciences, it serves also as source for deriving, ordering and providing the knowledge for other disciplines, whether in thoughts or in written processed form.

Remember that the construction of the General Design Science (i.e. on the generic plane of technical systems) was stated as first goal (see part II). As next goal, this knowledge should be transformed for different application regions. In a deductive process we can obtain (Specialized) Design Sciences for many individual kinds or types of TS and/or various additional features.

Derivations from Design Science can be useful especially from two characteristics according to our morphological diagram, figure 5--1. One derivation can be made from characteristic 3 -- recipients. Completely different knowledge systems are needed for:

- students -- as design-educational instruction, in which the design knowledge is selected and ordered according to didactic and pedagogic concepts;
- teachers or researchers/scholars -- as scientific treatments with all reasoning, justifications, substantiations and foundations;
- practitioners -- e.g., as design manuals in which especially the prescriptive statements find their place and the ordering must occur from pragmatic criteria (compare design situations, Section 7.2.2.3.5).

If today we still remark about the small practical application of Design Science, then important causes are that such derivations do not exist for practitioners to the extent as they would be needed. Some indications are given in Chapter 9.

A second derivation can be performed from characteristic 5 -- type of object. For each type of TS the corresponding knowledge can be derived and concretized. Sources are the knowledge of Design Science and the corresponding technical (branch, domain) knowledge of the area. The present knowledge, e.g., about machine elements, is joined with that of Design Science to form the Specialized Design Science for machine elements. Let us first treat the second of these classes of Design Science.

# 8 DESIGN SCIENCE FOR THE TS-TYPES

In the chapters of part II (Chapters 4 -- 7) we discussed the General Design Science, which is associated with the generic plane of technical systems. A general design process was also found (figure 7--13). However, each actual design process depends particularly on the object to be designed -- the technical system. Therefore all kinds of technical systems, as they were discussed in the Theory of Technical Systems (Section 7.1, and [214,219]), display specific design processes and demand beside the general knowledge also concrete and specific technical (branch, domain) knowledge.

The first two classes of technical systems are process systems and object (real) systems, and these influence the essence of the design processes.

In the earlier chapters we have demonstrated many classification characteristics, see figures 5--1, 7--4, 7--9 and Section 7.1. In the area of object systems, to which we pay more attention, the particulars and peculiarities of an individual design process are caused especially by the following classification characteristics:

- function and mode of action (mode of operation, action principle) of the technical system (TS), features of individual specialties or industry branches;
- degrees of complexity (plant to individual component part);
- level of originality of the TS (novel design, adaptation, variant, re-use);
- number of items of the manufactured TS (mass production to single item "one-of-a-kind" production);
- size and other features of organization where TS are manufactured (TS manufactured in small, medium or large organizations).

Some examples should illustrate these characteristics:

- when designing internal combustion engines or Kaplan turbines, and many other TS-families, the phase of conceptualizing will rarely be needed, because these TS almost always use the same mode of operation. Their principle and overall organ structure is pre-determined. The situation may be different at the level of the individual partial organs, which can certainly also be newly conceptualized;
- for material processing TS and special machines, conceptualizing forms the decisive phase of the design procedure in which the quality of the TS is established;
- the structure of plant exists predominantly of available and partially of purchasable elements. Thus for designing of plant ("project engineering") the main task is the selection of elements and establishing their relationships;
- the design process for a car in mass or series production, where hundreds to thousands

of pieces are manufactured, differs substantially from designing for single-item "one-of-a-kind" manufacture;

- only a part of the general design process, namely the last form-giving phase, is applied if a variant is to be designed. In this case only some of the data of an available product are changed;
- large enterprises mostly possess better technical and financial possibilities (including computer systems). The division of labor also reaches deeper, and thus also the knowledge level.

If each of these different classes of designing is taken as an example of a potential "Specialized Design Science," different suggestions for classifying characteristics logically emerge, depending on which superior criteria are chosen. One thing is certain, namely that each of these classes can allow setting up a specialized science, and therefore a specialized knowledge system. This statement is supported by knowing that in individual areas the "know-how"-information has become familiar (even if we can hardly speak of superior knowledge systems).

We recommend that the Specialized Design Sciences should be defined as completely as the General Design Science, so that they serve as many as possible of the branch experts, and can also possibly present a meaningful unit for engineering education.

From this point of view, the following hierarchy of ordering characteristics appears appropriate:

- complexity of technical systems: the difference between project engineering of plant and designing of components is large;
- branch, field of expertise: TS-families often differ greatly from each other in reference to the technical knowledge systems. The complexity level of "machines" is most applicable for this consideration;
- level of originality of technical systems: each technical system at any arbitrary complexity level and in any arbitrary branch can appear as a novel system or an adaptation (variant). According to kind of adaptation the changes can be different -- changes only in dimensions (within a given component structure), changes to the component structure (e.g., arrangement, configuration), to changes in higher (more abstract) structures (compare figure 7--3 and Section 7.4);
- kind of production: each technical system can be manufactured in mass production, as single-item production, or in other suitable quantity ranges (for instance, a plant, for which several of the elements may be mass-produced).

In strict logic, therefore, a "branch Design Science" can be considered as an element of the partial quantity of sciences within each complexity level; and also an "originality Design Science" is an element of the partial quantity of each branch, and a "kind of production Design Science" is an element of the partial quantity of each originality level.

This statement can be illustrated (see figure 8--1), by first subdividing the machine systems (crudely) according to the degree of complexity, and then associating the TS-families with the individual levels and indicating examples.

Regarding the completeness of Design Science in individual areas it must be clear that the individual states of embodiment of the ordering characteristics still make possible a general knowledge system which is broadly applicable and differs only in certain special aspects from the other knowledge systems in the region. Knowledge systems which belong as elements of several levels and kinds of sciences can emerge in this way. As example we mention the classes of "design for properties" (compare figures 7--4 and 7--5). Thus a

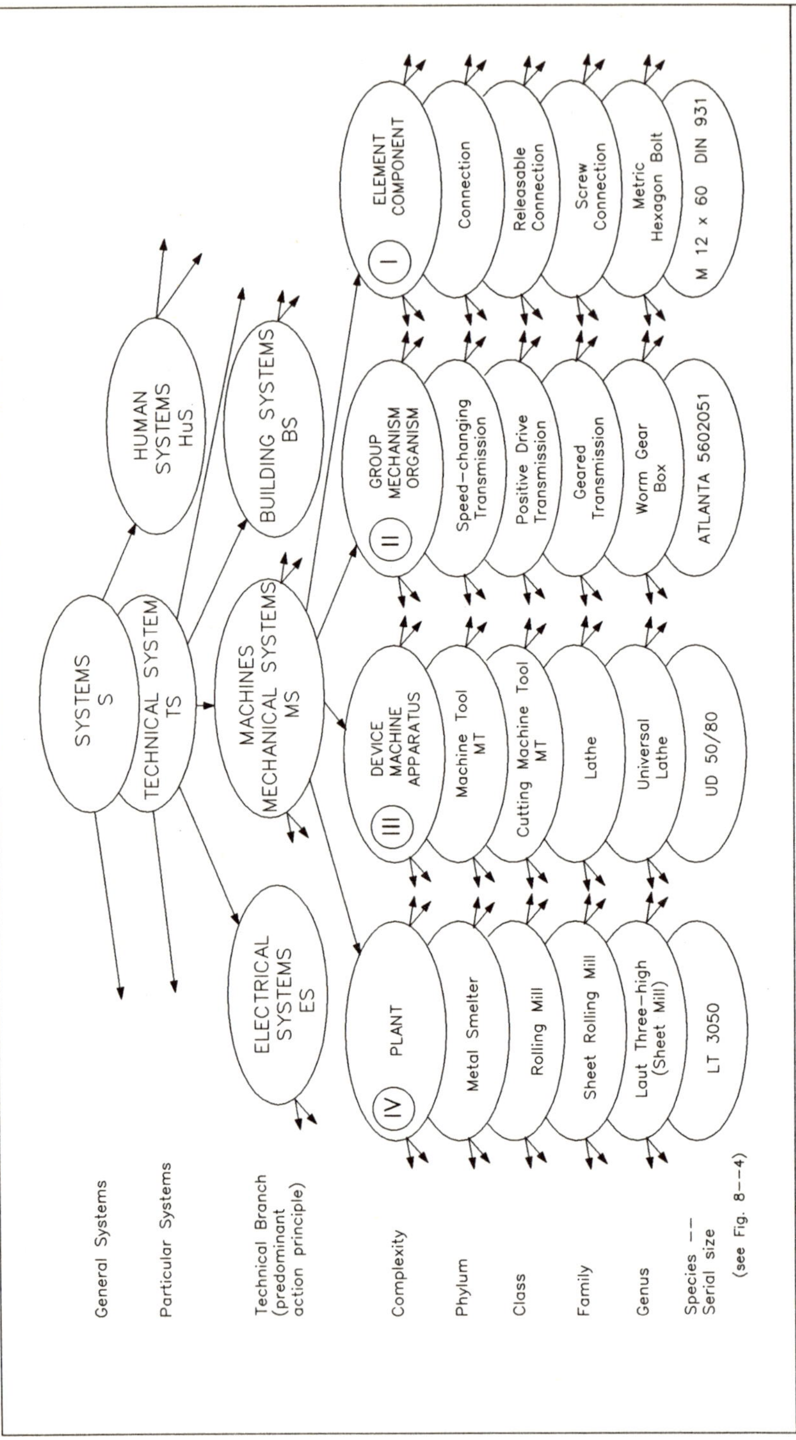


Figure 8--1 Hierarchy of Technical Systems (TS-Hierarchy)

complete Specialized Design Science can be assembled also from several rather "independent" elements, which are joined to the kernel of the branch (domain) knowledge (compare figure 8--5).

## 8.1 Design Science for Individual Levels of Complexity of Technical Systems

The hierarchy of technical systems can be erected according to the degree of complexity on four basic planes, even if each of these sub-divisions (units) can consist of additional levels. We are dealing with level IV -- plant, level III -- machines (instruments, devices), level II -- assembly groups and level I -- elements, components.

The best known unit of this hierarchy is presented by the machine, instrument, device and apparatus unit, which deliver to the user a complete complex of effects (outputs), and form a material unit, e.g., locomotive, vacuum cleaner, television set, house or measuring instrument. These machines can be decomposed, depending on the complexity of their structure, according to different criteria into many partial groups which display ill-defined boundaries. These partial groups again consist of assembly groups, down to elements, components.

We will now treat the specific features of the individual levels and thereby ask the question of how advantageous a Specialized Design Science can be, and at which level of the concretization it could be optimally situated.

### 8.1.1 Design Science for Plant

Plant, major projects and infrastructure are technical system of the highest level of complexity (level IV). They consist of machines down to elements, components. This kind of complex technical system includes manufacturing plant, energy generating plant, chemical plant, traffic and utilities infrastructure, according to which purpose the complex (or very complex) function (capability) of the plant should serve. The modern kitchen, which prepares, stores and performs other functions for a family or a social unit, could be called a plant.

#### 8.1.1.1 Peculiarities of Plant

Question:
What characterizes plant (especially for designing)?

Compared with other technical systems, a plant displays several characteristic features:

- it is technically complex;
- it is costly to very costly;
- it contains several kinds of technical systems: building, machine, electrical systems, etc.;
- it combines many purchasable elements into a unity (only a small part of any plant is manufactured to order);
- in planning -- designing -- the emphasis lies clearly in conceptualizing and establishing the processes, and the arrangement (configuration) of the chosen sub-units -- machines;

- the technology of the transformation is the dominating element in reference to quality;
- the operator (factor) environment plays an unusually important role, not only technically but also socially and, in large plants, also politically;
- it is realized predominantly (or almost exclusively) in single-item (made-to-order, one-of-a-kind) production, although many of its elements can be manufactured in larger numbers of pieces.

#### 8.1.1.2 Planning Plant, and Demands on Planning

Questions:
1. How does designing proceed? Does experience about designing/planning exist?
2. Do particular demands exist with reference to designing?

Referring to the above characteristic features, design planning or project engineering of plant is a particular process. It can appear surprising, that today the planning process is already formalized very broadly and reacts to particular demands. For instance:

- questions about potential costs are asked very early in the process;
- the location of the proposed plant must be established immediately, and the solution must react to the location and its conditions;
- because a plant influences the environment, plant building is subject to several regulations and permissions, e.g., building inspection, fire department approval, trade supervision, safety regulations, work protection;
- the planning engineer is often at the same time the realizer of the project, supervising its construction.

Planning proceeds as a rule in several stages. The general procedure usually occurs in following phases:

- pre-planning, investigative project, feasibility study, in which an approximate idea of the process, the plant structure, but also the initial costs should be established;
- elaboration of the project; this delivers accurate statements about processes (technology) of all corresponding streams (of energy, material, information; forward working and feedback control); layouts of machines, apparatus; plans for electric installation, piping, heating, ventilating, roadways, maintenance areas, etc.; a complete estimate of all costs;
- granting of building permits, and of construction contracts and building project sub-contracting (including some more detailed designing).

It is clear that plant engineering deals with a very demanding engineer work, which requires much specialized technical knowledge and is concerned with many risks (possibly including the financial risks).

Therefore it is not surprising that in this area very early attempts were made to streamline project engineering. The example of the method SLP (Systematic Layout Planning) [313] documents this effort; some of the details of this method (figure 8--2) show also the breadth of this attempt.

'Systematic Layout Planning' -- SLP -- investigates:

- A five basic influencing factors for manufacturing the product (P,Q,R,S,T)
- B in four basic phases of project planning (I to IV)
- C with the help of a block schematic of the procedure
- D using sets of agreed symbols (literal, numeric and graphical)

A **The Five Basic Influence Factors**

- P Product -- the items and systems to be produced
- Q Quantity -- production volume, numbers of each items to be made
- R Routing -- manufacturing process, operations and their sequence
- S Supporting services -- other supplies and tasks for manufacturing
- T Time -- manufacturing time

B **The Four Basic Phases**

- I Location (of the area to be laid out) -- situation of the plant (factory, workshop, machine complex)
- II General overall layout -- general solution (project) for the plant
- III Detailed layout plan -- detailed spatial layout of the plant
- IV Installation (planning and moves) -- realization and start-up

The phases lead logically into each other. Between the individual phases an approval process takes place.

C **Block Schematic of the Project Procedure**

Represents in graphical form the individual activities, their sequence and relationships.

D **Agreed Symbols**

These consist of graphical, alphanumeric and colored symbols, classifying and numbering systems for the tables, scales, etc.

Example: the symbols for the evaluation scale

- A -- absolutely necessary
- E -- especially necessary
- I -- important
- O -- ordinary
- U -- unimportant
- X -- not desirable

Literature

Muther, R., Systematic Layout Planning, Boston: Industrial Education Institute, 1961

Figure 8--2 Systematic Layout Planning -- SLP

### 8.1.1.3 Special Technical Knowledge

Questions:
1. Is special technical knowledge needed, and in which areas?
2. Does this knowledge already exist? If so, in which form does it exist?

To be able to fulfill the demands, a special range of knowledge had to be developed. Cost calculating and the characteristics connected with costs and calculation can serve as typical example:

The inquiry necessary for judging the economics of a project, especially of the cost emerging in the first project engineering phases, is extremely difficult. In the past, methods which allowed rough cost estimates had to be developed (compare figure 7--15). One of the developed methods consists of creating a simple model to obtain a first idea of the dimensions of the buildings, apparatus, etc. Thereby one can establish the extent of the enclosed space, and multiply this by experience numbers (e.g., cost per unit volume). Through suitable additions, the costs can be obtained for further equipment (e.g., for electrical, heating and sanitary installations). The method and the corresponding numbers form the technical knowledge of the project engineer.

To this area belongs also the knowledge about the mean values, which refer to proportions of individual types of cost contributing to the total costs. From a larger number of manufacturing plant, the following estimation values were obtained:

proportion of machines and apparatus costs about 33%, costs of buildings about 16%, engineering work about 16%, conductors and fittings about 10%, measuring and control equipment about 6,5%.

Such technical knowledge must be determined also in all other partial areas.

#### 8.1.1.4 Position and Content of the Specialized Design Science for Plant

Questions:

1. On which plant level can the Specialized Design Science be meaningfully and efficiently set up, and serve optimally?
2. Do special demands on contents exist?
3. What is the state of the knowledge?

In the taxonomy of technical systems (as part of the Theory of Technical Systems) the related kind of systems used are summarized in hierarchically graduated higher sets (taxa). Thereby we can define species, genus, families, classes, phylums. This ordering of technical systems, analogous to the classification used botany and zoology, is suitable for all complexity levels of technical systems, as shown in the examples in figure 8--1, and defined in figure 8--4.

For plant in figure 8--1 it is shown that a concrete series size LT-3050, of the genus "Lauth three-high," belongs to the family of sheet rolling mills, that these (and others) belong to the class of rolling mills and these in turn (and others) again to the phylum of metal processing plant. We have set the question, on which level of this hierarchy it would be meaningful to situate the fundamental Specialized Design Science.

In plant construction, predominantly single pieces are manufactured, and it would surely be doubtful to construct the Specialized Design Science on the level of "species." From this point of view, and with regard to the effectiveness of such information systems as goal, we must set the selected level as high as possibly. Here the context to concreteness (e.g., the representation of a plant) tends to be lost, and perhaps also too much of the similarity aspects. But also with these reservations, the highest plant level appears favorable for situating the Specialized Design Science, at least for a start. The knowledge system for designing can then be completed with the broadest special knowledge for individual levels (compare figure 8--5).

The state of knowledge for constructing the Specialized Design Science presents itself on this level as very favorable. Apart from the already mentioned knowledge (figure 8--2), extensive further knowledge is available in the published literature, and especially in the documentation and the basic information of known project engineering organizations.

### 8.1.2 Design Science for Instruments, Machines, Apparatus

The complexity level III of technical systems -- machine, apparatus, instrument, device (also house, bridge) -- belongs to the best known on the part of users, but also on the part of manufacturers and designers. Generally this level is supplied with the designation "product" -- something manufactured which is to be sold. Mostly these are products in commercial trading, because this unit is a useful and profitable object. For an enterprise (organization) these products form an output, which necessitates the existence of the firm. General explanations about design processes in Section 7.4 are valid as analogs for all complexity levels.

#### 8.1.2.1 Peculiarities of Machine

Question:
What characterizes (especially for designers) the complexity level III?

The description of this complexity level of technical systems can be established with following features:

- a technical system (TS) with capabilities which are demanded by the consumer (technical unit);
- a TS, which forms a spatial unit;
- TS with extensive diversity of types, each with very different demands;
- TS with continuously increasing requirements in depth and breadth;
- TS which are manufactured mostly in large to very large numbers of pieces;
- TS which are found mostly in a development series, have many predecessors, and at the same time are accompanied by members of a size range and/or by competing products.

#### 8.1.2.2 Designing of Machines and the Particular Demands on Designing

Questions:
1. How does the designing of machines proceed?
2. Do certain demands on designing exist?

When designing this complexity class, the phase of conceptualizing (compare figure 7--13) is rarely needed, because mostly the new products do not change the existing mode of action (i.e. the organ structure remains unchanged). The goal is mostly a further development, therefore adaptation or transformation designing, as well as variant designing. The emphasis lies accordingly at the highest abstraction of the preliminary layout, in which the individual organs and their forms are established.

In the layout stage, designers strive on one hand to realize the new requirements, to become marketable, on the other hand to increase the fitness for manufacture, and to lower the manufacturing costs.

Designing is also adjusted to the quantity of the product to be manufactured, and must consider the high demands of production and assembly. In most cases a prototype (or a set of prototypes) is first prepared, realized (executed, manufactured), and tested, after which the documentation for series and mass production is executed. The degree of automation of the manufacturing plant must also be taken into consideration, because the questions of raw material costs, and rational production and assembly (including robot assistance) play an especially important role. Despite the implied sequence of prototype to mass production, design and production strive towards the common and simultaneous processing of the range of problems, by using processes and methods such as "simultaneous" or "concurrent engineering," TQM and QFD (see also Chapter 1, and Section 7.1.3.6), so that problems of manufacturing are already avoided in the prototype.

A characteristic feature is that the design time is shortened as much as possible, all work proceeds under time pressure.

The design process in its structure and procedure depends on the size of the enterprise (organization) and thereby also on the available design technology.

An important element which influences designing, is the list of requirements (the requirements or design specification), which establishes the complexity of the products to be designed. Some further elements can be seen from figure 7--14.

### 8.1.2.3 Special Technical Knowledge

Question:
1. Is a special technical knowledge needed, and where is it needed?
2. Does this knowledge exist, and in which form does it exist?

From the description of the characteristic features of the machine and their design process in our last paragraphs it is clear that with respect to the emphasis of Design Science there are few methodical problems in this class. The biggest concern is the branch (domain) knowledge and, connected with the abundance of properties of the future technical system to be achieved, the knowledge about the individual properties. We refer to the knowledge in the north-western quadrant of figure 5--2 and Section 7.3, which describes this area in general. A different situation can emerge in designing and developing new products.

In the individual specialties (branches) particular demands emerge, which also require a special knowledge as peculiarity of this area. Let us take as example the food and pharmaceutical industries and their machines. For all these machines, a particular demand that occurs in no other area is made -- the possibility and need for ease of cleaning. Due to the severe hygienic regulations, no machine can reach the necessary quality without this capability. The technical knowledge to achieving friendliness for cleaning must be extensive, to produce the following capabilities of the system:

- a certain state and condition of the effector surfaces, which make cleaning possible;
- disassembly capability of the machines;
- accessibility of the areas to be cleaned and disinfected;
- a certain surface quality (no porosity which could support the growth of microorganisms);
- if possible, no complicated forms, but only simple forms;
- suitable coverage (enclosure) from outside, so that disturbances are avoided;
- no touching of sensitive metals by aggressive materials (especially foods);
- application of special sealing and containment organs.

This knowledge is mainly available, otherwise such machines with the desired qualities simply could not be built and used. However, the range of knowledge varies from maker to maker, and the form of this knowledge is not always the most favorable for designers. Its necessary revision for Design Science is no mere formal exercise. Systematic questioning reveals not only many possibilities, but also large gaps. Such a process lets the old question emerge again, namely on which taxonomic level the range of problems should be processed. Where can the information system -- the Specialized Design Science -- be set up, to not only efficiently, but also economically fulfill its function. This is discussed in the next section.

### 8.1.2.4 Situating the Specialized Design Science for Machines

Question:
On which taxonomic plane can the fundamental Specialized Design Science be meaningfully situated, so that an optimal information system emerges?

In engineering practice, the task is frequently set to design a concrete object -- a technical system -- so that it can then be manufactured. This object can be a weaving loom, a tractor, a press, a street, a high voltage electrical transmission line, or a lathe. Figure 8--3 shows some further areas and machines to illustrate the existing diversity. In addition, the complexity levels III and IV are connected with individual branches of the economy.

The central demand for us is to prepare an information system that can be made available for designers. We proceed on the assumption that all designers must acquire a certain knowledge system. The most important element of this knowledge should be formed by the Specialized Design Science, with which we are now concerned. The Basic Specialized Design Science (BSDS -- see below) should be built on one of the planes of the taxonomy (figure 8--4). Should this happen on the plane of the "phylum" or directly on the "species" plane, where the designers work? Or should several Specialized Design Sciences (SDS) be set up, or even on all levels? And how should the specialties of the other levels be incorporated into the system, if only one or a few SDS exist?

Figure 8--5 shows a possible construction of the technical knowledge system with such structure (on a BSDS). The conventional circle (according to figure 5--2) representing the General Design Science (GDS) on the plane of the TS is projected not only onto the structure in the Specialized Design Sciences (SDS), but already contains the general knowledge and transmits it. The indicated segment in the north-west quadrant shows as an example "design for manufacture." The next level drawn in this diagram, the "phylum" level (example machine tools), is chosen as "Basic SDS." This science should be set up as completely as possible (e.g., in a fundamental book or a data base). All knowledge systems below this level, on the levels "families" to "species," expand and concretize the knowledge through corresponding "completions." The figure shows not only the completions from the TS-plane, but also those originating from the concrete organization and even the concrete case (series size, special wishes). The latter elements can already be considered on the higher planes of the hierarchy.

### 8.1.2.5 Contents of Basic Specialized Design Sciences (BSDS)

Questions:
1. Which are the contents areas?
2. What is the state of knowledge in individual areas?

We can assume at this stage that the structure of contents is known for the General Design Science (GDS). Figure 5--2 reflects the four fundamental quadrants. Each Specialized Design Science can be derived from this model and will contain the same parts, which are, of course, filled with the knowledge covered in the specialty.

| Branch of Economy | Technical System TS | |
|---|---|---|
| | Equipment for (purpose) | Typical Machine |
| Mining | Accessing<br>Delivering<br>Preparing | Cutting machine<br>Conveyor<br>Screening machine |
| Energy generation | Steam raising<br><br>Electric generating | Steam boiler<br>Water conditioner<br>Steam turbine<br>Gas turbine<br>Water turbine<br>Generator |
| Smelting | Pig iron smelting<br>Steel smelting | Blast furnace<br>Bessemer converter<br>Rolling mill |
| Chemical industry | Coal scrubbing<br>Pigment & die producing<br>Explosives producing | Pressure vessel<br>Piping<br>Destillation column |
| Pharmaceuticals | Medicines producing | Press |
| Papermaking | Papermaking | Calander roll |
| Metalworking industry | Chipless forming<br><br>Chip–forming<br>Heat treatment<br>Foundry<br>Assembly | Press<br>Forging hammer<br>Machine tool<br>Furnace<br>Die casting machine<br>Jigs and fixtures |
| Constructional industry | Oil exploration<br>Building<br>Roadworks<br>Hydro–construction<br>Materials manufacture | Drill rig<br>Personnel lift<br>Scraper<br>Concrete mixer<br>Block press |
| Transportation | Railway<br><br>Shipping<br>Space travel | Locomotive<br>Wagon<br>Passenger liner<br>Rocket |
| Textile industry | Textile manufacture<br><br>Dressmaking | Spinning machine<br>Weaving loom<br>Sewing machine |
| Food industry | Sugar refining<br>Cheese production<br>Milk processing | Concentrator<br>Press<br>Centrifuge |
| Medicine | Diagnosis<br>Therapy | X–ray apparatus<br>Artificial heart<br>Prosthesis |
| Printing, offices | Printing<br>Office work | Printing machine<br>Typewriter<br>Calculator |
| Agriculture | Transporting<br>Harvesting<br>Lumbering | Tractor<br>Combine<br>Chain saw |
| Distribution, trade | Self–service<br>Packing | Check–out<br>Wrapping machine |

Figure 8--3 Examples of Technical Systems for Important Branches of the Economy

| General Hierarchy of Technical Systems | | | | Example | |
|---|---|---|---|---|---|
| Level of Concretization | Designation of level | Definition by ... (graphical model) | Established Design Characteristics | Designation of level | Graphical representation |
| 0 | Technical System (TS) | 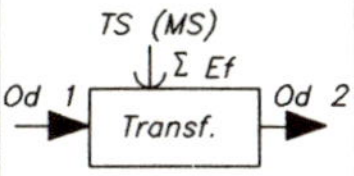  | TS general;<br>MS (as special case) exhibits mainly mechanical mode of action | | |
| 0.2 | Phylum of TS | General 'black box'<br>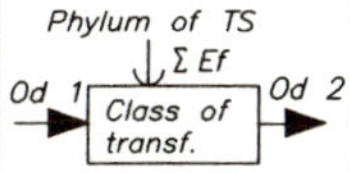  | – Phylum of operands<br>– Class of operations | Machine tools | 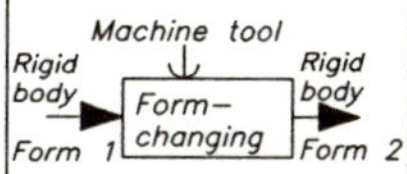  |
| 0.4 | Class of TS | Detailed 'black box'<br>Sketch of technological principle<br>Basic function structure | – Family of operand<br>– Technological principle of transformation<br>– Necessary input effects and from these the basic TS–internal processes and functions | Lathe (metal–working) | 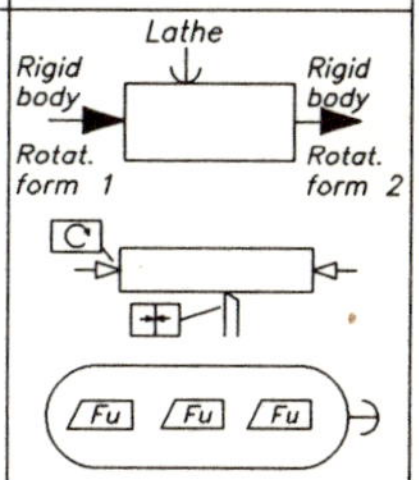  |
| 0.6 | Family of TS | Detailed function structure<br>Rough component structure<br>Concept sketch | – Species of operand<br>– Function structure (functions and their sequence)<br>– Inputs to the TS<br>– Families of function carriers (organs)<br>– Combination and basic arrangement of organs | Universal lathe | 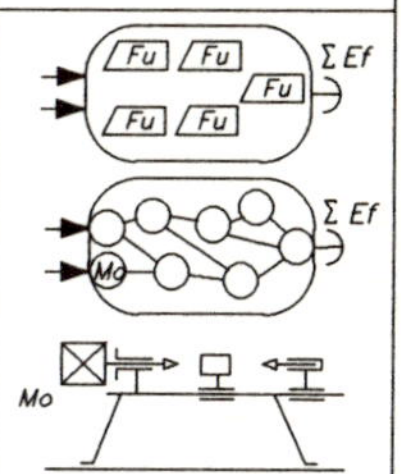  |
| 0.8 | Genus of TS | Drawing of component structure (conditions of similarity will differ according to type series)<br>Drawings of common (re–used) components and subassemblies | – Complete parts<br>– Arrangement<br>– Partial form<br>– Some dimensions<br>– Types of materials<br>– Some tolerances and surface properties | Lathe type SUR | 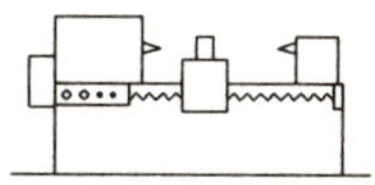 |
| 1 | Species of TS, or Serial size | Complete set of workshop documentation | Complete and definitive specifications for parts and arrangements<br>For all components:<br>Forms<br>Dimensions<br>Materials<br>Manufacturing methods<br>Tolerances<br>Surface finish | Lathe SUR 5 |   |

| Figure 8--4 | Taxonomy of Technical Systems (TS), with Example of Machine Tools (Machine System -- MS) |
|---|---|

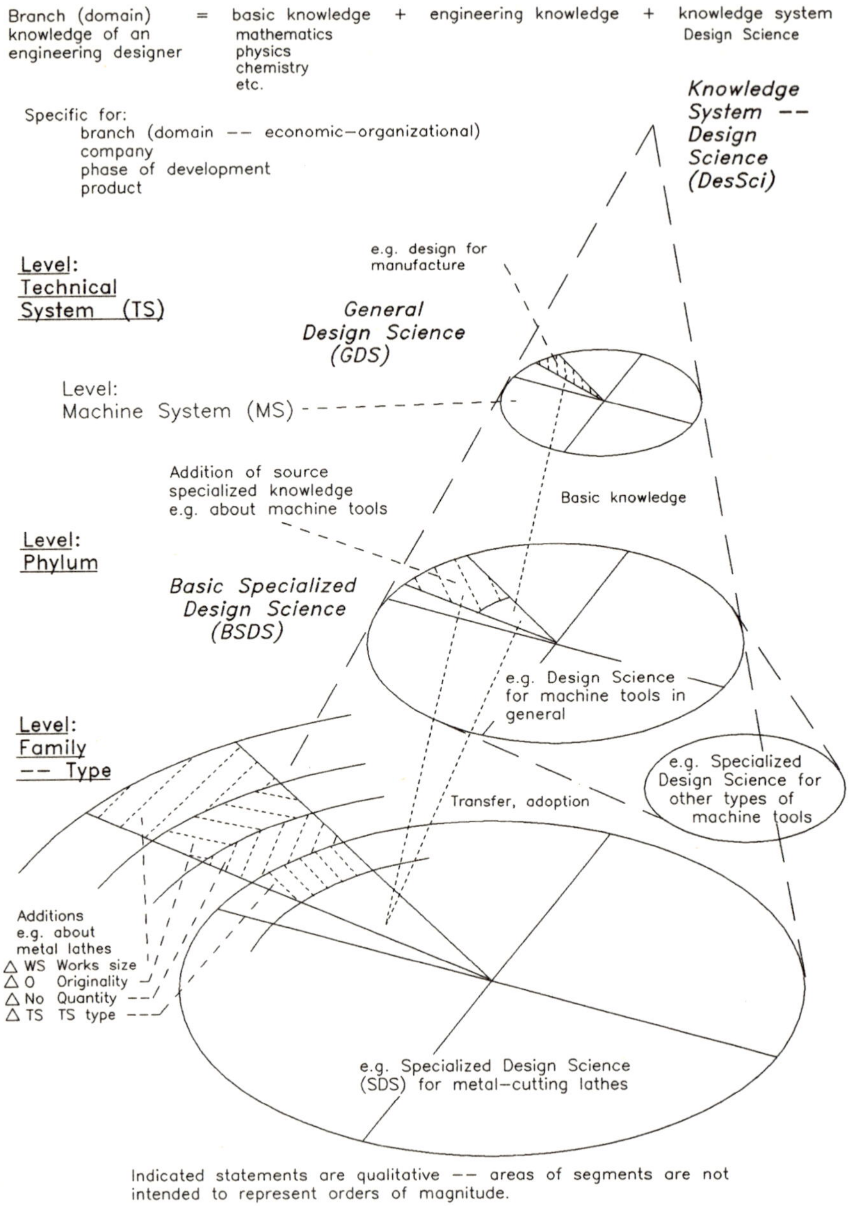


| Figure 8--5 Part 1 of 2 | Constitution of Branch (Domain) Knowledge for Design |
|---|---|

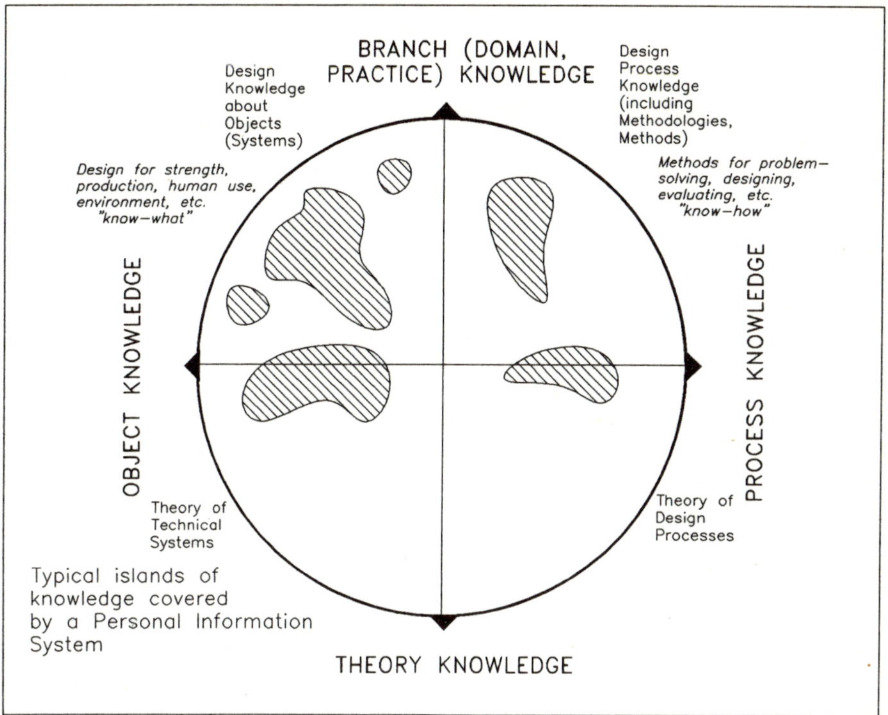


| Figure 8--5 Part 2 of 2 | Personal Information System (PIS) for Engineering Design |
|---|---|

To develop an accurate idea of the contents, we briefly describe these fundamental knowledge elements, and machine tools should serve as example of the product families.

A further remark is needed to the terminology: we will use the term "Basic Specialized Design Science (BSDS)" consistently as a technical term (*terminus technicus*), it designates a knowledge system for designing at a specific complexity level and in a branch (specialty).

### -- *Theory of Phylum of Technical Systems (TS-phyla, e.g., Machine Tools -- south-west)*

Starting from the "map" in figure 5--2, we ask about the contents of the south-west quadrant. In accordance with the Theory of Technical Systems, the quadrant contents can be specified through those part segments which are shown in figure 7--9. The machine tools are analyzed and described from all these points of view. Thereby the typical functions and features of this phylum are found and all solutions cataloged. This analysis is understandably further deepened, if we consider and process the classes within the phylum, in our case the lathes, milling machines or planing/shaping machines. For these, the new knowledge is captured in the completions, or if this would be beneficial, the basic plane could be put on one of the deeper planes. Thereby we could generate the Basic Specialized Design Science (BSDS) for lathes, milling machines and further classes, instead of only for machine tools. Particular attention should be given to the specific properties of the phylum (or the class), for instance in our case to precision, stiffness or stability. The Basic Specialized Design Science could also be set up at two levels, e.g., phylum and classes.

### -- *Theory of Design Processes for TS-Phylum (south-east)*

The theory in this sector should describe the special design processes for creating the phylum and the classes. It will lean largely on the general treatment in the General Design Science (GDS) and emphasize only the specific conditions of the field, be it in the area of the designers, their working methods, representation methods, working means, management or working conditions in designing a product phylum or a class.

The statements in the two northern prescriptive quadrants are considerably more concrete in comparison with the two southern theoretical quadrants.

### -- *Technical Knowledge about Objects of designing (north-west)*

Whereas in the theory of the TS-phylum the structures, laws and forms of construction, properties and all other laws are described, the contents of this quadrant are directed towards the provision of practical knowledge. The question to be answered is: "How do I achieve or realize this structure, property and further features?"

Especially the fitness of the proposed TS for different life phases should be supported through suitable instructions which directly serve the creating designers. The instructions as "know-how" of the designers are generated as individual, almost independent knowledge segments. Some of these are fairly well known, such as design for manufacture and assembly (e.g. [48,49,80,81]), and may even to some extent be computer-supported [80,81]. Others wait for a first formalization, such as design for maintenance, the environment, among others. If we say, that they wait for a first formalization, this does not mean that no relevant knowledge exists. It is usually found in other areas, or in a different form that is not so appropriate, directly usable or understandable for the designers. The large step will be fully done if (for instance) the "design for manufacture" knowledge is derived from manufacturing technology.

Concerning the distribution of this kind of knowledge on the individual hierarchical planes, this is formed unlike the relevant theory. A much more concrete statement can already be made on the upper, more general levels, because the majority of the knowledge is universally valid. On the lower planes only specialties are added, such as instructions for machine tool beds made of reinforced concrete as part of the "design for material and manufacturing" area, which are certainly not treated in any other specialty.

### -- *Technical Knowledge about Designing (north-east)*

In this north-east quadrant we expect to find the practical knowledge about designing, the processes. At present, the predominant share of this knowledge is formed by methodical references: the procedural models recommended to be followed, and also the methods which support the completion of individual design steps. Additional partial segments should give practical references for application of the working means (e.g., computers), management or working conditions.

This knowledge area, in comparison with the others, has been filled in recent years mostly through design research. Unfortunately, this happened mainly on a general level, where (because of attempted universality) no sufficiently concrete methodologies can be formulated. The general references and rules of conduct (compare figure 7--14) are to be adjusted not only for the branches, domains and specialties, but also for well defined conditions of a certain enterprise organization, or even for one particular design group or certain designers. As the practice shows, this adjustment is not as simple or trivial as we may initially suppose. This is certainly one of the elements, which can explain the present lack of broad application the design methodology in engineering practice.

The close relationship of the working methods with the object and/or with the level of the branch (domain) is already evident in figure 5--4. For instance many design characteristics are already defined for an individual branch at the start of a design process, i.e. designing begins from the organ structure (or even only within the component structure) which is determined for the TS-class (compare Section 7.4). This fact, well known in the practice, has not up to now been clearly and explicitly expressed in the theory.

-- *Summary*

- *State of the Knowledge*

The presented model of the Design Sciences at different levels of abstraction, which appear as elements of the holistic knowledge basis for designers, shows clearly the advantages of such an information system in comparison with the present practice. The impression could almost emerge that no large difference exists between these systems. That is deceiving, because many of the knowledge elements are certainly available. Their displacement and/or transformation brings order, which is indispensable not only for the theory but also for educational instruction and practice.

- *Form of the Knowledge*

In constructing the Specialized Design Science, we should not forget its form. We have already repeatedly underlined this aspect and validated it with examples (figures 7--21, 7--22, 7--23). With the example of the knowledge about representation (engineering drawing) in figure 8--6 we again want to show the importance of a clear formulation and presentation of a knowledge complex, to aid judgment and to copy.

### 8.1.3 Design Science for Assembly Groups and Elements

The complexity level of lowest technical systems is formed by assembly groups, mechanisms, organisms (II) and elements, components (I), i.e. the component structure. The functions of these technical systems are often problematic, and an exact coincidence "elementary function = elementary technical system" does not and cannot exist. Especially for the formation of assembly groups, organizational aspects (e.g., sequence of assembly) can predominate over functional aspects. We treat both of these complexity levels at the same time, because their range of problems is closely related and they are both represented in engineering by the discipline of "machine elements."

#### 8.1.3.1 Peculiarities of Assembly Groups and Elements

Question:
What is characteristic (especially with regard to designing) for these technical systems?

The characteristic features in the class for complexity levels I (elements) and II (assembly

## 19. The Detail Drawing: Basics, Phases, Elements

**Representation basics**

Projections, axonometric representations, views, sections, sectioning arrangements, break lines, details, features, etc.

**Geometric basics**

3-, 4-, 5-, 6-, 8-sided figure, rectangle, square, ellipse, oval, parabola, involute, curve, chord, tangent, etc.

**General basics**

Printing, formats, lines, scales, title block, parts lists, standards numbers, tolerances, limits and fits, abbreviations, etc.

**Detail Drawing**

Phases of generation

Representation part 1
Dimensioning part 2
Completion part 3

**Drawing technique**

| | |
|---|---|
| Coordinates Dimensions | 1. X length<br>2. Y height<br>3. Z depth/thickn. |
| Views | 1. front view<br>2. top view<br>3. side view |
| Auxiliary views | 1. 3 additional views<br>2. sections<br>3. detail enlargements |

**General technique**

| | |
|---|---|
| Portions for completion | 1. materials<br>2. surface and treatment data<br>3. title block completion |
| Type of material | 1. basic (casting)<br>2. semi-finished<br>3. raw part |
| Data for title block | 1. number, test class<br>2. origin data<br>3. indicators, changes, replacements |

**Dimensioning technique**

| | |
|---|---|
| Parts | 1. auxiliary line<br>2. dimension line<br>3. dimension |
| Measurement types | 1. toleranced<br>2. untoleranced<br>3. reference sizes |
| Toleranced dimension types | 1. with symbol and quality<br>2. with allowances<br>3. with special data & notes |
| Dimensioning styles | 1. serial (chain) dimensioning<br>2. parallel (datum) dimensioning<br>3. combined dimensioning |
| Dimension types | 1. external<br>2. internal<br>3. hole spacing |
| Type of hole spacing | Tolerance<br>1. positive<br>2. negative<br>3. symmetrical |
| Auxiliary dimension types | 1. internal radii & chamfers<br>2. external radii & chamfers<br>3. angles |
| 'Parts theory' | 'triangle-theory' by M. Weber |

| | |
|---|---|
| Dimensioning types | 1. functional<br>2. production-<br>3. testing-related methods |
| Requirements | 1. function<br>2. manufacture<br>3. testing |
| Dimension groups | 1. functional<br>2. non-functional<br>3. auxiliary dimn. |
| Function groups | function dimns.<br>1. 1st priority<br>2. 2nd priority<br>3. 3rd priority |
| Function sub-groups | 1. 4th priority<br>2. 5th priority<br>3. last priority |
| Groups in general -- tolerances | 1. group fine<br>2. group interm.<br>3. group coarse |
| Groups in general -- tolerances for metal cutting manufacture | 1. shaft based tolerance -<br>2. hole based tolerance +<br>3. bilateral tolerance |
| General advice for use of tolerances | 1. shaft based system<br>2. hole based system<br>3. with symmetry symbols |

| | |
|---|---|
| Data for form and position tol. | 1. ISO symbols<br>2. tolerance<br>3. datum reference |
| Tolerance groups | 1. dimension tol.<br>2. form tolerance<br>3. position tol. |
| Types of form tolerance -- line-based | 1. straightness<br>2. roundness<br>3. linear form |
| Types of form tolerance -- surface-based | 1. flatness<br>2. cylindricity<br>3. plane form |
| Types of position tolerance | 1. deviation of direction<br>2. place deviation<br>3. motion deviation |
| Deviation of direction | 1. parallelism<br>2. perpendicularity<br>3. slope |
| Deviation of position | 1. position<br>2. symmetry<br>3. concentricity |
| Deviation of motion | 1. motion<br>2. circular run-out<br>3. flatness run-out |

Literature

Krist, Th., Technisches Zeichnen (Technical Drawing), Darmstadt: Tabellenverlag, 1975, p. 55

**Figure 8--6** Survey of Knowledge about Production of Detail Drawings

groups) can be listed as follows, whereby the varying and abundant forms of embodiment of individual features are typical:

- technical systems (TS) for which the task that they have to fulfill in many different conditions is simple, but that are not exactly designed for a specified block of functions (therefore not as organs, function carriers, or functional units);
- TS which occur as "species" in many different machines (i.e. gear wheels, bearings, shafts, levers);
- TS which occur either in mass production (screws, rolling bearings, seals) or in single-item or small series production (special casings or shafts);
- TS that are the originators of all properties of the higher technical systems. Their form, raw materials, dimensions, surface qualities and arrangement in space (configuration) are those design properties which determine all other properties.

#### 8.1.3.2 Designing of Assembly Groups and Elements, and their particular demands

Questions:
1. How does designing proceed?
2. Do special demands on designing exist?

Designing the assembly groups and elements (components) contains concrete decisions about all design properties. Therefore this phase has an especially large importance for the quality of the products (compare figure 8--7). Design methodology should therefore offer exact procedures for their individual (different) classes. A very concrete set of technical knowledge must accompany the concrete methodical references. We will return to this aspect later.

Specific procedural plans can be derived (compare figure 7--14) with ever more concrete contents on different taxonometric planes. On the upper plane of the machine elements -- components, figure 8--8 shows a procedural plan as example, which on one hand is founded on the general procedural model (figure 7--13), on the other hand defines how the individual design properties and characteristics can be progressively established. The assignment of the individual steps to the generally defined stages is obvious here. The elements and components become definitive and complete after a set of repetitions and iterations (at least twice: in the preliminary layout and in the definitive layout), by repetition of the steps (iteration of values of properties) in the flow chart. The necessity to approach the goal by repetitions (iterative procedure), is conditioned by the complex relationships among the design properties. They are not only in a complex relationship to the demands, but are also mutually in different relationships. Designers must work first with assumptions, which can only be defined in the second (or third) process. The flow chart shows a possibility for optimization, namely the work divided into certain blocks with internal feedback (return) loops, to avoid too many repetitions of the full procedure.

Figure 8--9 presents another flow chart, which represents a procedural plan for laying out of hydrodynamic bearings (a family of connections). We do not want to enter into details of the procedure in this context. If we now compare this process with that in figure 8--8, or even with the general procedure in figure 7--13, the difference can clearly be seen: on the general plane of generic TS, mostly extensive and generalized processes appear; in contrast, the flow chart of figure 8--9 specifies rather simple operations, such as "determine the Sommerfeld number." (Note and caution: the definition of Sommerfeld number in figure 8--9 is conventional for most of Europe. The inverse definition is used in other world regions. The scales and shapes of the value diagrams in the literature may therefore be different.) The references are

| Influenced Properties and Characteristics | | | Constituent of Layout — Selected Design Properties: Form | | Dimen–sioning | Material |
|---|---|---|---|---|---|---|
| | | | Shape, Geometry | Surface | | |
| Fulfilling of Functions | Realizing the Technical Process | | ■ | ◧ | ■ | ◧ |
| Fulfilling of Functions | Safety against: | non–permissible operating conditions | ◧ | □ | ◧ | □ |
| Fulfilling of Functions | Safety against: | mechanical failure by static or dynamic forces | ■ | ◧ | ■ | ◧ |
| Fulfilling of Functions | Safety against: | long–time loss of strength | ◧ | □ | □ | ■ |
| Fulfilling of Functions | Safety against: | non–permissible deformation: force | ■ | □ | ■ | ◧ |
| Fulfilling of Functions | Safety against: | non–permissible deformation: thermal expansion | ■ | □ | ■ | ■ |
| Fulfilling of Functions | Safety against: | relaxation & settling | ◧ | ◧ | ◧ | ■ |
| Fulfilling of Functions | Safety against: | wear | ◧ | ■ | ◧ | ■ |
| Fulfilling of Functions | Safety against: | corrosion | □ | ◧ | □ | ■ |
| Fulfilling of Functions | Safety against: | endangering personnel | ◧ | □ | ◧ | □ |
| Manufacture | manufacture | | ■ | ◧ | ■ | ■ |
| Manufacture | quality control | | ■ | ◧ | ◧ | □ |
| Assembly | assembly | | ■ | □ | ◧ | □ |
| Assembly | quality control | | ◧ | □ | ◧ | □ |
| Transportation, packaging | | | ■ | □ | ■ | □ |
| Maintenance, repair | | | ■ | ◧ | ◧ | □ |
| Environmental loading | | | ◧ | □ | □ | □ |
| Disposal, liquidation, recycling | | | □ | □ | ◧ | ■ |
| Deadlines | | | ◧ | ◧ | ◧ | ◧ |
| Conformity to standards | | | ■ | ◧ | ■ | ◧ |
| Costs | | | ■ | ■ | ■ | ◧ |

Legends: ■ strong influence ◧ intermediate influence □ weak influence

Figure 8––7 Influence of Design Properties on the Properties of the Designed Technical System

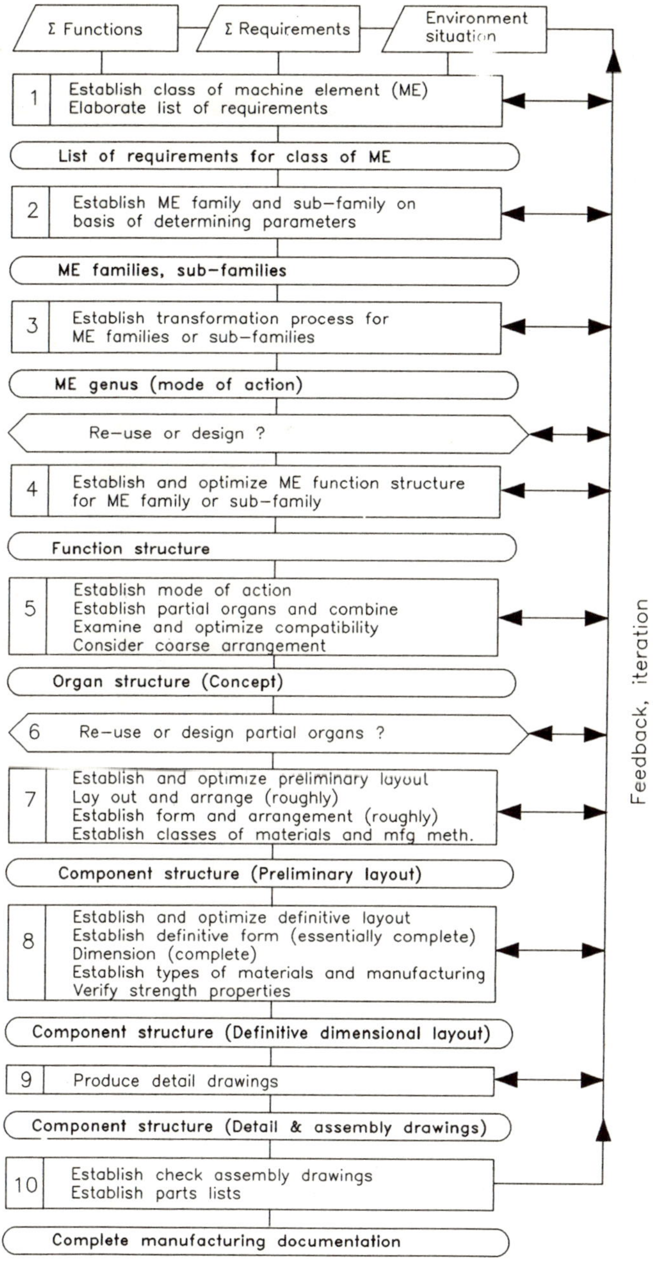


Figure 8--8 General Procedural Model for Designing of Novel Machine Elements

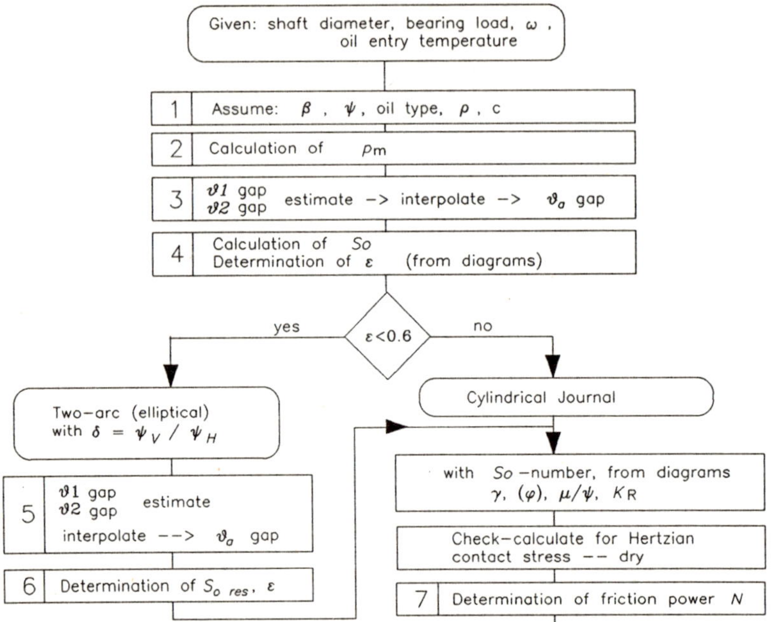


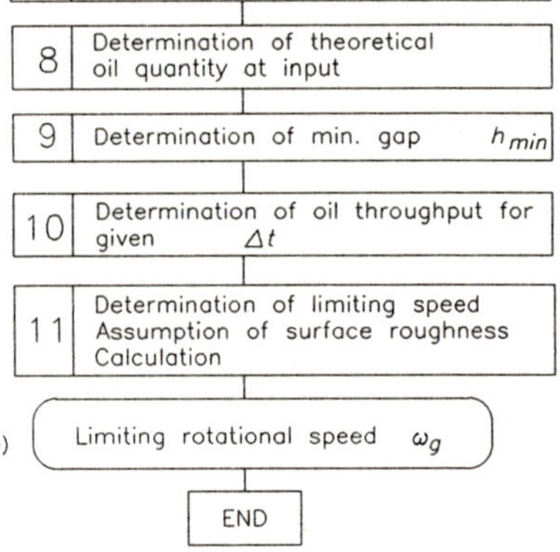


Legends:

| | |
|---|---|
| R | shaft radius |
| $D=2.R$ | shaft diameter |
| B | bearing width |
| $\Delta R$ | radial increase of bearing liner |
| $\Delta D$ | diametral increase of bearing liner |
| e | excentricity |
| $h_{min}$ | smallest gap |
| h | gap thickness |
| $\beta=B/D$ | bearing aspect ratio |
| $\psi=\Delta R/R=\Delta D/D$ | relative bearing clearance |
| $\varepsilon=e/\Delta R$ | relative excentricity |
| $\omega$ | angular velocity |
| $\omega_G$ | limiting angular velocity |
| P | load |
| $\mu$ | friction coefficient |
| $P_R$ | friction force |
| N | friction power |
| $S_o$ | Sommerfeld number |
| $S_{o\ res}$ | Sommerfeld number (horizontal clearance) |
| $\rho$ | oil density |
| $\eta$ | dynamic viscosity |
| $\nu=\eta/\rho$ | kinematic viscosity |
| c | specific heat |
| $t_E$ | oil entry temperature |
| t | maximum oil temperature in gap |
| $p_m=P/(B.D)$ | mean bearing pressure |
| $p_{max}$ | maximum bearing stress (Hertz contact) |
| $\varphi$ | angle to oil inlet |
| $\gamma$ | angle to smallest gap |
| $\varphi_a$ $\vartheta_a$ | angle to oil pressure relief |
| E | Young's modulus of elasticity |
| Q | oil quantity (throughput) |
| H | -- horizontal |
| V | -- vertical |
| W | -- shaft |
| L | -- bearing liner |
| R | -- theoretical |
| E | -- required |

$\omega_G \propto \eta.\psi^{1/2}/(\rho.R.\Delta R)$ $\quad p_{max} \propto \psi.E_W.E_L/(E_W+E_L)$

$N=\mu.P.R.\omega=\rho.Q_R.c.(t-t_E)$

$S_o=p_m.\psi^2/(\eta.\omega)$ $\quad S_{o\ res}=p_m.\psi_H^2/(\eta.\omega)$

Literature:


Barwell, F.T., Lubrication of Bearings, London: Butterworths, 1956

Doughtie, V.L., & Vallance, A., Design of Machine Members, New York: McGraw-Hill, 1964

Freeman, P., Lubrication and Friction, London: Pitman, 1962

Juvinall, R.C., & Marshek, K.M., Fundamentals of Machine Component Design (2 ed), New York: Wiley, 1991

Ott, H., Vorlesungen: Maschinenelemente (Lecture Notes: Machine Elements), Zürich: ETH, 1980


| Figure 8--9 Part 1 of 2 | Procedure for Design Calculations of Hydrodynamic Journal Bearings |
|---|---|

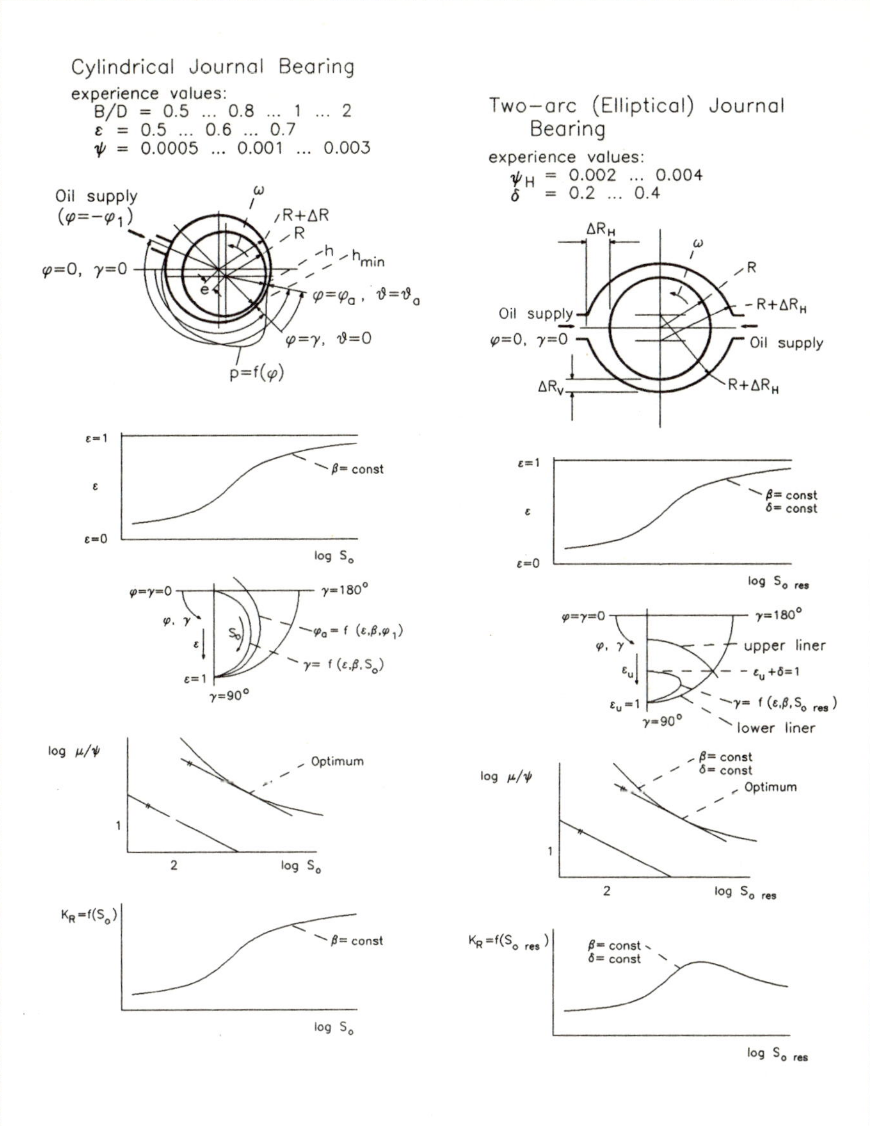


| Figure 8--9<br>Part 2 of 2 | Procedure for Design Calculations of<br>Hydrodynamic Journal Bearings |
|---|---|

much more concretized, and therefore also the objectivity of the decision, because the decision criteria are concrete and simple. For instance, "if the dimension ***e*** is numerically smaller than 0.6, then ... ." It is clear, that such an instruction could already be executed by the computer, for this statement is algorithmizable.

With these examples we have shown the particular situation at this complexity level, and the analysis has demonstrated the differences between the procedural models and a procedural plan by juxtaposing them. We could also recognize how important the design methodology for this area. The almost exclusive concentration on the level of machines (complexity level III) of the existing design methodology, as reported in the literature, has only little justification. This is probably another facet of the big picture of reasons why design methodology has lacked broad applications in engineering design practice.

#### 8.1.3.3 Special Technical Knowledge

Questions:
1. Is a special technical knowledge needed, and where is it needed?
2. Does this knowledge exist, and in which form?

The example of procedure for laying out of hydrodynamic bearings (figure 8--9) is not only instructive for the methodology but also for the relation with knowledge. We learn from this example in which quantity special knowledge is needed for the procedure and for each individual step.

Two different kinds of the necessary organized knowledge exist:

1. The "engineering knowledge," referring to physics, chemistry, and others areas of knowledge, which must be available for the acquisition of the function, reliability, strength and all further properties of the technical system. Beside the relationships of individual sizes, e.g., in scientific and empirical formulae, other data and values (both science-based and heuristic) must exist as individual characteristic magnitudes so that the parts can be laid out. These are for instance the permissible stress values, the possible wall thickness of a casting, the normally used maximum flow rate, and others.
2. A further kind of knowledge emerges especially in the connection with considerations about procedure and CAD. This should give designers and the computer additional items of information. These can be catalogs of the solutions of individual problems emerging in the procedure of designing, but also masters (compare Section 8.4) of individual documents of the methodical series, such as a list of requirements (design specification), function structure, sketch of principles (organ structure, concept), among others. The master list of requirements for bearings (part 1), and masters of TS-structures (part2), as shown in figure 8--10 can serve as an example of this form of process knowledge.

The "engineering knowledge" for individual families was explored, researched, published and taught as part of the discipline "machine elements" for almost hundred years. It is just over 100 years since C. Bach (*Die Maschinenelemente* -- The Machine Elements, Stuttgart 1891) published a new view about the calculation of machine elements (and wrote: "... the method of relationship numbers has had its foundation removed.") The contents of machine elements has since then remained quite stable in its structure, but in newer works we can find methods and values for individual areas and individual operations. Altogether, knowledge is amply available for this area -- even at times in a form suitable for designers. Many computer programs exist, and gradually also knowledge-based systems are being developed which are capable of form-giving for whole organs in dialogue with the designer. The program CADOBS for "rotating shafts and bearing arrangements" is quoted here as an example [152].

The situation of the second kind of knowledge is not nearly as satisfactory. Some areas have been processed, but more in the form of examples and not as systematic knowledge.

### 8.1.3.4 Situating the Specialized Science for Elements, Components

Question:
On which taxonomic planes can the Basic Specialized Design Science be meaningfully situated, so that an optimal information system emerges at this level of complexity?

In the treatment of the higher complexity levels, the principle of a taxonomy has been presented and explained using the example of machine tools. The order at the complexity levels I and II discussed here can be set up on similar way. As an example for this we show one of the best known groups of the components (machine elements) -- connections. Let us follow the origin of this hierarchy in figure 8--11, but first we present a terminological remark: figure 8--11 uses the term "organ" as the "means to solve a problem" (compare figure 7--3), something that fulfills a certain function -- a function carrier. Although the terms "organs" and "components" (elements) are not identical, the diagram is valid for both.

Now we can discuss figure 8--11, and the scheme and examples of connections. "Connections" in general are placed on the phylum-plane, with other organs that carry different functions. The question about the degrees of freedom of the connection gives rise to the origin of several classes: the first with no degrees of freedom (the fixed connections), then those with 1 or 2 degrees of freedom, among others. According to the direction of the degree of freedom, a relative axial motion (guidance) or rotational motion (bearing) of a part can exist. The latter needs also the prevailing force direction as radial or axial (thrust) bearing. These lower classes of connections are then arranged according to the kind of friction: the family of rolling bearings with rolling friction, the family of sliding bearings with sliding friction. Consideration of lubrication of the lower classes yields dry bearings, hydrodynamic and hydrostatic bearings. The way in which lubricant is supplied bring pressurized or self-supplying bearings, to which all the further lower classes belong. The form the lubrication gap defines the species of the sliding bearing connections: circular bearing, two-arc bearing, or segmented bearing (fixed or pivoted segments). These can be associated again with individual sizes or concrete type sizes. Let us remember here the flow chart, figure 8--8, which partly algorithmizes this step.

The hierarchy described above has many applications. This hierarchy is not only an ordering, it lets us also recognize that such a diagram demarcates the possible step-wise solution path for designing of connections. We again refer to the possibility that such a representation, accompanied by precise conditions for the steps downward (compare figure 8--9), forms the basis for a computer program.

If we now put the question where the basic specialized design knowledge should be situated in this hierarchy, then the experience in the area of the machine elements can help us. The published literature has in practice been divided into two main streams. One presents works about machine elements (e.g. [110,146,236,377,395,406]), i.e. the highest phylum of general technical systems of the first complexity level. The other treats the connecting elements, bearings [66,158,185], i.e. the phylum, class or family plane. The knowledge on the lower planes is concretized through the corresponding completions (TS). We consider it expedient to take over this practice for the construction of the knowledge system for designing, which brings many advantages.

That would mean, in such a model, that we should process the machine elements into a Basic Specialized Design Science (BSDS), and to plan and work out the TS-completions systematically. With regard to contents we would be concerned about working out the part about designing, because the currently available publications treat this problem very sparsely, and about completing further parts about machine elements.

Harmonizing the view about the systematics and other areas of machine elements with the Theory of Technical Systems will produce no revolutionary changes, contrary to initial

| Requirements | Units, Values, Tolerances | Commitment (Fixed requirement, desirable) |
|---|---|---|
| *Requirements conditional on function* | | |
| Loading | | |
| – magnitude radial | | |
| – magnitude axial | | |
| – time variation | | |
| – frequency of shock | | |
| – possible exceptional loading peaks | | |
| Motion | | |
| – rotational speed | | |
| – time variation | | |
| Damping | | |
| Shaft | | |
| – dimensions | | |
| – material | | |
| – positional accuracy | | |
| – elasticity | | |
| Neighboring TS | | |
| – interface locations | | |
| – dimensions | | |
| – behavior | | |
| – space restrictions | | |
| Energy losses | | |
| Environment situation | | |
| – temperature | | |
| – humidity | | |
| *Requirements conditional on operation* | | |
| Reliability | | |
| Lifespan | | |
| Maintenance | | |
| – extent (time, cost) | | |
| – replacement possibilities | | |
| *Ergonomic requirements* | | |
| Noise emission | | |
| Safety | | |
| Possibility of monitoring | | |
| *Requirements regarding manufacturing properties* | | |
| Production technologies | | |
| Batch size / order quantity | | |
| *Economic requirements* | | |
| Costs | | |
| – manufacturing costs | | |
| – running costs | | |

Figure 8--10 Part 1 of 2 | Master List of Requirements for Bearing Arrangements

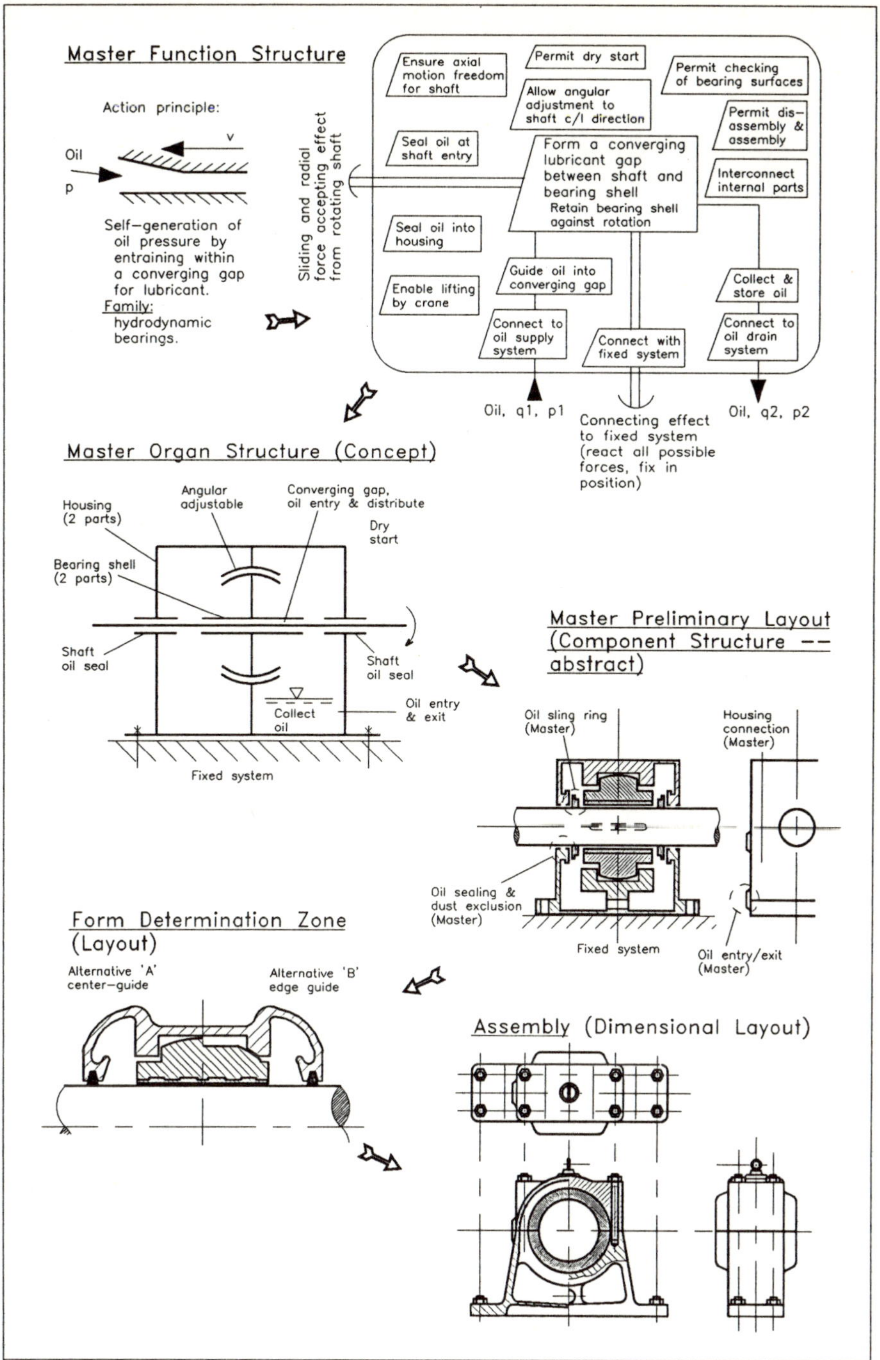


Figure 8--10 Part 2 of 2 | Masters for Hydrodynamic Journal Bearings

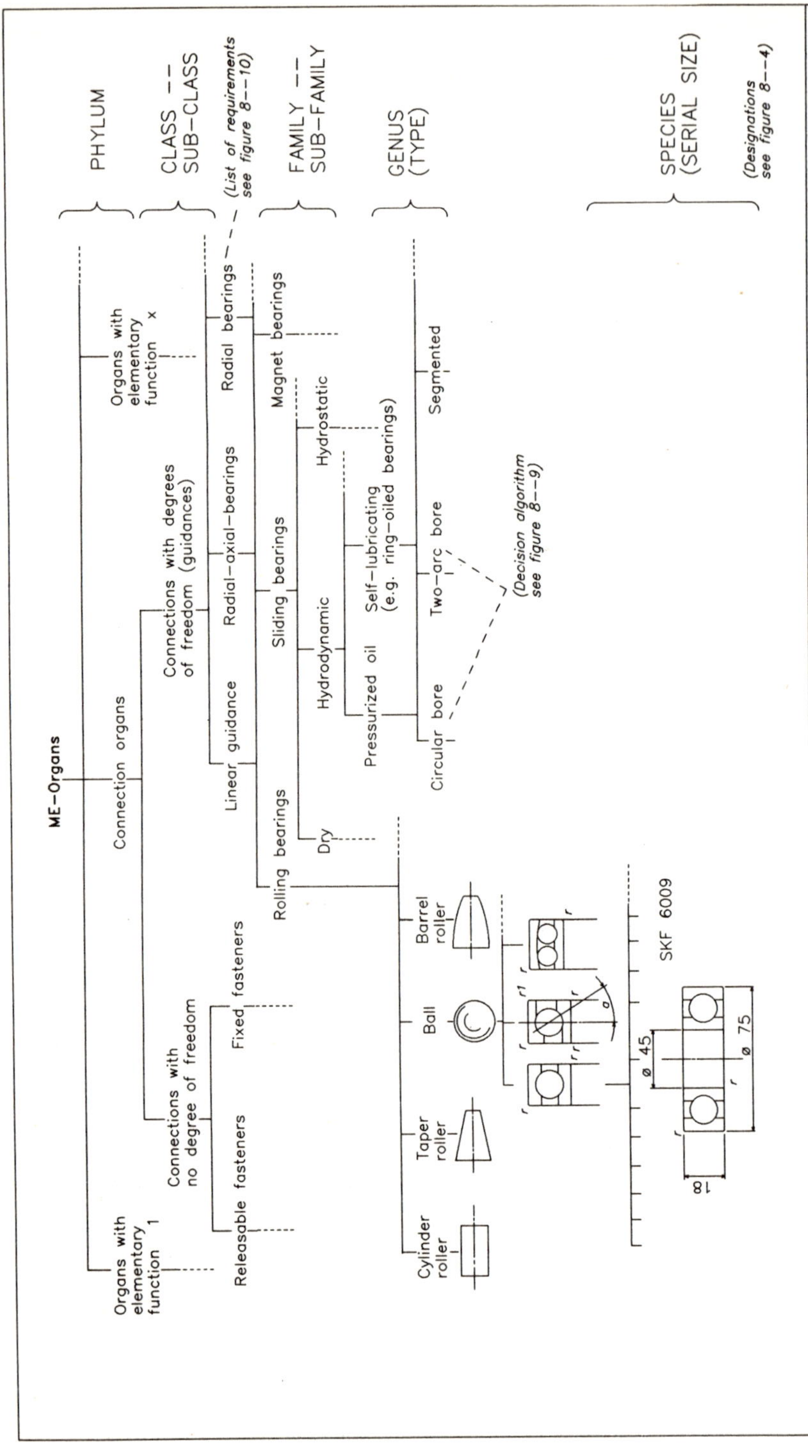


Figure 8--11 Machine element (ME) Organs -- Example: Connective Organs

expectations. It is important as introduction for a uniform set of ordering systems, whereby the general order is facilitated.

## 8.2 Design Science for Individual Branches -- TS Families

The treatment of Design Science with regard to the complexity levels was useful, but it has not led us to an information system for designers.

In engineering practice, designers always have the task to design a concrete object -- a product, an artifact -- we use the inclusive term "technical system" (TS). This may not only be a machine in general, but possibly a certain boiler, or a particular steam turbine, or a bridge, or a house, a transformer, a high voltage instrument, a crane or a personnel elevator. The enumeration of these particular objects (TS) would certainly be very long, even endless. This enumeration would be different if we consider more generally bridges, cranes, elevators, or even lifting equipment in general: here we are already on the plane of the product families (compare figure 8--4), namely the branches or specialties. The kinship of individual products in a product family can be different. It begins mostly with the common function (or functions), and especially with partial functions, and continues with the application of similar modes of action, manufacturing technologies, raw materials, ways of constructing, forms and a series of further historically conditional features. This is the reason why the knowledge of these areas can be developed in similar ways, can be jointly set up, and why also the transfer of experiences can be at all possible.

For designers an expedient field of activity emerges with a product family, on which they bring their experiences to bear. The broader and deeper the kinship of the products, the more favorable can be the exploitation of the available technical knowledge in individual families.

The present structure of the branches, specialties and/or product families was generated pragmatically both in the theory (discipline books, compare the titles in figure 7--1), as well as in practice. In the theory, the function (purpose) or the action principle, and possibly other features, have prevailed as ordering principles. The diagram in figure 2--1 indicates some ordering possibilities in a branch (specialty). The formation of the branches in practice was influenced sometimes by production kinship (e.g., instrumentation mechanics), which for Design Science is not a favorable kinship.

In any case the existing order is not homogeneous, and cannot be taken over without correction. On the other hand, otherwise introduced information systems exist, as well as a certain order among the quantity of the addressees to which Design Science turns, and which have a clear task field.

It cannot become the goal of Design Science to set up and recommend a certain branch order. The goal can only be to collect the necessary knowledge of the specialty for the design process, to organize and to process it, so that it is available in favorable form.

If we remain at the complexity level III -- machines, then the analyses of designing and technical knowledge in Section 8.1.2 have been comprehensively exercised, and can serve as basis for all specialties on this plane. A new element will likely be the level of originality, i.e. possibly new problems will emerge in the development of novel products, which are unknown for the daily cases. Also the number of items to be manufactured can be an influential factor.

At the complexity levels I and II, a series of specialties emerges beside the connections (see figure 8--11), but also in the lower levels of the family. The fact that mass production is possible, and very demanding manufacturing technologies are necessary (e.g., for rolling bearings), allows a deeper specialization. At a factory for rolling bearings, the technical and economic parameters now depend more on the construction of the manufacturing processes

| GENERAL DESIGN SCIENCE | Examples | SPECIALIZED DESIGN SCIENCES | Examples |
|---|---|---|---|
| <u>Transformation Process</u> (TrfP):<br>The transformation of state of an operand demands a series of operations (partial transformations) in a certain sequence. Individual operations and their sequence depend on and are determined by the selected technology (Tg) for the transformation, based on insights from various sciences (e.g. physics, chemistry, material science). Each transformation can be realized by several technologies, and these can be categorized by technological principles. The technologies are implemented by the operators of the transformation.<br><u>Model</u>: -- general transformation process. | DesSci<br>Fg. 7--2<br>TTS<br>Fg. 3.2 | A certain class of operands in a given state are to be transformed into a different state (narrower or broader options) using a defined technology.<br><u>Masters</u> (paradigm patterns, stencils, proforma, exemplars, guidelines, prototypes, archetypes):<br>-- states of the operands;<br>-- suitable technologies and their allocation to transformations;<br>-- concrete classes of transformation processes. | TTS<br>Fg. 2.4<br>& 3.2 (ct)<br>ED<br>Fg 8--3<br>for<br>family of<br>"positioner" |
| <u>Technical System</u> (TS):<br>Apart from its effects (they cause the transformation), the technical system (TS) is carrier of its properties. These enable all phases of the life of a TS, and allow them to proceed effectively and efficiently. The properties for any TS can be classified into a few general classes, preferably according to life phases and their operators.<br><u>Models</u>: -- Life cycle (see below);<br>-- Classes of properties of the TS that cover the individual life situations of the TS. | DesSci<br>Fg. 7--4<br>TTS<br>Fg. 7.1<br>ED<br>Fg. 4--3 | A certain type of TS is carrier of the capabilities to exert a set of effects to cause the transformation, and of other properties that ensure advantageous behavior in individual life phases. Properties are adapted to the operators of individual phases.<br><u>Masters</u>: -- typical life cycle of the TS-type;<br>-- specification of needed effects;<br>-- specification of other properties (e.g. defined in a design specification). | ED<br>Fg. 8--4<br>(positioner)<br>DesSci<br>Fg. 8--10<br>ED<br>Fg. 8--2 |
| <u>TS Life Cycle</u>:<br>The life cycle of a TS is a system of processes (with operators), consisting of planning, originating (designing, production planning, producing), distributing, using (exploiting), and disposing. Each phase contains further partial phases, e.g. maintaining, modernizing, etc.<br><u>Model</u>: -- general life cycle with all possible phases. | DesSci<br>Fg. 7--8<br>TTS<br>Fg. 10.1<br>ED<br>Fg. 4--10 | Every TS-type passes through specific processes and stages which place particular requirements on the TS, e.g. for hygiene in food-processing machinery, for transport of larger units.<br><u>Masters</u>: -- specific life cycle processes and stages for individual TS-types and sub-types. | TTS<br>Fg. 10.4<br>for quantity<br>manufacture |
| <u>TS Development in Time</u>:<br>TS develop over any period of time, depending on their usage, available technical knowledge, existing materials, economic situation, and other factors. The total development consists of elementary changes of individual properties of the TS (i.e. successively produced or upgraded TS)<br><u>Descriptions</u>: -- theses about development;<br>-- general impulses that drive development;<br>-- general factors that influence developments. | DesSci<br>sect.<br>7.1.3.3<br>TTS<br>Ch 11 | Development in time of a TS-type is specific, and influenced by a series of specific factors. Knowledge of the development contributes to understanding the accepted solutions, which may not always be rational.<br><u>Reports</u>: -- about the development of the TS-type;<br>-- analysis of development factors;<br>-- possible trends of the development. | TTS<br>Fg. 11.1--<br>11.4 for<br>different<br>families |
| <u>Design Process</u> (DesP):<br>Concerning structure, every design process (DesP) can be divided into a finite number of partial processes (phases, operations).<br>-- guiding pictures of "design phases";<br>-- guiding pictures of "design operations";<br>-- guiding pictures of "basic operations" (esp. open-ended problem solving). | DesSci<br>Fg. 7--12<br>ED<br>Fg. 5--4<br>ThKoP<br>Fg. 2.4<br>& 2.6 | The design process for a particular TS-type has certain peculiarities that influence the individual phases and operations. According to its position in the hierarchy, regulations or recommendations can be worked out for individual phases or operations. | ThKoP<br>Fg. 4.7,<br>4.3 & 4.4 |

Figure 8--12 Part 1 of 2 — Comparisons of Statements from General and Specialized Design Sciences

| GENERAL DESIGN SCIENCE (cont.) | Examples | SPECIALIZED DESIGN SCIENCES (cont.) | Examples |
|---|---|---|---|
| The course of progress of designing (DesP) can be supported and controlled with the help of methods. The total progress flows according to a general procedural model (design strategy, design methodology). Progress is iterative and recursive, and involves work on one partial system at a time.<br>-- guiding pictures of 'generalized procedure';<br>-- guiding pictures of several 'specialized procedural plans' | DesSci<br>Fg. 7--13<br>ThKoP<br>Fg. 5.12<br>ED<br>Fg. 6--1<br>& 6--2 | The course of progress of designing depends strongly on the TS-type, and therefore we can prescribe or recommend<br>-- specific procedure for a TS-type,<br>-- specific methodologies and methods as the strategic tools. | DesSci<br>Fg. 7--14,<br>8--8<br>& 8--9<br>TTS<br>Fg. 12.7 |
| The individual phases and operations of the DesP can be performed with the help of certain methodologies, methods and principles.<br>-- guiding pictures of design methodologies, methods, principles;<br>-- allocation of these to phases and operations. | DesSci<br>Fg. 8--13<br>ED<br>Fg. 5--6<br>& 5--7<br>ThKoP<br>Fg. 5.14 | As the existing operational breadth is limited, so also the tactical tools are limited.<br>Masters: -- individual methods or principles can be allocated to individual operations, and can be accompanied by many concrete recommendations and knowledge. | DesSci<br>Fg. 8--13,<br>8--11<br>8--6 &<br>7--23<br>TTS<br>Fg. 12.8 |
| As with all other processes, various recurring specific situations occur in the design process (DesP) that demand specific information in order to overcome them. Solving these frequently recurring situations can be acquired (learned) up to the state of 'intuitive action', ability, or stereotype.<br>Models: -- guiding model of the situation -- dependent on ....;<br>-- catalogs of general design situations;<br>-- guiding picture of special design situations. | DesSci<br>Fg. 7--14 | Many specific design situations frequently recur, they show very concrete solutions with limited numbers of alternatives, and demand certain definite information.<br>Masters: -- catalog of specific situations;<br>-- procedural models for individual situations;<br>-- specification of relevant information for individual situations. | DesSci<br>Fg. 8--11<br>ED<br>Fg. 6--4<br>TTS<br>Fg. 8.2,<br>12.7, 12.8<br>& 12.9 |
| Design processes are realized in design systems (i.e. design divisions). The potential of a design system depends on the operators (see model of the DesP):<br>-- designer, or design team;<br>-- working means (equipment, tools, computers, etc.);<br>branch object and design process knowledge;<br>-- leadership, management of the DesP;<br>-- environment, working conditions.<br>Many relationships exist among the operators.<br>The work within the design process should be regulated and controlled with the help of suitable<br>-- procedural organization,<br>-- compositional organization,<br>considering the special nature of design work and its coordination into the enterprise organization. | DesSci<br>Fg. 7--10<br>ED Fg. 5--2<br>& 5--3<br>ThKoP<br>complete<br>book<br><br>TTS<br>Fg. 10.9<br>& 10.5<br>ThKoP<br>Fg. 8.4 | The design system for designing particular TS-types has specific requirements and solutions that depend on several variables.<br>Masters: -- masters of the design system (versions, options);<br>-- requirements for individual designer-types;<br>-- necessary information systems;<br>-- listing of information sources;<br>-- listing of information carriers (media);<br>-- guiding picture of the procedural organization (versions);<br>-- guiding picture of the compositional organization (versions). | DesSci<br>Fg. 7--20<br>(quoted<br>figures)<br>ThKoP<br>Append. 3 |
| The designer is the most important operator of the design process. Designers can appear in various forms (individuals, design group, team) and requirements for designers depend on these forms.<br>Models: -- guiding picture for individual design types;<br>-- guiding picture for the design team -- composition, behaviour, procedural manner, etc. | DesSci<br>Fg. 7--17<br>ThKoP<br>Fg. 3.2,<br>3.3, 3.7,<br>8.6 & 8.7 | Special requirements are set for designers of particular TS-types. The general statements can be converted into concrete guiding pictures.<br>Masters: -- concrete guiding picture of individual design-types;<br>-- concrete guiding picture for design teams -- composition, behaviour, procedural manner, etc. | ThKoP<br>Fg. 3.4<br>& 3.6 |

DesSci ... Hubka, V. & Eder, W.E., Design Science: Introduction to Needs, Scope and Organization of Engineering Design Knowledge (this book)
TTS ... Hubka, V. & Eder, W.E., Theory of Technical Systems, New York: Springer-Verlag 1988
ED ... Hubka, V. & Eder, W.E., Engineering Design, Zürich: Heurista, 1992
ThKoP ... Hubka, V., Theorie der Konstruktionsprozesse (Theory of Design Processes), Berlin/Heidelberg: Springer-Verlag, 1976

Figure 8--12 Part 2 of 2 — Comparisons of Statements from General and Specialized Design Sciences

than on the construction of the bearing as product. The considerations about design process and technical knowledge, which were treated in Section 8.1.3, form also the basis for all components-related specialties, just as for the machines.

## 8.3 Design Science for Other TS-types

The most important ordering features which have caused us to treat the Specialized Design Sciences were degrees of complexity and product families. Of those classification characteristics enumerated in the beginning, three still remain: level of originality, number of pieces and organization size.

When considering the degree of complexity and the product families we mentioned everywhere further influencing factors, and figure 8--5 indicates also the corresponding completions in the whole system. Therefore we will not enter into further details.

One type of technical system which is very special, and has up to now remained without explicit treatment, although we have spoken repeatedly about its range of problems and task. It deals with *processes* as systems of operations, and with the range of problems of designing of processes, as distinct from designing TS-object systems. Especially in connection with plant we mentioned the importance of the process and the technology.

Some areas exist in which this range of problems already exhibits independent works, such as *Technologische Betriebsprojektierung* (Technological Works Project Engineering) by W. Rockstroh [368]. The boundaries are very fuzzy, and often the whole plant project engineering task is covered. Also in the book SLP [313] referred to above, the process is presented as "R-routing" and arranged among the five influencing factors (compare figure 8--2).

The combination of designing with some specialties such as chemical engineering appears at least unusual for a mechanical engineer. But also in such specialities, the task of the chemical engineer, namely "designing the process," has been acknowledged and integrated into education. In the research towards Design Science, chemical engineers have also participated, e.g., S.A. Gregory [175,177].

## 8.4 Design Sciences in Comparison

In the previous sections of this chapter we have analyzed the formation of the Specialized Design Sciences at various levels of complexity of the TS, in several branches (TS-families), and also at various levels of abstraction.

In order to increase understanding of the differences between the various kinds of SDS, figure 8--12 contrasts the statements of the General and the Specialized Design Sciences. The selection of statements should, where possible, reflect the whole breadth of the knowledge, and thus contribute to better understanding. The individual statements are augmented by references to suitable examples, mostly diagrams. These figures are in part published within this book, or were published in the Theory of Technical Systems [214,219], Engineering Design [228], or Theorie der Konstruktionsprozesse (Theory of Design Processes) [202].

# 9 SPECIALIZED DESIGN SCIENCE FOR INDIVIDUAL CLASSES OF RECIPIENTS

*As you stand here, remember those who have enlarged the horizon of your mind and pray for those who teach.*
*Inscription on wall plaque, Salisbury Cathedral, St. Edmunds Chapel*

*Knowledge is of two kinds. We know a subject ourselves, or we know where we can find information upon it.*
***Samuel Johnson**, Letter to Lord Chesterfield, February 1755*

Question:
For whom is the stated knowledge system intended?

Starting from the statement morphology in figure 5--1 we deal now with the third characteristic -- the recipient. The knowledge about designing can be addressed to different recipient classes. The recipients and their needs, goals, tasks and state of knowledge, are the cause for important decisions on contents and form of the knowledge systems, and thereby also the division of recipients into individual classes.

It would however be wrong when designing the knowledge systems to already connect these knowledge systems with a certain formal idea, e.g., to believe that each knowledge system demands written processing such as a book. A system can possibly remain formulated only in an intellectual form and still provide good services. In such a state it can only with difficulty be copied and transmitted to others for further applications, e.g., for initial and continued education.

The individual Specialized Design Sciences (SDS) for recipient classes can be derived either from the General Design Science (GDS), or from the types of technical systems (TS-types) as described in Chapter 8. In this way a General Design Science can emerge for students *per se*, or a Specialized Design Science for students of turbine construction, or other branches. Often such a knowledge system is no longer acknowledged as a Design Science, but is called a special formation (an educational presentation, design for students).

The following classes should specify only most the frequently occurring recipient groups. Other groups can be attached to this series, or the groups subdivided or combined in different ways, as happens in practice. The most frequently occurring groups are:

- design researchers;
- design practitioners;
- design students -- inexperienced people, novices -- and design teachers -- educators;
- design lay persons -- including the general public;
- computer program logicians, who look for a “logic” for a program.
- others

**Learning About Design** -- Learning (formally or informally) about the processes of engineering design, but also about products and processes that exist, involves four basic kinds of contents:

- knowledge and advice for new learners, potential engineering designers and support people in education and training, who should learn sufficient object and design process knowledge based on some fundamental concepts (preferably with theories), with examples from a typical field of future experience, but from a limited knowledge background;
- knowledge for older and experienced practitioners, re-learners in further education or self-study, who should upgrade and modernize their range of knowledge;
- knowledge for managers, who should become aware of the object and design process knowledge needed by the practitioners;
- knowledge for teachers of engineering design, who should transmit the required knowledge and supervise the initial forms of practice.

Learning about designing involves learning to understand, find, recall, interpret, transfer, relate and use several kinds of information, some of which should also be transformed into (internalized) knowledge. Such information includes:

- hard information -- facts, rules, analytical models related to phenomena, as a part of object knowledge;
- soft information -- principles, guidelines, suggestions, partly related to phenomena and object knowledge, but also partly about processes, procedures, techniques, methods, etc.;
- heuristics, hints, tips -- advice and short-cuts based on knowledge, e.g., about failures and their (historic, societal and physical) causes;
- personal, individualistic and idiosyncratic behavior expectations, learning styles;
- team building (bonding) -- social interaction, teamwork, cooperation, communication, coping with stress, time management, record-keeping, assessments, and so on;
- information storage and retrieval, presentation forms -- including systematic "bringing onto paper" of ideas to free the mind of organizational tasks, which supports the needed interaction between perception of external records and mental action to develop ideas;
- methodology -- linkage of methods based on some connective theory that provides a logic of connectivity, but also including iterative (learn more and revise), recursive (decompose, restructure, repeat process on smaller units, re-combine), and opportunistic actions.

Related topics for education (teaching and learning) are needed for enhancing the psychological and sociological factors, the *cognitive* domain [73]; changing attitudes and learning styles, the *affective* domain [260]; and improving abilities and skills, including those in the *psycho-motor* domain.

Learning to design can be achieved in various ways [128,129,132,137]. The oldest is a "school of hard knocks," as was traditional in the past. This has produced some excellent engineering designers, but many potentially good designers have been turned off, discouraged by lacking success and recognition. An improvement was the "school of apprenticeship," although it depended on each master providing adequate explanation and guidance. Masters (e.g., of designing) are not necessarily good instructors. For the learner it was not easy to find the essentials, to abstract processes, and to explain why and how to improve the designing process. Nor was it easy to find good knowledge about the objects being designed, or to transfer that knowledge among different types (species) of objects. At times it was even difficult to find several or unusual ways of solving the design task. The "school of scientific approach" is usual in engineering science, but not yet usual for engineering design. It involves the foundation of a scientifically logical theory, such as our Design Science, coupled with observing, abstracting, classifying, analyzing, defining, checking, intuitive and logical understanding, and guided practice [130,135,136]. Whilst there has been extensive research on some aspects of Design Science, the coverage is incomplete, lacking overview and connectivity. The potential exists, but with no guarantees at present.

## 9.1 Design Science for Design Researchers

For the recipients "design researchers" the most extensive and most comprehensive formulation of Design Science should be presented; it should be very comprehensive in the theories (see southern quadrants in figure 5--2), justifications and substantiations, as well as in the considerations about conceptualizing, but contain less details in the prescriptive area (north of the map).

We can associate the considerations contained in this book with this kind of Design Science. This Design Science should also form the basis for the derivation of other kinds.

## 9.2 Design Science for Design Practitioners

As we declared the support of practical engineering design work as one of our goals, we thereby set the Design Science for practitioners as a goal of our efforts.

It is logical that this class must be arranged into further kinds, so that the emphases of the knowledge can be set according to the user. A certain amount of knowledge is needed from conceptualizing engineers, layout engineers, or development engineers, a different form and set of knowledge is expected from detail designers, although both can be offered some parts of similar contents.

Engineering practice demands above all prescriptive knowledge, in the object region and (currently to a lesser extent) in the design process area. This must be accompanied by the knowledge ordered according to design situations. Adjusting the form to the questions demands the application of special forms of the presentation of knowledge, such as models, matrices, catalogs, guidelines (compare Section 7.3.3, and figures 7--21, 7--22, 7--23).

It can be said that a part of the prescriptive knowledge already exists in the published handbooks, allegedly processed into a suitable form and with corresponding contents; still we miss the complete handbook in exemplary form within the meaning of the ideas developed here.

In order for a concrete "personal information system" (PIS) to emerge from this knowledge system, much additional information about the object (material Design Science), as well as about a concrete enterprise and designer is necessary. Additions (completions) could correspond to the model in figure 8--5 in their construction. A designers' personal information system (PIS) need not provide full coverage of the complete "map," it is more likely to consist of islands of knowledge as indicated in figure 8--5, part 2.

The help to be achieved for practicing designers could lead to an under-valuation of the abstract kinds of knowledge found at the extremes of the individual ordering features. Even if these kinds do not serve the engineering practice directly, they have the important task of transferring experiences from one area into the others.

We have set the justified question about the construction of these knowledge systems at the level of TS-types. Where should the Basic Specialized Design Science (BSDS) be situated, and which are the completions to individual hierarchical planes? The range of problems of the object (material) and design process knowledge does not make it easy to set priorities. Experience shows that a basis lying relatively high on the abstraction axis can be profitable and that individual completions can refer to individual functions (specialist).

## 9.3 Design Science for Inexperienced People, Students, Novices

An important variation of Design Science relates to the knowledge system for students. Traditionally such a treatment is called "design instruction." Numerous publications with suitable titles have existed for some time. The objectives of individual books were adapted to further demands, so that the title "design instruction" displays rather different contents.

We learn the intents of authors in the area of design instruction from studying some of the well-known German books. Already F. Reuleaux expressed some fundamental thoughts in his *Construktionslehre* (co-authored by C. Moll) [364] (1854). He saw the task and object of design instruction in the arrangement and explanation of the rules, coupled with practical demands, which were to be based on the principles of mechanics, and should be easily applicable. He pleaded for the application of a "regulated design procedure," which he presented and explained as valid for all of mechanical engineering (see Section 7.2.1). It is certainly understandable that these ideas about design processes do not agree with the present view of design methodology, but the harmony with the objectives of Design Science is exciting.

H. Tschochner, in his book *Konstruieren und Gestalten---Abriß einer Konstruktions- und Gestaltungslehre* (Designing and Form-giving -- Sketch of a Study of Design and Form-giving) [426] (1954), intended to clarify the nature of designing and form-giving, and to investigate the questions of relationships. The main part of the book consists of form-giving (60%), in which the basic guidelines for form-giving are summarized for different production process. Tschochner's intent was that his book should be "for the entering and young designer a kind of branch experience, to give him references for the practice, and finally to try to give the outline of a design instruction to be further expanded."

The *Konstruktionslehre des allgemeinen Maschinenbaus* (Design Instruction for General Mechanical Engineering) by R. Matousek [288] (1957) belongs to the valued works of this area. It was translated into English as the first German design instruction (Blackie, 1963), under the title *Engineering Design -- A Systematic Approach* [289], which points to an uncertainty (in both languages) about the explanation of the German word "Lehre" (does it mean apprenticeship? guidelines to learning? gauge to measure achievement?) and its possible and appropriate English equivalents. The book was addressed to "earnestly striving young engineers, not to such people who are satisfied with tracing from existing patterns without getting to know the reasoning and considerations which lead to the final product."

The author undertook the experiment to show in "easily comprehensible and organized form" a methodical procedure, which makes possible for the beginner a rational path towards a solution. The author expressed his conviction that a design instruction which exists of an aggregation of guidelines and rules cannot deliver the technology of designing for the beginner. Because only practice makes the perfect master, the book presented many exercises.

At the peak point of the efforts around rationalizing design work, the book *Konstruktionslehre* (Design Instruction) by G. Pahl and W. Beitz appeared, in 1977 in the first edition, 1988 in the second revised and expanded edition, and 1995 in its third edition

[333]. The book should address "in similar ways the student of machine apparatus and gearing construction and the designer in practice active in these areas, independent of specialty or branch." The book is also addressed to teachers and researchers, for which it wants to be a basis and inspiration to their own further development. It delivers a strategy of designing. The presented procedures, methods and resources for designing technical systems refer to conceptualizing, laying out (embodying, arranging, configuring) and elaborating (detailing, parametrizing), and the descriptions are formulated in a language understandable for the practice. This book is set up on the basis of mechanical systems and methodical (systematic) procedure. It is divided according to the individual design stages (conceptualizing, laying out, detailing) and the design steps as part of these stages.

The design instruction by Pahl and Beitz is without doubt the best existing doctrine, not only in contents (it contains a very broad spectrum of design knowledge), but also in the suitable and practice-related examples, and in the understandable form and presentation.

Already this small study has shown, and verified with quotations, that publications about design instruction were and are used as school books, but that they are not composed as special educational textbooks. All address broader circles (also practitioners), and (at least some of) their contents give the impression as if they were intended as applied Design Sciences rather than as design instructions. An applied Design Science can certainly serve as a design instruction, but only in part, it cannot satisfy all the demands that need to be respected.

### 9.3.1 Requirements for a Textbook for Design Education

A good textbook should contain a relevant and systematically founded and expanded presentation of a knowledge area (up to this definition, the demands are identically with those of science), and use pedagogic, didactic and learning-psychological knowledge. Such a textbook should be oriented towards the appropriate level of education.

It is clear that the already discussed (and other) design instructions are at least defective with respect to pedagogics and didactics. A different question refers to whether the offered presentation of the knowledge area is systematic and complete.

A three-dimensional picture of the demands on design instruction can be shown, if we use the model of the teaching and learning process (see figure 9--1). We see that this process model is directly derived from the general transformation process (compare figures 7--2, 7--10, 8--8). In the case of the teaching process, the person -- the student -- appears as operand with certain abilities for designing, and these should be brought from a state of relative ignorance of designing into a state of adequate knowledge. As operators of this transformation, the teachers work with their technical means and their knowledge, coordinated by the management of the process. The obligatory question about "technology" for these transformation processes (operand -- student) shows the necessity of special knowledge for the transformation of the state of knowledge and ability of a person. We are now proceeding in the area of pedagogics and didactics.

The didactic knowledge can in this connection be expressed in very simplified and condensed ways, as represented by the "principles of education" assembled in figure 9--2. If the application of these principles is explored in the existing design instructions, we come to the conclusion that they have not been considered, and that these books are hardly suitable as textbooks within the meaning of the definition.

Several problems are still to be overcome before we can arrive at the right knowledge system, and therefore also at a textbook. Above all this concerns the last demand of the above definition, namely to "... be oriented towards the appropriate level of education." This

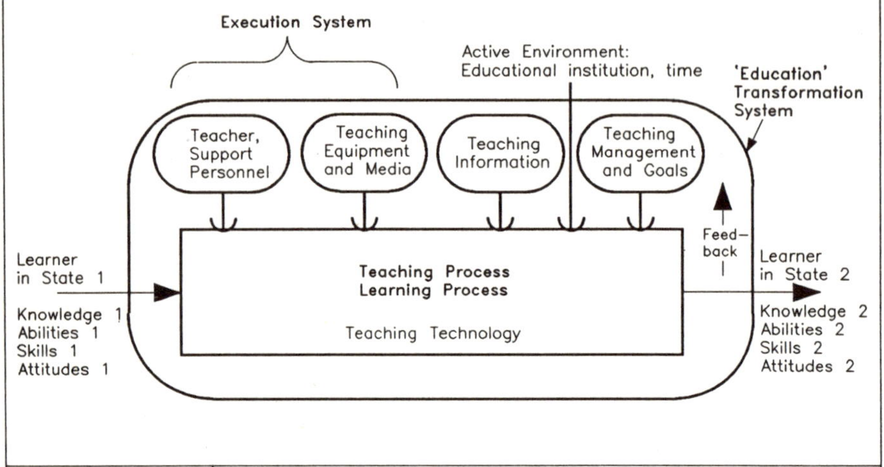


Figure 9--1 Model of the Teaching and Learning Process

demand postulates, surely with justification, that textbooks should be written for different levels. The four design instructions which have been analyzed here (Section 9.3), do not mention the level explicitly, because they are generally used at technical universities. In addition, they also address practitioners.

The literature also contains other works, e.g., *Methodisches Konstruieren* (Methodical Designing) by K. Hohmann [193], which originates from the area of the technical colleges and intends to give an incentive for self-study. To achieve simple handling, "extensive explanations were avoided." A second example is *Entwickeln und Konstruieren mit System* (Developing and Designing with System) by E. Gerhard [165], which is above all addressed to those who are "in the profession."

We can say that the mentioned books have been addressed at different recipient groups. A judgment of the didactic value needs a further study.

Another problem emerges in connection with the different types of teaching presentations which can be used for design education. The possibilities and formation elements of individual kinds are analyzed in Hubka [203]. For each of these types, a special knowledge system would, in principle, have to be prepared -- e.g., for lectures, group instruction, testing of technical knowledge, exercises, projects, case studies, programmed instruction, practical experience, excursions, research, self-study, etc.

Corresponding references can be found in the literature for some of these types, but hardly a complete knowledge system processed according to didactic principles.

### 9.3.2 Knowledge System for Design Education

Our considerations up to this point have discovered especially the defects and the deviations from a theoretically ideal condition.

The question must now be: How can the knowledge system practically be set up for design education? Certainly not -- as an extreme case -- to include all the types mentioned above as elements of the system, and still further elements considering the principles of

1) The principle of TOTALITY demands that for every detail of the subject matter (contents for learning) the context relative to the totality of subjects becomes active. Every subject should be observed and developed from all aspects, should be supported by prior experience (of staff and student), and should indicate the scope of new tasks.

2) The principle of ACTIVITY should bring the students (the learners) into an internalized relationship with the subject matter, such that they are actively confronted by their task with their whole inner unity.

3) The principle of INDIVIDUALITY requires that the educational experience is fitted to the individuality of each student (learner), and should consider the uniqueness in time and situation of the active participation of the student.

4) The principle of CURRENCY is predicated on the fact that the formative action on the student (learner) depends on the current educational situation. The sitation must be considered in light of the subject matter (learning content and form of presentation) that appeals to the student, and under what subjective conditions that student approaches the situation.

5) The principle of RELEVANCE is satisfied when all details are brought into a conclusive and sensible sequence that reveals the truth of the subject matter. A logical closure can also be postulated that tries to achieve a measure of completeness of the subject matter to provide a surveyable orienation and order.

6) The ACQUISITION OF KNOWLEDGE should develop the ability to approach new tasks, and to independently find ways of solving (open– and closed–ended) problems in a planned (methodical, systematic) fashion.

| Figure 9--2 | Principles of Education |
|---|---|

individuality and currency (see figure 9--2).

In this connection, the role of the teachers must be considered as the decisive operator in education (model figure 9--1). They build the instruction in at least two levels -- in the teaching concept which is anchored in the curriculum (overall teaching plan), and in the preparation of individual teaching presentations (time periods). Didactics always asks the questions: "what? how? where? for what? with what? when? and who?" (but for each of them also "why?") Judged strictly according to these criteria, no teaching period would be able to be repeated, because of the constantly changing circumstances.

It is therefore the role of the teachers, and their task, always to prepare a new knowledge system for each course and for each time period. This is an individual performance, for which a teacher should be provided with good documents and good knowledge elements, to be able to maintain the quality of education. For this the didactic knowledge (how? with what? for what?) must therefore also be available in suitable form. An important role is played by the articulation of each instruction form, it places special demands on the knowledge system (details may be found in Hubka [203]).

If this idea is projected onto the present knowledge systems, it is clear that the books on design instructions considered above are not directly usable as textbooks, but serve as material for the preparations of the teachers. Missing is a branch didactics for design education, which should provide support to the teachers.

In future the perfection of the knowledge system should be an aim in addition to the branch didactics. The new contents according to Design Science must be prepared for the teachers and as much as possible be differentiated for the distinctive levels of education. These new design instructions, derived from Design Science and considering the didactics, can improve the teaching considerably. Nevertheless, good teachers are the main condition.

## 9.4 Design Science for Lay Persons

Such a Design Science appears almost as a contradiction. Yet the knowledge about the roles of technology -- particularly its origin through planning ahead, designing and manufacturing -- is an important element in the general knowledge for each population. Ignorance about technology increases the fear and aversion to technology.

The design knowledge should also be prepared for these recipients, in different kinds, according to classes of lay people, such as:

- buyers and salespeople -- retail and wholesale;
- organizational economists -- managers;
- politicians -- legislators, who determine the social organization and its directions of development and philosophies;
- school children;
- teachers in general educational schools;
- the general population;
- etc.

The knowledge about the design process (designing), as important process in the development and realization of a technical system, should serve at least as enlightenment and entertainment, as is usual today for the "pure sciences" -- e.g., programs on television about nature and atomic physics. For other lay groups, the preparation is much more necessary, because the national and organizational economists need products, and these are manufactured or prepared using technical means (systems). Ignorance about the development and realization of the products, about their manufacture by means of technical system (which must themselves be designed and realized), and about the tasks of designing can be very injurious for the long-term health of the economy.

These discussions about Specialized Design Sciences for classes of recipients should suffice to outline the range of problems.

# 10 QUALITY OF DESIGN SCIENCE

*Looking and seeing are different things which rarely coincide.*
***Alexander Blok***
*-- additional comment by the authors:*
*hearing and listening follow the same rule.*

*A new product will only be accepted by the engineering community and the market if it shows both technical and economic advantages.*
***E.W. Constant, II**, in **The Origins of the Turbojet Revolution**, Johns Hopkins Studies in the History of Technology, Baltimore: Johns Hopkins U.P., 1980*

Design Science, as a system of statements, a knowledge or information system, can be good or bad, yet it has a certain level of quality as a whole and in its individual parts and elements. The quality of the knowledge -- the information -- can be judged from its properties. Of these, the most important are:

- contents: correct -- true, provable? (explanations, sources)
- contents: complete -- organized, classified, retrievable? (not only separate sections or statements without an added statement of relationships and a range of application)
- contents: expedient, structured (ordered)?
- form: clear, understandable representation?
- form: clear, quick orientation possible?

Further properties offer additional characteristics which can support the first group:

- age of the information;
- source of the information (author, publisher);
- availability of the information (not protected, not only personal experience).

The quality of the contents of Design Science is the determining factor for designers, so that the results -- the technical systems being designed -- will function correctly and can carry all the desired properties. The structure and form influence especially the efficiency and effectiveness of the design process. But the quality of the contents is important also for other classes of recipients (see Chapter 9).

Accuracy and truth are fragile ideas according to philosophy. In technology, the agreement of a statement with the real fact is not so problematic, especially not in the knowledge area of the object, in the branch knowledge. There each statement is examined practically in the light of reality. We must still reckon with error in the application of the knowledge, and consider it in judging the quality of the knowledge.

More difficult is the examination of the truth in the area of designing -- process knowledge. Is the recommendation given by a method right or wrong? It can help some designers in some situations very well, but others may receive little or no assistance.

The *empirical support* (compare figure 5--1) for the statements can enhance this consideration. After Stegmüller [410], we can distinguish the following classes:

- Pre-scientific (A) is the term used for statements that appear less systematic and consistent, or that cannot be (or to date have not been) tested and verified. Many declarations from the practice for the practice fall into this category of statements.
- Scientific: singular "understanding" (B) is the term used for the social-scientific approach. It leads to acceptable practice-based hypotheses of similar object-fields by means of case descriptions and in-depth case analyses (typically for historical science and leadership theories). The action-research concept is a more recent research method in this area.
- Scientific: inductive -- statistical (C) is the term applied to the method of substantiation, and the research concept, used in the "empirical (hard) sciences." According to the most recent scientific understanding, all knowledge is hypothetical in character, and must be continually re-examined and substantiated, especially for application [254,255,388]. The statistical (inductive) support of practice-based hypotheses is widely used, but cannot deliver "proofs" (in the formal logical sense) for the "truth" of statements [254,255,412].

## 10.1 The Quality of the Current Knowledge System of Design Science

The question about measuring the quality of the knowledge that belongs into the sphere of Design Science and that is available today can only be answered with difficulty in a total judgment. To be able to make a more precise appraisal statement, we must first gain suitable evaluation means (evaluation elements). This means that we must first select a set of criteria which determine the quality, we must further establish the structural parts (classes) of the knowledge, and finally the type of the knowledge system (its hierarchical level).

One such attempt is shown in figure 10--1. As criteria we have used some properties of the information, such as completeness, availability (accessibility), among others. These properties must be determined for the four main areas of Design Science (compare figure 5--2), in addition also for the General Design Science (therefore for technical system) and for the Specialized Design Sciences. The last of these areas is certainly very problematic, for the state of SDS in individual branches and fields (compare Chapter 8) is not everywhere the same. Because we are not concerned with establishing exact single statements, but rather a crude total characteristic which makes possible the comparison of the individual areas, we accept these inaccuracies.

For the evaluation we accept a four-stage scale: very good, good, deficient, poor. The evaluation is therefore executed in eight areas, shown as the rows of the matrix in figure 10--1, and in six properties that appear as the columns B -- G of the matrix.

In many cases, the knowledge system which is being evaluated does not exist as a formalized "unit." The books *Theory of Technical Systems* [214,219] and *Theorie der Konstruktionsprozesse* (Theory of Design Processes) [202] make available a description of the entire areas in the general plane of technical systems. But the systems for individual TS-families, which contain mainly the currently usual knowledge in the available information carriers (see for example figure 7--1), are missing. Many additional elements which belong to this area are scattered in other areas or exist only as personal experiences.

| Categories of Design Knowledge (*see figure 5--2*) | | | Completeness | Availability (information medium, carrier) | Empirical support (methodological category) | Form (direct or indirect answer) | Representation | Structural classification | Important tasks for the future, with the aim of increasing quality |
|---|---|---|---|---|---|---|---|---|---|
| A | | | B | C | D | E | F | G | H |
| Statements about technical objects -- systems (object knowledge) | Theory of Technical Systems (TTS) | general | deficient | good | deficient | good | deficient | good | – completing, unifying, homogenizing, etc.<br>– increasing the empirical support |
| | | special | good | good | good | deficient | deficient | deficient | – improving accessibility form representation |
| | Design object knowledge | general | deficient | deficient | deficient | deficient | deficient | good | – completing<br>– improving accessibility, form, representation<br>– increasing empirical support |
| | | special | good | deficient | good | deficient | deficient | deficient | – increasing accessibility<br>– improving form and representation |
| Statements about the design process (process knowledge) | Theory of Design Processes | general | deficient | good | deficient | deficient | deficient | good | – improving with respect to all criteria, especially unifying and homogenizing |
| | | special | deficient | deficient | deficient | poor | poor | deficient | – improving with respect to all criteria, especially the structural classification |
| | Design process knowledge | general | good | good | deficient | good | good | deficient | – increasing empirical support |
| | | special | deficient | deficient | deficient | poor | poor | deficient | – completing<br>– improving accessibility, form, representation<br>– increasing empirical support |

Legend: very good, good, deficient, poor

Figure 10--1 Quality of Design Knowledge

That makes the evaluation complicated, especially regarding completeness.

Despite all of these difficulties, important knowledge can be obtained from the matrix. Seen globally, the branch knowledge can be considered from almost all criteria as the qualitatively better area, compared with the process knowledge. Especially better is the situation in the theory. This difference can be explained by considering the general development of the knowledge, and the priority of the branch knowledge for engineering practice.

A comparison of the theoretical knowledge with the branch (domain, technical) knowledge shows that the theory has advanced further in the object knowledge regions. Exactly the reverse is the situation in the process knowledge, where the branch technical knowledge has a better total quality.

If the knowledge systems are contrasted for the general and the specialized areas, the situation in the branch knowledge is different where specialized areas are rated more highly. Contrary to this, the situation in the general process knowledge is more favorable than in specialized areas. The broader validity of the process knowledge compared to the branch knowledge certainly plays an important role.

# 11 FUTURE TASKS IN DESIGN SCIENCE

*One of the principal objects of theoretical research in any department of knowledge is to find the point of view from which the subject appears in its greatest simplicity.*
***Josiah Willard Gibbs****, founder of the geometric approach in classical thermodynamics, 1881*

*Only an inventor knows how to borrow, and every man is or should be an inventor.*
***Ralph Waldo Emerson***

Finally we need to look at the future, after we have dealt comprehensively with the past and present in the previous chapters. The "childhood years" of our discipline, in which we are today, already set clear tasks regarding the maturation process of Design Science: completion, unification, homogenization of the available knowledge.

The tasks can be derived more accurately, and differentiated for individual areas, by using the evaluation results presented in Chapter 10. Therefore we have used the evaluation matrix to formulate the future tasks from the recognized "weaknesses" (see figure 10--1, column H). As a result of the evaluation, some essential tasks are shown in the figure. The sequence of these tasks should now come under discussion, i.e. the strategy for the further development of Design Science. Because the General Design Science has now assumed completely clear forms, the structure model and the terminology are worked out, and further areas have been treated in detail (witness this book), attempts would be expedient to expand some Specialized Design Sciences according to this model. The experiences will then serve to examine the expediency of the proposed models and to possibly correct them in the second round.

A second direction should serve for constructing the technical knowledge areas. Much knowledge exists in these areas, but its formalization is qualitatively insufficient. This task is closely connected with the construction of knowledge-based systems. Thereby it has gained high attractiveness and urgency.

The third task complex has the completion of the theory as its goal, so that a balanced relationship to experience and engineering practice is obtained.

May we offer a closing suggestion for the reader, that you now return to Chapter 1 and review the initial descriptions of designing. In this way we hope that you will be able to relate and holistically recombine the elements; to bring the individual sections into context and focus; to complete your appreciation of the theory of Design Science as presented in this book.

# Bibliography

This compilation, placed in alphabetical order, and for each author in time sequence, includes all references listed in the figures


[1] -- , **Brockhaus Enzyklopädie** (Brockhaus Encyclopedia), Wiesbaden: Brockhaus, 1975.

[2] -- , **The Concise Oxford Dictionary**, (7 ed), Oxford: Univ. Press, 1982

[3] -- , **Curriculum for Design: Engineering Undergraduate Courses**, Proc. of Working Party, Loughborough: SEED Publ., 1985

[4] -- , **Curriculum for Design: Compendium of Engineering Design Projects**, Loughborough: SEED Publ., and London: Design Council, 1988

[5] -- , **Education and Training of Chartered Engineers for the 21st Century**, London: Fellowship of Engineering, 1989

[6] -- , **ESDU Index** (annual edition), Engineering Sciences Data Unit, London, England

[7] -- , **ESDU Catalogue of Engineering Software** (annual edition), Engineering Sciences Data Unit, London, England

[8] -- , **Great Soviet Encyclopedia** (25 vols.), New York: Macmillan, 1973--81.

[9] -- , **Improving Engineering Design**: Designing for Competitive Advantage, National Research Council, Washington: National Academy Press, 1991

[10] -- , **Instructions for Consumer Products**, Dept. of Trade and Industry, London: HMSO, 1988

[11] -- , "ISO 9000 and Quality" (Task Force for Standardization Technical Committee ISO/TC 176), *Automotive Engineering*, **99** No. 11, Nov 1992, p. 19--23

[12] -- , **Lueger Lexikon der Technik** (Lueger Lexicon of Technology), Reinbek: Rowohlt, 1971.

[13] -- , **Meyers Enzyklopädisches Lexikon** (Meyers' Encyclopedic Lexicon), Mannheim: Bibl. Institut, 1974

[14] -- , **Meyers neues Lexikon** (Myers' New Lexicon), Leipzig: VEB Verlag, 1977.

[15] -- , National Science Foundation: Program Announcement "Design Theory and Methodology", Nr. OMB 3145--0058 12/31/85

[16] -- , **The New Encyclopedia Britannica**, Chicago: Enc. Britannica, 1973

[17] -- , **The Practice of and Education for Engineering Design** Conference, *Proc.I.Mech.E.* **178** 1964, Pt. 3B, p. 34--35 and 51--52

[18] -- , **Timber Design Manual**, Ottawa: Laminated Timber Institute of Canada, 1972 (Imperial Edition), 1980 (Metric Edition)

[19] -- , **Tomorrow's Engineering Designers: the real needs of industry** Conference, Loughborough University, 25--26 Sept. 1979, Worthing: Publ. Serv., 1979

[20] -- , VDI Richtlinie 2220: **Produktplanung - Ablauf, Begriffe und Organisation** (Product Planning - Procedure, Terminology and Organization), Düsseldorf: VDI, 1980

[21] -- , VDI Richtlinie 2221: **Methodik zum Entwickeln und Konstruieren technischer Systeme und Produkte** (Methodology for Developing and Designing Technical Systems and Products), Düsseldorf: VDI, 1985

[22] -- , VDI Guideline 2221: **Systematic Approach to the Design of Technical Systems and Products**, Düsseldorf: VDI, 1987 (edited by K.M. Wallace)

[23] -- , VDI Richtlinie 2222 (1): **Konstruktionsmethodik: Konzipieren technischer Produkte** (Design Methodology: Conceptualizing Technical Products), Düsseldorf: VDI, 1977

[24] -- , VDI Richtlinie 2222 (2): **Konstruktionsmethodik: Erstellung und Anwendung von Konstruktionskatalogen** (Design Methodology: Generating and Using Design Catalogs), Düsseldorf: VDI, 1982

[25] -- , VDI Richtlinie 2223: **Begriffe und Bezeichnungen im Konstruktionsbereich -- Empfehlungen** (Terms and Designations in the Design Region -- Recommendations), Düsseldorf: VDI, 1973

[26] -- , VDI Richtlinie 2225: **Technisch-wirtschaftliches Konstruieren** (Technical-Economic Designing), Düsseldorf: VDI, 1964

[27] -- , VDI Richtlinie 2242 (1) and (2): **Ergonomiegerechtes Konstruieren** (Design for Ergonomics), Düsseldorf: VDI, 1983

[28] -- , VDI Richtlinie 2802: **Kostenvergleichsrechnung bei Wertanalysen** (Cost Comparison Calculations in Value Analyses), Düsseldorf: VDI, 1976

[29] -- , **WDK 18: Proceedings of the Institution of Mechanical Engineers, International Conference on Engineering Design, ICED 89 Harrogate** (2 vols.), London: I.Mech.E., 1989

[30] -- , **XII. Internationales Wissenschaftliches Kolloquium der Technischen Hochschule Ilmenau** (12th International Scientific Colloquium of the Technical University Ilmenau), Section L -- Konstruktion (Design), p. 17--25, Ilmenau, DDR: T.H., 1967

[31] Abbott, H., **Safe Enough to Sell? Design and Product Liability**, London: Design Council, 1980

[32] Abbott, H., **Safer by Design: The Management of Product Design under Strict Liability**, London: Design Council, 1988

[33] Ackoff, R.L., **Scientific Methods: Optimizing Applied Research Decisions**, New York: Wiley, 1962

[34] Ackoff, R.L., **Redesigning the Future: A Systems Approach to Societal Problems**, New York: Wiley, 1974

[35] Ackoff, R.L., **The Art of Problem Solving**, New York: Wiley, 1987 (Reprint of 1978 edition)

[36] Ackoff, R.L., **Creating the Corporate Future: Plan or Be Planned For**, New York: Wiley, 1981

[37] Adams, J.L., **Conceptual Blockbusting** (2. ed.), San Francisco: Freeman, 1980

[38] Akin, Ö., "Necessary Conditions for Design Expertise and Creativity", *Design Studies*, **11** 1990, p. 107--113

[39] Alexander, C., "The Determination of Components for an Indian Village", in [237], p. 83--114

[40] Alexander, E.R., "The Design of Alternatives in Organizational Contexts: A Pilot Study", *Admin. Sci. Quart.* **24** 1979, p. 382--404

[41] Alger, J.R.M., and Hays, C.V., **Creative Synthesis in Design**, Englewood Cliffs, NJ: Prentice-Hall, 1964

[42] Altschuller, G.S., **Erfindungen -- (K)ein Problem in Russland** (Inventing -- (Not) A Problem in Russia), Berlin: Verlag Tribüne, 1973

[43] Altschuller, G.S., **Creativity as an Exact Science**: Theory of the Solution of Inventive Problems (2 ed.), Gordon & Breach, 1987

[44] Anderson, J.R. (ed), **Cognitive Skills and their Acquisition**, Hillsdale, NJ: L. Erlbaum Assoc., 1981

[45] Anderson, J.R., **The Architecture of Cognition**, Cambridge, MA: Harvard Univ. Press, 1983

[46] Andreasen, M.M., **Synthesemethoder på Systemgrundlag -- Bidrag til en Konstruktionsteori** (Synthesis Methods as Basis for System -- Contribution to a Design Theory), Lund: T.H. (Doctor Thesis), 1980

[47] Andreasen, M.M., & Hein, L. (1987) **Integrated Product Development**, London: IFS Publ. and Berlin/Heidelberg: Springer-Verlag

[48] Andreasen, M.M., Kähler, S., Lund, T., and Swift, K.G., **Design for Assembly** (2. ed.), London: IFS Publ., and Berlin/Heidelberg: Springer-Verlag, 1988

[49] Andreasen, M.M., and Ahm, T., **Flexible Assembly Systems**, London: IFS, 1988

[50] Andreasen, M.M., and Olesen, J., "The Concept of Dispositions", *Jnl. Engng. Design*, **1** No. 1 1990, p. 17--36

[51] Archer, L.B., **Systematic Method for Designers**, London: Council for Industrial Design, reprint 1964

[52] Archer, L.B., "The Structure of the Design Process", in G. Broadbent and A. Wade (ed.), **Design Methods in Architecture**, New York: Wittenbern, 1969

[53] Archer, L.B., **Technological Innovation -- a Methodology**, Frimley, Surrey: Inforlink, 1971

[54] Archer, L.B., "Systematic Method for Designers", in [96]

[55] Archer, L.B., "The Implications for the Study of Design Methods of Recent Developments in Neighbouring Disciplines", in [215], p. 833--840

[56] Arciszewski, T., "Design Theory and Methodology in Eastern Europe", in [367], p. 209--217

[57] Argyris, C., and Schön, D.A., **Organizational Learning: A Theory of Action Perspective**, Reading, MA: Addison-Wesley, 1978

[58] Argyris, C., **Action Science: Concepts, Methods and Skills for Research and Intervention**, San Francisco, CA: Jossey-Bass, 1985

[59] Armstrong, L.V. & Hartman, J.B., **The Diesel Engine**, New York: Macmillan, 1959

[60] Artobolevskii, I.I., "General Problems in the Theory of Machines and Mechanisms", *Mech. & Mach. Th.* **10** 1975, Nr. 2/3

[61] Artobolevskii, I.I., "Past, Present and Future of the Theory of Machines and Mechanisms", *Mech. & Mach. Th.* **11** 1976 Nr. 6, p. 353--61

[62] Ashby, W.R., **An Introduction to Cybernetics**, London: Methuen Univ. Paperbacks, 1968

[63] Ashford, F., **The Aesthetics of Engineering Design**, London: Business Books, 1969

[64] Asimow, M., **Introduction to Design**, Englewood Cliffs, NJ: Prentice-Hall, 1962

[65] Bailey, R.W., **Human Performance Engineering**: A Guide for Systems Designers, New York: Prentice-Hall, 1982

[66] Barwell, F.T., **Lubrication of Bearings**, London: Butterworths, 1956

[67] Beitz, W., "Engpass Konstruieren -- Situation in der Bundesrepublik Deutschland" (Bottleneck Designing -- Situation in the Federal Republic of Germany), in [213], p. 7--15

[68] Berne, E., **Games People Play: The Psychology of Human Relationships**, Harmondsworth: Penguin, 1964

[69] Bertalanffy, L. von, **General Systems Theory** (revised ed.), New York: Braziller, 1968

[70] Billington, D.P., **Thin Shell Concrete Structures**, New York: McGraw-Hill, 1982

[71] Birkhofer, H., private communication

[72] Birolini, A., & Hubka, V. (eds), **WDK 15: Zuverlässigkeit technischer Systeme**, Zürich: Heurista, 1988

[73] Bloom, B.S. (ed) et al., **Taxonomy of Educational Objectives: Cognitive Domain**, New York: McKay, 1956

[74] Bodily, S.E., **Modern Decision Making: A Guide to Modeling with Decision Support Systems**, New York: McGraw-Hill, 1985

[75] Bolotin, V.V., **Prediction of Service Life for Machines and Structures**, New York: ASME Press, 1989

[76] Booker, P.J., "Principles and Precedents in Engineering Design", London: Inst. Engng. Designers, 1962

[77] Booker, P.J. (ed.), **Conference on the Teaching of Engineering Design**, London: Inst. Engng. Designers, 1964

[78] Booker, P.J. (ed.), **Teaching of Engineering Design** (Proc. of Second Conference), London: Inst. Engng. Designers, 1966

[79] Booker, P.J., Contribution of the Editor in [79], p. 326--327

[80] Boothroyd, G., Dewhurst, P., & Knight, W., **Product Design for Manufacture and Assembly**, New York: Marcel Dekker, 1991

[81] Boothroyd, G., **Assembly Automation and Product Design**, New York: Marcel Dekker, 1991

[82] Bossert, J.L., **Quality Function Deployment**, New York: Marcel Dekker, 1990

[83] Briggs-Myers, I., **Introduction to Type**, Gainesville, FL: Center for Applications of Psychological Type, 1979

[84] Bromme, R., and Hömberg, E., **Psychologie und Heuristik** (Psychology and Heuristics), Darmstadt: Steinkopff, 1977

[85] Brunner, K., **Einführung in den Umgang mit Geschichte** (Introduction to the Approach to History), Wien: Literas-Verlag, 1985

[86] Buhl, R., **Creative Engineering Design**, Ames, IA: Iowa S.U. Press, 1960

[87] Carlson, S.E., Shonkwiler, R., & Ingrim, M.E., "Comparison of Three Nonderivative Optimization Methods with a Genetic Algorithm for Component Selection", *Jnl. Eng. Design*, **5** No. 4, 1994

[88] Checkland, B., **Systems Thinking, Systems Practice**, Chichester: Wiley, 1984

[89] Christianson, L.L., & Rohrbach, R.P., **Design in Agricultural Engineering**, St, Joseph, MI: ASAE, 1986

[90] Churchman, C.W., **Systems Approach**, New York: Dell, 1983 (reprint of 1969 ed.)

[91] Clausing, D.P., **Total Quality Development**, New York: ASME Press, 1994

[92] Corfield, K.G., **Product Design**, London: NEDC, 1979

[93] Crawford, R.P., **The Technique of Creative Thinking**, Burlington, VT: Fraser, 1954 (reprint 1964)

[94] Crosby, P.B., **Quality is Free: The Art of Making Certain**, New York: McGraw-Hill, 1979

[95] Cross, N., **Design Participation**, London: Academy Editions, 1972

[96] Cross, N. (ed.), **Developments in Design Methodology**, Chichester: Wiley, 1984

[97] Cross, N., Kees, D., & Roozenburg, N., **Research in Design Thinking**, Delft: University Press, 1992

[98] Cross, N., **Engineering Design Methods** (2 ed: Strategies for Product Design), Chichester: Wiley, 1994

[99] Cross, N., "Editorial", *Design Studies*, **16** No. 1, Jan 1995, p. 2--3

[100] Davis, L.E., & Taylor, J.C. (eds), **Design of Jobs**, Harmondsworth: Penguin, 1972

[101] Dawes, R.M., **Rational Choice in an Uncertain World**, San Diego, CA: Harcourt Brace Jovanovich, 1988

[102] DeBono, E., **PO: Beyond Yes and No**, Harmondsworth: Penguin, 1973

[103] DeBono, E., **Eureka! Illustrated History of Inventions from the Wheel to the Computer**, London: Thames & H., 1979

[104] DeBono, E., **Tactics: The Art and Science of Success**, London: Collins, 1985

[105] Deming, W.E., **Out of the Crisis**, Cambridge, MA: MIT, 1986

[106] Descartes, René, **Discours de la méthode pour bien conduire sa raison et chercher la vérité dans les sciences**, Verlag Girardet

[107] Descartes, René, **A Discourse on Method etc.**, transl. J. Veitch (London 1912)

[108] DeSimone, D.V., **Education for Innovation**, Oxford: Pergamon, 1968

[109] Dieter, G., **Engineering Design**, New York: McGraw-Hill, 1983

[110] Dimarogonas, A., **Computer Aided Machine Design**, Englewood Cliffs, NJ: Prentice-Hall, 1989

[111] Dixon, J.R., **Design Engineering: Inventiveness, Analysis and Decision Making**, New York: McGraw-Hill, 1966

[112] Dixon, J.R., "Engineering Design Science: The State of Education", *Mechanical Engng.*, Feb 1991, p. 64--67

[113] Dixon, J.R., "Engineering Design Science: New Goals for Engineering Education", *Mechanical Engng.*, March 1991, p. 56--62

[114] Dorsch, F., **Psychologisches Wörterbuch** (Dictionary of Psychology), Hamburg: Felix Meiner, 1971

[115] Doughtie, V.L., & Vallance, A., **Design of Machine Members**, New York: McGraw-Hill, 1964

[116] Driggs, I.H., & Lancaster, O.E., **Gas Turbines for Aircraft**, New York: Ronald Press, 1955

[117] Drucker, P.F., **The Effective Executive**, New York: Harper & Row, 1964

[118] Drucker, P., **Post-Capitalist Society**, New York: Harper Business, 1993

[119] Dym, C.L., **Engineering Design: A Synthesis of Views**, Cambridge: Univ. Press, 1994

[120] Eder, W.E., and Gosling, W., **Mechanical Systems Design**, Commonwealth and International Library # 308, Oxford: Pergamon Press, 1965

[121] Eder, W.E., "The Design Method -- Definitions and Methodologies", in [175], Chap. 3, p. 19--31

[122] Eder, W.E., "Design Research -- Technologies and Varieties of Design", in [175], Chap. 33, p. 311--315

[123] Eder, W.E., "Systematisierung und Rationalisierung der Konstruktionsarbeit in Grossbritannien" (Systematizing and Rationalizing Design Work in Great Britain), in [30], p. 17--25

[124] Eder, W.E., "Design Engineering Methodologies: Comparisons of English- and German-Language Regions and Influences of Culture", in [207], p. 41--53

[125] Eder, W.E. (ed.), **WDK 13: Proceedings of the 1987 International Conference on Engineering Design** (2 vols.), New York: ASME Press, 1987

[126] Eder, W.E., "Information Systems for Designers", in [29], p. 1307--1319

[127] Eder, W.E., "Engineering Design -- a Perspective on U.K. and Swiss Developments", in Proc. Conference on **Design Theory and Methodology, DTM 90**, p. 327--336

[128] Eder, W.E., Hubka, V., Melezinek, A., & Hosnedl, S., (eds), **WDK 21: Engineering Design Education -- Ausbildung der Konstrukteure -- Reading**, Zürich: Heurista, 1992.

[129] Eder, W.E., "A European Outlook on Engineering Design in Education", *Jnl. Engng. Education* (ASEE), **82** No. 2, April 1993, p. 118--122

[130] Eder, W.E., "Developments in Education for Engineering Design -- Some results of 15 Years of WDK Activity in the Context of Design Research", *Jnl. Eng. Design*, Carfax, **5** No. 2, 1994, p. 135--144

[131] Eder, W.E., "Engineering Design -- Art, Science and Relationships", *Design Studies* **16**, 1995, p. 117--127

[132] Eder, W.E., "Comparisons -- Learning Theories, Design Theory, Science", *Jnl. Engng. Education* (ASEE) **83** No. 2, April 1994, p. 111--119

[133] Eder, W.E., "Bekannte Methodiken in den U.S.A. und Kanada" (Design Methods in the U.S.A. and Canada), *Konstruktion*, Springer-Verlag, **46** Heft 5, Mai 1994, p. 190--194

[134] Eder, W.E., "Methode QFD -- Bindeglied zwischen Produktplanung und Konstruktion" (Method QFD -- Connective Link between Product Planning and Design), *Konstruktion*, Springer-Verlag, **47** Heft 1/2, Feb 1995, p. 1--9

[135] Eder, W.E., "Learning Design -- Advantages for Procedures", in **Proc. 1995 ASEE Annual Conference**, Design in Engineering Education Division, Washington: ASEE, 1995 (in press)

[136] Eder, W.E., "Learning Processes -- Learning About Procedures", in **Proc. 1995 ASEE Annual Conference**, Design in Engineering Education Division, Washington: ASEE, 1995 (in press)

[137] Eder, W.E., "Context for Creative and Systematic Design in Engineering Education", in [230] (in press)

[138] Edholm, O.G., **The Biology of Work**, New York: McGraw-Hill World University Library, 1967

[139] Edison, T.A., Interview in *Life*, 1932

[140] Ehrlenspiel, K., **Kostengünstiges Konstruieren** (Designing for Favorable Cost), Berlin/Heidelberg: Springer-Verlag, 1985

[141] Ehrlenspiel, K., & John, T., "Inventing by Design Methodology", in [125] p. 29--38

[142] Eisenberger, M., "The Mechanical Strength Critic in CASE", Report EDRC 12--36--90, Pittsburgh, PA: Eng. Des. Res. Center, Carnegie-Mellon Univ., 1990

[143] Ellinger, J.H., **Design Synthesis**, London: Wiley, 1968

[144] Erasmus, Desiderius, of Rotterdam, **Praise of Folly**, (translated by B. Radic, 1971, revised by A.H. Levi), New York: Viking Penguin, 1994

[145] Ertas, A., & Jones, J.C., **The Engineering Design Process**, New York: Wiley, 1993

[146] Faires, V.M., **Design of Machine Elements** (4. ed.), New York: Macmillan, 1965

[147] Farr, M., **Design Management**, London: Hutchinson, 1966

[148] Faupel, J.H., **Engineering Design: a Synthesis of Stress Analysis and Materials Engineering**, New York: Wiley, 1964

[149] Feilden, G.B.R., **Engineering Design**, Report of Royal Commission, London: HMSO, 1963

[150] Ferreirinha, P., "Berechnung von Funktionskosten" (Calculation of Function Costs), in [209], p. 73--79

[151] Ferreirinha, P., "Kalkulationssystem für Neuentwicklungen mit Anwendung von Funktionskosten" (Cost Calculation System for New Developments Using Function Costs), in [213], p. 166--173

[152] Ferreirinha, P., Hubka, V., Andreasen, M.M., and Rosenberg, R., "Rechnerunterstütztes Entwerfen und Detaillieren von Konstruktionsgruppen" (Computer Aided Laying Out and Detailing of Assembly Groups), in [221], p. 105--116

[153] Ferreirinha, P., Hubka, V. and Eder, W.E., "Early Cost Calculation -- Reliable Calculation, not just Estimation", in **Design for Manufacturability 1993 Conference Proc.** (P.J. Guichelaar, ed), New York: ASME Press, DE-Vol 52, p. 97--104

[154] Finger, S., & Dixon, J.R., "A Review of Research in Mechanical Engineering Design. Part I: Descriptive, Prescriptive and Computer-Based Models of Design Processes", *Res. Engng. Design*, **1**, 1989, p. 51--67

[155] Finger, S., & Dixon, J.R., "A Review of Research in Mechanical Engineering Design. Part II: Representations, Analysis, and Design for the Life Cycle", *Res. Engng. Design*, **1**, 1989, p. 121--137

[156] Finniston, Sir M.F. (chairm.), **Engineering Our Future: Committee of Inquiry Report into the Engineering Profession**, Cmnd. 7794, London: HMSO, 1980

[157] Franke, H-J., "Konstruktionsmethodik und Konstruktionspraxis -- Eine Kritische Betrachtung" (Design Methodology and Design Practice -- A Critical Assessment), in [215], p. 910--924

[158] Freeman, P., **Lubrication and Friction**, London: Pitman, 1962

[159] French, M.J., **Engineering Design: The Conceptual Stage**, London: Heinemann Educ., 1971

[160] French, M.J., **Conceptual Design for Engineers** (2. ed. of [159], London: Design Council and Berlin/Heidelberg: Springer-Verlag, 1985

[161] French, M.J., **Invention and Evolution; Design in Nature and Engineering**, Cambridge: University Press, 1988

[162] Frost, M., **Value for Money -- the techniques of cost-benefit analysis**, Aldershot: Gower, 1971

[163] Gaffert, G.A., **Steam Power Stations**, New York: McGraw-Hill, 1952

[164] Gasparski, W., **Understanding Design: The Praxiological-Systemic Perspective**, Seaside, CA: Intersystems Publ., 1984

[165] Gerhard, E., **Entwickeln und Konstruieren mit System** (Developing and Designing with System), Grafenau: Expert Verlag, 1979

[166] Gershman, M., **Getting It Right the Second Time**, Reading, MA: Addison-Wesley, 1990

[167] Gibson, J.E., **Introduction to Engineering Design**, New York: Holt Rinehart and Winston, 1968

[168] Glegg, G.L., **The Design of Design**, Cambridge: Univ. Press, 1969

[169] Glegg, G.L., **The Selection of Design**, Cambridge: Univ. Press, 1972

[170] Glegg, G.L., **The Science of Design**, Cambridge: Univ. Press, 1973

[171] Goldberg, P., **The Intuitive Edge**, Los Angeles, CA: Tarcher (New York: St. Martin Press), 1983

[172] Goralski, A., "Exploration Matrices", Problem, Method, Solution -- Techniques of Creative Thinking, Vol. 3, Warsaw: WNT Publ., 1980

[173] Gordon, W.J.J., **Synectics: the development of creative capacity**, New York: Collier, 1961

[174] Gosling, W., **The Design of Engineering Systems**, London: Heywood, 1962

[175] Gregory, S.A. (ed.), **The Design Method**, London: Butterworths, 1966

[176] Gregory, S.A., "Design Science", in [175] p. 323--330

[177] Gregory, S.A., **Creativity and Innovation in Engineering**, London: Butterworths, 1972

[178] Guilford, J.P., and Hoepfner, R., **The Analysis of Intelligence**, New York: McGraw-Hill, 1971

[179] Hadamard, J.S., **Essai sur la psychologie de l'invention dans le domaine mathematique**, Paris: Gauthier-Villers, 1975

[180] Hales, C., "Designer as Chameleon", *Design Studies*, **6** 1985, No. 2, p. 111--114

[181] Hales, C., **Analysis of the Engineering Design Process in an Industrial Context** (2 ed), Eastleigh, Hants: Gants Hill Pub., 1991

[182] Hales, C., **Managing Engineering Design**, Harlow (Essex): Longman Scientific, 1993

[183] Hall, A.D., **A Methodology for Systems Engineering**, Princeton, NJ: Van Nostrand, 1962

[184] Halling, J. (ed.), **Principles of Tribology**, London: Macmillan, 1975

[185] Hamilton, P.H., Rhodes, R.G., and Wells, C.S., **Communication in Design**, Curriculum for Design: Preparation Material for Design Teaching, Loughborough: SEED Publ., 1988

[186] Hansen, F., **Konstruktionssystematik** (Design Systematics) (2. ed.), Berlin: VEB Verl. Technik, 1966

[187] Hansen, F., **Konstruktionswissenschaft -- Grundlagen und Methoden** (Design Science -- Basis and Methods), München: Carl Hanser, 1974

[188] Harrisberger, L., **Engineersmanship: a philosophy of design**, Belmont, CA: Brooks/Cole, 1966

[189] Hauser, J.R., & Clausing, D., "The House of Quality", *Harvard Business Review*, May-June 1988, p. 63--73

[190] Heller, E.D., **Value Management: Value Engineering and Cost Reduction**, Reading, MA: Addison-Wesley, 1971

[191] Hobbes, R.G., **Reminiscences of Seventy Years' Life, Travel and Adventure**, London: E. Stock, 1893-95

[192] Hofmeister, K., "Quality Function Deployment: Market Success through Customer-Driven Products", in **Food Products Development from Concept to the Marketplace**, E. Graf & I.S. Saguay (eds.), New York: Van Nostrand Reinhold, 1990

[193] Hohmann, K., **Methodisches Konstruieren** (Methodical Designing), Essen: Girardet, 1977

[194] Hollins, B., and Pugh, S., **Successful Product Design**, London: Butterworths, 1989

[195] Holt, K., **Product Innovation**, London: Newnes-Butterworths, 1977

[196] Horlock, J.H., **Combined Power Plants: including combined cycle gas turbine (CCGT) plants**, Oxford: Pergamon, 1992

[197] Horne, D.F., **Dividing, Ruling and Mask Making**, New York: Crane, Russak, 1974

[198] Hubka, V., "Grundlegender Algorithmus für die Lösung von Konstruktionsaufgaben" (Fundamental Algorithm for the Solution of Design Problems), in [30], p. 69--74

[199] Hubka, V., **Theorie der Maschinensysteme** (Theory of Machine Systems), Berlin/ Heidelberg: Springer-Verlag, 1974

[200] Hubka, V., "Konstruktionswissenschaft" (Design Science), *VDI-Zeitschr.* **116**, 1974, Nr. 11, p. 899--905, and No. 13, p. 1087--1094

[201] Hubka, V., "Intuition und Konstruktionsgefühl" (Intuition and Feel for Design), *Schw. Masch.* **75** Nr. 50, p. 42--44

[202] Hubka, V., **Theorie der Konstruktionsprozesse** (Theory of Design Processes), Berlin/ Heidelberg: Springer-Verlag, 1976

[203] Hubka, V., **Konstruktionsunterricht an Technischen Hochschulen** (Design Education at Technical Universities), Konstanz: Leuchtturm, 1978

[204] Hubka, V., **WDK 1: Allgemeines Vorgehensmodell des Konstruierens** (General Procedural Model of Designing), Goldach: Fachpresse, 1980

[205] Hubka, V. (ed), **WDK 2 a & b : Bibliographie der Konstruktionswissenschaft** (Bibliography of Design Science), Zürich: Heurista, 1980 (an occasional serial publication, Bibliography of Design Science)

[206] Hubka, V., Andreasen, M.M., Eder, W.E., Pighini, U., Schlesinger, A., & Wyss, M., **WDK 3: Fachbegriffe der wissenschaftlichen Konstruktionslehre in 6 Sprachen** (Terminology of the Science of Design Engineering in 6 languages), Zürich: Heurista, 1981

[207] Hubka, V. (ed.), **WDK 5: Konstruktionsmethoden in Übersicht: Proc. ICED 81 Rome** (Design Methods in Review), Milano: tecniche nuove, 1981

[208] Hubka, V. (ed.), **WDK 6: Konstruktionsunterricht in Übersicht** (Design Education in Review): Proc. ICED 81 Rome, Milano: tecniche nuove, 1981

[209] Hubka, V., and Eder, W.E. (ed.), **WDK 7: Ergebnisse von ICED 81** (Results of ICED 81 Rome), Zürich: Heurista, 1981

[210] Hubka, V. (ed.), **WDK 9: Dietrych zum Konstruieren** (Dietrych about Designing), Zürich: Heurista, 1982

[211] Hubka, V. (ed), **WDK 8 : Konstruktionsmethoden - Reading** (Design Methods - Reprints), Zürich: Heurista, 1983

[212] Hubka, V., "Systematische Heuristik" (Systematic Heuristics), *Schw. Masch.* **83** Nr. 38, p. 33--35

[213] Hubka, V. (ed.), **WDK 10: CAD, Design Methods, Konstruktionsmethoden**: Proc. ICED 83 Copenhagen (2 vols.), Zürich: Heurista, 1983

[214] Hubka, V., **Theorie Technischer Systeme** (Theory of Technical Systems, 2nd revised and expanded ed. of [199], Berlin/Heidelberg: Springer-Verlag, 1984

[215] Hubka, V. (ed.), **WDK 12: Theory and Practice of Engineering Design in International Comparison: Proc. ICED 85 Hamburg** (2 vols.), Zürich: Heurista, 1985

[216] Hubka, V. (ed), **WDK 11: Führung im Konstruktionsprozess - Reading** (Leadership/ Management in the Design Process - Reprints), Zürich: Heurista, 1985

[217] Hubka, V., **WDK 1: Principles of Engineering Design**, Zürich: Heurista, 1987 (translated and edited by W.E. Eder, first edition by London: Butterworths, 1982)

[218] Hubka, V. & Schregenberger, J.W., "Paths Towards Design Science", in [125] p. 3--14

[219] Hubka, V., and Eder, W.E., **Theory of Technical Systems**, New York: Springer-Verlag 1988 (revised translation of [214]

[220] Hubka, V., Andreasen, M.M., and Eder, W.E., **Practical Studies in Systematic Design**, London: Butterworths, 1988

[221] Hubka, V., Baratossy, J., and Pighini, U. (ed.), **WDK 16: Proceedings of ICED 88** (3 vols.), Budapest: GTE and Zürich: Heurista, 1988

[222] Hubka, V. & Schregenberger, J.W., "Eine neue Systematik konstruktionswissenschaftlicher Aussagen -- Ihre Struktur und Funktion" (A New Systematic Order for Design-Scientific Statements -- its Structure and Function), in [221] p. 103--117

[223] Hubka, V., "Design for Quality", in [29], p. 1321--1333

[224] Hubka, V. & Schregenberger, J.W., "Eine Ordnung Konstruktionswissenschaftlicher Aussagen" (A Classifying Arrangement for Design-Scientific Statements), *VDI-Zeitschr.* **131** (1989) no.3.

[225] Hubka, V., and Eder, W.E., "Design Education and Design Science", in Proc. of the S. Neaman International Workshop on **The Place of Design in the Engineering School**, Haifa, Israel: S. Neaman Press, 1989, p. 31--55

[226] Hubka, V., and Kostelic, A., **WDK 19: Proc. of the 1990 International Conference on Engineering Design, ICED 90 Dubrovnik** (4 vols.), Zürich: Heurista and Judeko, 1990

[227] Hubka, V. (ed.), **WDK 20: Proceedings of ICED 91 Zürich** (2 vols.), Zürich: Heurista, 1991

[228] Hubka, V. & Eder, W.E., **Engineering Design**, Zürich: Heurista, 1992 (revised and expanded ed. of [217]

[229] Hubka, V. & Eder, W.E., **Einführung in die Konstruktionswissenschaft** (Introduction to Design Science), Berlin/Heidelberg & New York: Springer-Verlag, 1992

[230] Hubka, V. (ed.), **WDK 23: Proc. ICED 95**, Zürich: Heurista, 1995 (in press)

[231] Hubka, V., "Specialized Design Science", in [230] (in press)

[232] Imai, M., **Kaizen**, New York: Random House, 1986

[233] Ishikawa, K., **What is Total Quality Control? The Japanese Way**, Englewood Cliffs, NJ: Prentice-Hall, 1985

[234] Jeffries, W.R., **Tool Design**, New York: Prentice-Hall, 1955

[235] Jewkes, J., Sawers, D., and Stillerman, R., **The Sources of Invention**, London: Norton, 1958

[236] Johnson, R.C., **Optimum Design of Mechanical Elements**, New York: Wiley, 1961

[237] Jones, J.Ch., and Thornley, D.G., **Conference on Design Methods**, Oxford: Pergamon, 1963

[238] Jones, J.C., "Design Methods Reviewed", in [175], Chap. 32, p. 295--310

[239] Jones, J.Ch., **Design Methods -- seeds of human futures** (2. rev. ed.), New York: Reinhold Van Nostrand, 1992

[240] Juran, J.M., **Management of Quality** (4 ed), New York: J.M. Juran, 1981

[241] Juvinall, R.C., & Marshek, K.M., **Fundamentals of Machine Component Design** (2 ed), New York: Wiley, 1991

[242] Kammel, D., "Kennen und Erkennen am Beispiel von Konstruktionsvergleichen" (Knowing and Recognizing on the Example of Design Comparisons), in [227], p. 1350--1353

[243] Katz, R.H., **Information Management for Engineering Design Applications**, New York: Springer-Verlag, 1985

[244] Kelley, H.H., **Attribution in Social Interaction**, New York: General Learning Press, 1971

[245] Kesselring, F., **Technische Kompositionslehre** (Study of Technical Composition), Berlin/Heidelberg: Springer-Verlag, 1954

[246] Kesselring, F., "Technisch-wirtschaftliches Konstruieren" (Technical-Economic Designing), *VDI-Zeitschr.*, **106** 1964, Nr. 30, p. 1533

[247] Kimber, M.D., and Smith, D.G., "SEED -- A Unique Formula for Grassroots Improvement in Design", in [29], p. 1111--1124

[248] Klaus, G., **Wörterbuch der Kybernetik** (Dictionary of Cybernetics), Frankfurt: Fischer, 1969

[249] Klaus, G., **Kybernetik in philosophischer Sicht** (Cybernetics in Philosophical View) 4th ed., Berlin: Dietz Verlag, 1965

[250] Klir, G.J., **An Approach to General Systems Theory**, New York: Van Nostrand, 1969

[251] Klir, G.J. (ed.), **Applied General Systems Research**, New York: Plenum, 1978

[252] Klir, G.J., **Architecture of Systems Problem Solving**, New York: Plenum, 1985

[253] Klir, G.J., "Developments in Uncertainty-Based Information", in Yovits, M.C. (ed), **Advances in Computers**, Vol. 36, San Diego, CA: Academic Press, 1993

[254] Koen, B.V., **Discussion of *The* Method**, Austin, TX: The University of Texas at Austin, College of Engineering (college ed.), 1984

[255] Koen, B.V., **Definition of the Engineering Method**, Washington, D.C.: ASEE, 1985

[256] Koller, R., **Konstruktionsmethode für Maschinen-, Geräte- und Apparatebau** (Design Method for Machine, Device and Apparatus Construction), Berlin/Heidelberg: Springer-Verlag, 1979

[257] Koller, R., **Konstruktionslehre für den Maschinenbau** (Study of Designing for Mechanical Engineering, 2. ed.), Berlin/Heidelberg: Springer-Verlag, 1985

[258] Kotarbinski, T., **Gnosiology: The Scientific Approach to the Theory of Knowledge**, Oxford: Pergamon, 1966 (translated from 1929 ed.)

[259] Kotarbinski, T., **Praxiology**, Oxford: Pergamon, 1965

[260] Krathwohl, D.R. (ed) et al., **Taxonomy of Educational Objectives: Affective Domain**, New York: McKay, 1965

[261] Krick, E.V., **An Introduction to Engineering and Engineering Design**, New York: Wiley, 1965

[262] Krick, E.V., **An Introduction to Engineering and Engineering Design** (2. ed.), New York: Wiley, 1969 (substantially changed from 1st ed.)

[263] Krick, E.V., **An Introduction to Engineering: methods, concepts and issues**, New York: Wiley, 1976

[264] Krist, Th., **Technisches Zeichnen** (Technical Drawing), Darmstadt: Tabellenverlag, 1975, p. 55

[265] Kuesel, T.R., "Whatever Happened to Long-Term Bridge Design?" *Civil Engng.*, **6**, No. 2, February 1990, p. 57--60

[266] Kuhn, T.S., **The Structure of Scientific Revolutions** (2 ed), Chicago: U. of Chicago Press, 1970

[267] Kuhn, T.S., **The Essential Tension: Selected Studies in Scientific Tradition and Change**, Chicago: Univ. of Chicago Press, 1977

[268] Kumar, K., **Prophecy and Progress: The Sociology of Industrial and Post-Industrial Society**, Harmondsworth: Penguin, 1991 (1st ed. 1978), p. 274

[269] Kusiak, A. (ed.), **Concurrent Engineering: Automation, Tools, and Techniques**, New York: Wiley, 1993

[270] Langdon, R., **Design Policy** (6 vols.), Proc. Design Policy Conference, London: Design Council, 1984

[271] Layard, R. (ed), **Cost-Benefit Analysis**, Harmondsworth: Penguin, 1972

[272] Leech, D.J., **Management of Engineering Design**, London: Wiley, 1972

[273] Lewis, W., & Samuel, A., **Fundamentals of Engineering Design**, New York: Prentice Hall, 1989

[274] Leyer, A., **Maschinenkonstruktionslehre** (Study of Machine Design), Heft 1 (1963), Heft 2 (1964), Heft 3 (1966), Heft 4 (1968), Basel: Birkhäuser

[275] Leyer, A., **Machine Design**, Glasgow: Blackie, 1974 (translation of [274]

[276] Lickley, R.L., **Report of the Engineering Design Working Party**, London: SERC, 1983

[277] Liggett, J.V., **Fundamentals of Position Tolerance**, Dearborn, MI: Soc. of Manuf. Eng., 1970

[278] Locascio, A., & Thurston, D., "Quantifying the House of Quality for Optimal Product Design", DE-Vol. 68, **Design Theory and Methodology - DTM '94**, New York: ASME Press, 1994, p. 43--54

[279] Lochner, R.H., & Matar, J.E., **Designing for Quality: an introduction to the best of Taguchi and western methods of statistical experimental design**, White Plains, NY: Quality Resources, 1990

[280] Lohmann, H., "Die Bedeutung der Methodologie für die Lehre der Technik" (The Importance of Methodology for Education in Technology), *Die Fachschule* 1957, p. 128

[281] Lohmann, H., "Zur Theorie und Praxis der Heuristik in der Ingenieurerziehung" (To the Theory and Practice of Heuristics in Engineering Formation), *Wiss. Zeitschr.* TH Dresden 1959/60, p. 1059 and 1281

[282] Love, S.F., **Design Methodology**, Waterloo, ONT: Designectics International Inc., 1973

[283] Lüscher, M., "Zum Verhältnis von Kreativität und Intuition" (To the Relationship of Creativity and Intuition), *Neue Züricher Zeitung*. Samstag/Sonntag 16./17. April 1988. Nr. 88. 25.

[284] Lüscher, M., **Das Harmoniegesetz in uns: Checkliste für die vier Selbstgefühle** (The Law of Harmony Within Us: Check-list for the Four Feelings),

[285] Marples, D.L., "The Decisions of Engineering Design", London: Inst. Engng. Designers, 1960 and *IRE Trans. Engng. Management*, EM--8 1961, p. 55--71

[286] Marshall, B., **Experiences in Being**, Belmont, CA: Brooks/Cole, 1971

[287] Matchett, E., and Briggs, A.H., "Practical Design Based on Method (Fundamental Design Method)", in [175], p. 183--200

[288] Matousek, R., **Konstruktionslehre des allgemeinen Maschinenbaues** (Study of Design in General Mechanical Engineering), Berlin/Heidelberg: Springer-Verlag, 1957

[289] Matousek, R., **Engineering Design**, London: Blackie, 1963 (translation of [288])

[290] Maunder, L., **Machines in Motion**, Cambridge: University Press, 1986

[291] Mayall, W.H., **Industrial Design for Engineers**, London: Iliffe, 1967

[292] Mayall, W.H., and Bishop, T., **Design Matters**, London: Design Council, 1985

[293] McCrory, R.J., "The Design Method -- A Scientific Approach to Valid Design", ASME-Paper 63-MD-4, March 1964

[294] McKim, R.H., **Experiences in Visual Thinking** (2. ed.), Monterey, CA: Brooks/Cole, 1980

[295] Melezinek, A., **Ingenieurpädagogik** (Engineering Pedagogics), Wien: Springer-Verlag, 1967

[296] Mellor, D., Report to the Design Council on the **Design of British Consumer Goods**, London: Design Council,

[297] Menon, H., **TQM in New Product Manufacturing**, New York: McGraw-Hill, 1992

[298] Middendorf, W.H., **Engineering Design**, Boston: Allyn and Bacon, 1969

[299] Miles, L.D., **Techniques of Value Analysis and Engineering** (2 ed), New York: McGraw-Hill, 1972

[300] Miller, D.W., and Starr, M.K., **The Structure of Human Decisions**, Englewood Cliffs, NJ: Prentice-Hall, 1967

[301] Miller, G.A., "The Magical Number Seven, plus or minus two: Some Limits on our Capacity for Processing Information", *Psychological Review*, **63** 1956, p. 81--97

[302] Miller, G.A., "Information and Memory", *Scientific American*, **195** 1956, p. 42--46

[303] Miller, G.A., "Human Memory and the Storage of Information", *IRE Trans. Information Theory*, **IT-2** No. 3, 1956, p. 128--137

[304] Miller, G.A., "Assessment of Psychotechnology", *American Psychologist*, **25** No. 11, 1970, p. 911--1001

[305] Moore, J.H., **Building Scientific Apparatus**, London: Addison-Wesley, 1983

[306] Morrison, D., **Engineering Design : the choice of favourable systems**, London: McGraw-Hill, 1968

[307] Moulton, A.E. (ed.), **Engineering Design Education**, London: The Design Council, 1976

[308] Müller, J., "Operationen und Verfahren des problemlösenden Denkens in der konstruktiven technischen Entwicklungsarbeit -- eine methodologische Studie" (Operations and Procedures of Problem Solving Thinking in Technical Designing and Developing -- a methodological study), *Wiss. Zeitschr.* d. T.H. Karl-Marx-Stadt, Jg. **IX**, 1967, Heft 1/2, p. 5--51

[309] Müller, J., "Denkpsychologie und Ingenieurmethodik -- Wege zur empirisch fundierten Methodikforschung" (Psychology of Thinking and Engineering Methodology -- Paths Towards Empirically Based Research into Methodology), in [215], p. 841--854

[310] Müller, J., **Arbeitsmethoden der Technikwissenschaften -- Systematik, Heuristik, Kreativität** (Working Methods of the Engineering Sciences -- Systematics, Heuristics, Creativity), Berlin/Heidelberg: Springer-Verlag, 1990

[311] Müller, J., "Akzeptanzbarrieren als berechtigte und ernst zu nehmende Notwehr kreativer Konstrukteure -- nicht immer nur böser Wille, Denkträgheit oder alter Zopf" (Barriers to Acceptance as Justified and Serious Self-defence Reaction of Creative Designers -- not always only ill will, thinking inertia or old hat), in [227], p. 769--776

[312] Munton, R., & Stott, J.R., **Refrigeration at Sea** (2 ed), London: Applied Science Publ., 1978

[313] Muther, R., **Systematic Layout Planning (SLP)**, Boston, MA: Industr. Educ. Inst., 1961

[314] Nadler, G., **Work Systems Design : The IDEALS Concept**, Homewood, IL: Irwin, 1967

[315] Nadler, G., "An Investigation of Design Methodology", *Management Science* **13** No. 10, 1967

[316] Nadler, G., **The Planning and Design Approach**, New York: Wiley, 1981

[317] Nadler, G., and Hibino, S., **Breakthrough Thinking**, Rocklin, CA: Prima Publ., 1989

[318] Newell, A., and Simon, H., **Human Problem Solving**, Englewood Cliffs, NJ: Prentice-Hall, 1974

[319] Nezel, I., **Allgemeine Didaktik der Erwachsenenbildung** (General Didactics of Adult Education), Bern: Haupt, 1992

[320] Nezel, I., "Standpunkt eines Psychologen zur Ausbildung (Förderung) der Kreativität" (Viewpoint of a Psychologist to the Formation (Advancement) of Creativity), in [230] (in press)

[321] Newsome, S.L., Spillers, W.R., and Finger, S., **Design Theory '88**, Proc. of 1988 Grantee Workshop on Design Theory and Methodology, New York: Springer-Verlag, 1989

[322] Norbye, J.P., **The Wankel Engine**, Philadelphia, PA: Chilton Books, 1971

[323] Norbye, J.P., **The Gas Turbine Engine**, Radnor, PA: Chilton Books, 1975

[324] Norman. D.A., **The Psychology of Everyday Things**, New York: Basic Books, 1988

[325] Norris, K.W., "The Morphological Approach to Engineering Design", in [237], p. 115--140

[326] Odrin, V.M., and Kartavov, S.S., **Selected Possibilities and Directions for Development for Morphological Analyses**, Kiev: Soviet Academy of Science, 1973 (in Russian)

[327] Odrin, V.M., **Morphological Synthesis of Technical Systems**, Kiev: Glushkov Cybernetic Institute, 1986 (in Russian)

[328] Odrin, V.M., **Morphological Synthesis of Technical Systems**, Moscow: VNIIPI, 1989 (in Russian)

[329] Orlov, P.I., **Foundations of Designing**, Osnovy konstruovania, 1977 (in Russian)

[330] Osborne, A.F., **Applied Imagination -- Principles and Procedure of Creative Thinking**, New York: Scribners, 1963

[331] Ott, H., **Vorlesungen: Maschinenelemente** (Lecture Notes: Machine Elements), Zürich: ETH, 1980

[332] Page, J.K., Contribution to **Building for People**, Conference Report, London: Min. of Public Buildings and Works, 1966

[333] Pahl, G., and Beitz, W., **Konstruktionslehre** (Study of Designing), Berlin/Heidelberg: Springer-Verlag, (2. ed.) 1988

[334] Pahl, G., and Beitz, W., **Engineering Design**, London: The Design Council, and Berlin/ Heidelberg: Springer-Verlag, 1984 (translated by A.J. Pomerans and K.M. Wallace, edited by K.M. Wallace)

[335] Papalambros, P.Y., and Wilde, D.J., **Principles of Optimal Design: Modeling and Computation**, Cambridge: Univ. Press, 1988

[336] Pare, E., Francis, L.L., and Kimbrell, J.T., **Introduction to Engineering Design**, New York: Holt Rinehart and Winston, 1963

[337] Pascal, Blaise, **Pensées**, transl. & ed. A.J. Krailsheimer (1966), Harmondsworth: Penguin Classics, 1970

[338] Parkinson, C.N., **Parkinson's Law**, Boston: Houghton Mifflin, 1957; & New York: Buccaneer, 1993

[339] Parkinson, C.N., **In-Laws and Outlaws**, New York: Buccaneer, 1993

[340] Parkinson, C.N., **The Law and the Profits**, Boston: Houghton Mifflin, 1960; & New York: Buccaneer, 1993

[341] Peter, L.J., & Hull, R., **The Peter Principle**, New York: Morrow, 1969; & New York: Buccaneer, 1993

[342] Peter, L.J., **Peter's Quotations: ideas for our times**, New York: Morrow, 1977; & New York: Buccaneer, 1993

[343] Petroski, H., **To Engineer is Human**, New York: St. Martin Press, 1985

[344] Petroski, H., **The Pencil: A History of Design and Circumstance**, New York: Knopf, 1989

[345] Petroski, H., **The Evolution of Useful Things**, New York: Knopf & New York: Vintage Books, 1992

[346] Petroski, H., **Design Paradigms**, Cambridge: University Press, 1994

[347] Phadke, M.S., **Quality Engineering Using Robust Design**, Englewood Cliffs, NJ: Prentice-Hall, 1989

[348] Phillips, L.N., **Design with Advanced Composite Materials**, London: Design Council, 1989

[349] Pitts, G., **Techniques in Engineering Design**, London: Butterworths, 1973

[350] Pitts, G. (ed.), **Information Systems for Designers**, Proc. Intern. Symposium 1974, 1977, 1979, 1981, 1988

[351] Poincaré, H., **The Foundations of Science**, Washington 1982 (original French 1905; includes **Science and Hypothesis**, New York: Dover, 1952; **Science and Method**, New York: Dover, 1952; **Value of Science**, New York: Dover, 1958)

[352] Polanyi, M., **The Tacit Domain**, New York: Doubleday, 1966

[353] Polya, G., **How To Solve It**, Princeton, NJ: Princeton U. P., 1945

[354] Polya, G., **Schule des Denkens -- vom Lösen mathematischer Probleme** (School of Thinking -- about Solving Mathematical Problems), Bern: Francke, 1980

[355] Popper, K., **The Logic of Scientific Discovery**, New York: Basic Books, 1959 (transl. of **Logik der Forschung**, Wien, 1935)

[356] Powilejko, R.P., "Classification of some Methods for Solving Design Problems", *Jnl. Informatics and its Problems* **5** 1972 (in Russian)

[357] Powilejko, R.P., **Technical Creativity**, Moskva: Science, 1987 (in Russian)

[358] Pugh, S., **Specification Phase**, Curriculum for Design: Preparation Material for Design Teaching, Loughborough: SEED Publ., 1986

[359] Pugh, S., **Total Design**: Integrated Methods for Successful Product Engineering, Wokingham: Addison-Wesley, 1991

[360] Putnam, H., "Computational Psychology and Interpretation Theory", in **Artificial Intelligence: The Case Against**, Reiner P. Born (ed), London & Sydney: Croom Helm, 1987 (paperback 1988), p. 1--17, quote from p. 13

[361] Rabins, M.J., **Goals and Priorities for Research in Engineering Design**, New York: ASME Press, 1986

[362] Rand, A., **Introduction to Objectivist Epistemology**, New York: Mentor, 1979

[363] Reswick, J.B., **Prospectus for an Engineering Design Center**, Cleveland, OH: Case Institute of Technology, 1965

[364] Reuleaux, F., and Moll, C.L., **Construktionslehre für den Maschinenbau** (Study of Design for Mechanical Engineering), Braunschweig: Vieweg, 1862

[365] Reuleaux, F., **The Constructor**, New York: Engineering Magazine, 1904 (translated by H.H. Suplee)

[366] Rinderle, J.R., and Watton, J.D., "Automatic Identification of Critical Design Relationships", in [125], p. 512--525

[367] Rinderle, J. (ed.), **Design Theory and Methodology DTM '90**, New York: ASME Press, 1990

[368] Rockstroh, W., **Die technologische Betriebsprojektierung** (Technological Project-Engineering for Enterprises), Berlin: VEB Verlag, 1977

[369] Rodenacker, W.G., **Methodisches Konstruieren** (Methodical Design), Berlin/Heidelberg: Springer-Verlag, 1970

[370] Roe, P.H., Soulis, G.N., and Handa, V.K., **The Discipline of Design**, Boston, MA: Allyn and Bacon, 1967

[371] Roozenburg, N., & Eekels, J. (eds), **WDK 17: EVAD -- Evaluation and Decision in Design -- Reading**, Zürich: Heurista, 1990

[372] Roozenburg, N.F.M. (ed), **WDK 22: Proc. of ICED 93** (3 vols.), Zürich: Heurista, 1993

[373] Roozenburg, N.F.M., & Eekels, J., **Product Design: Fundamentals and Methods**, Chichester: Wiley, 1995

[374] Ropohl, G., **Eine Systemtheorie der Technik** (A Systems Theory of Technology), München: Carl Hanser Verlag, 1979

[375] Ross, P.R., **Taguchi Techniques for Quality Engineering**, New York: McGraw-Hill, 1988

[376] Roth, K., **Konstruieren mit Konstruktionskatalogen** (Designing with Design Catalogs) (2 ed., 2 vols.), Berlin/Heidelberg: Springer-Verlag, 1994

[377] Rovida, E. (ed.), **WDK 14: Methodisches Konstruieren der Maschinenelemente** (Methodical Designing of Machine elements), Zürich: Heurista, 1987

[378] Runes, D.D. (ed.), **The Diary and Sundry Observations of Thomas Alva Edison**, New York: Philosophical Library, 1948

[379] Salvendy, G. (ed), **Handbook of Human Factors**, Wiley-Interscience, 1987

[380] Sandor, G.N., "The Seven Stages of Engineering Design", *Mechanical Engng.*, April 1964, p. 21--25

[381] Sapiro, J.S., **Organisatia i Effektivnoss Techniceskogo** (Organisation and Effectiveness of Technical Development), Moskva: Razvitija Moskva Ekonomika, 1980

[382] Sater-Black, K., & Iverson, N., "How to Conduct a Design Review", *Mechanical Engng.*, March 1994, p. 89-92

[383] Schmidt, H., **Lexikon der Philosophie** (Lexicon of Philosophy), Stuttgart: Kröner, 1974

[384] Schmidt, H., **Lexicon der Philosophie** (Lexicon of Philosophy) (16. ed.), Zürich: Buchklub Ex Libris, 1974

[385] Schön, D.A., **The Reflective Practitioner: How Professionals Think in Action**, New York: Basic Books, 1983

[386] Schregenberger, J.W., **Methodenbewusstes Problemlösen** (Method-Conscious Problem Solving), Bern/Stuttgart: Haupt, 1982

[387] Schregenberger, J.W., "Erfolgreicher konstruieren - aber wie?" (Designing More Successfully -- But How?), *Schw. Masch.* (1986), no.23, p. 46-49.

[388] Schurrman, E., **Reflections on the Technological Society**, Toronto: Wedge Publ., 1977

[389] Sedlmayr, H., **Verlust der Mitte** (Loss of the Middle), Salzburg: Otto Müller-Verlag, 1948

[390] Seifert, H., *et al.*, **Rechnerunterstütztes Konstruieren mit PROREN** (Computer Supported Designing with PROREN) (2 vols.), Bochum: Ruhr-Universität, 1986

[391] Seligman, M., **Helplessness: On Depression, Development and Death**, San Francisco, CA: Freeman, 1975

[392] Sereda, I.N., **Arbeiter -- Erfinder** (Worker -- Inventor), Rabocij -- Izobretatel, 1961

[393] Severns, W.H., & Fellows, J.R., **Heating, Ventilating and Air Conditioning Fundamentals** (2 ed), New York: Wiley, 1949

[394] Shapiro, H.I., **Cranes and Derricks**, New York: McGraw-Hill, 1980

[395] Shigley, J., and Mischke, C.R., **Mechanical Engineering Design** (5. ed.), New York: McGraw-Hill, 1989

[396] Shuster, H.D., **Teaming for Quality Improvement**, Englewood Cliffs, NJ: Prentice Hall, 1990

[397] Simon, H.A., **Models of Discovery**, Dordrecht: Reidel, 1977

[398] Simon, H.A., **The Sciences of the Artificial** (2. ed.), Cambridge, MA: MIT Press, 1981

[399] Singleton, W.T., **Man-Machine Systems**, Baltimore: Penguin, 1974

[400] Smith, C.O., "Design of a Saw Table", ASME Paper 87-WA/DE-9, Winter Annual Meeting, 1987

[401] Smith, C.O., "Problems in Machine Guarding", ASME Paper 87-WA/DE-6, Winter Annual Meeting, 1987

[402] Smith, C.O., "Coffee Grinder: Safe or Not?", ASME Paper 88-WA/DE-6, Winter Annual Meeting, 1988

[403] Smith, C.O., & Talbot, T.F., "Effects of Product Liability on Design", ASME Paper 92-WA/DE-8, Winter Annual Meeting, 1992

[404] Snow, C.P., **The Two Cultures & A Second Look**, Cambridge: University Press, 1969

[405] Spotts, M.F., **Design Engineering Projects**, Englewood Cliffs, NJ: Prentice-Hall, 1968

[406] Spotts, M.F., **Design of Machine Elements** (6. ed.), Englewood Cliffs, NJ: Prentice-Hall, 1985

[407] Sprague, R.H., & Carlson, E.D., **Building Effective Decision Support Systems**, Englewood Cliffs: Prentice-Hall, 1982

[408] Starfield, A.M., Smith, K.A., & Bleloch, A.L., **How to Model It**: Problem Solving for the Computer Age, New York: McGraw-Hill, 1990

[409] Starr, M.K., **Product Design and Decision Theory**, Englewood Cliffs, NJ: Prentice-Hall, 1963

[410] Stegmüller, W., **Theorie und Erfahrung** (Theory and Experience), Berlin/Heidelberg: Springer-Verlag, 1973

[411] Stegmüller, W., **Wissenschaftliche Erklärung und Begründung** (Scientific Explanation and Substantiation), Berlin: Springer, 1973

[412] Stegmüller, W., **Personelle und statistische Wahrscheinlichkeit** (Personal and Statistical Probability), Berlin: Springer, 1973

[413] Steuer, K., **Untersuchungen der Struktur der Konstruktionslehre im Maschinenbau als Voraussetzung der Konstrukteurausbildung** (Investigation of the Structure of Design Study in Mechanical Engineering as Prerequisite for Engineering Education), Dresden: 1963

[414] Subramanian, E., Podnar, G., and Westerberg, A., "A Shared Computational Environment for Concurrent Engineering", Report EDRC 05--48--90, Pittsburgh, PA: Eng. Des. Res. Center, Carnegie-Mellon Univ., 1990

[415] Suh, N.P., **Principles of Design**, Oxford: University Press, 1989

[416] Taguchi, G., **Introduction to Quality Engineering: Designing Quality into Products and Processes**, White Plains, NY: Quality Resources (Asian Productivity Organisation), 1986

[417] Taguchi, G., **Taguchi on Robust Technology Development: Bringing Quality Engineering Upstream**, New York: ASME Press, 1993

[418] Talukdar, S., and Fenves, S., "Towards a Framework for Concurrent Design", Report EDRC 18--14--90, Pittsburgh, PA: Eng. Des. Res. Center, Carnegie-Mellon Univ., 1990

[419] Talukdar, S., Sapossnek, M., Hou, L., Woodbury, R., Sedas, S., and Saigal, S., "Autonomous Critics", Report EDRC 18--13--90, Pittsburgh, PA: Eng. Des. Res. Center, Carnegie-Mellon Univ., 1990

[420] Talukdar, S., and Christie, R., "An Extended Framework for Security Assessment", Report EDRC 18--15--90, Pittsburgh, PA: Eng. Des. Res. Center, Carnegie-Mellon Univ., 1990

[421] Taylor, E.S., M.I.T. Report, Cambridge, MA: MIT Press, 1959

[422] Tillman, F.A., Eggeman, G.W., Hwang, C.L., & Fitzgerald, A.L., "The Application of Multiple Criteria Decision Making Methods to Design", in [215], p. 113--120

[423] Tjalve, E., Andreasen, M.M., and Schmidt, F.F., **Engineering Graphic Modelling: workbook for design engineers**, London: Butterworths, 1979

[424] Tjalve, E., **A Short Course in Industrial Design**, London: Newnes-Butterworths, 1979

[425] Tomiyama, T., and Yoshikawa, H., "Extended General Design Theory", in [464], p. 95--124

[426] Tschochner, H., **Konstruieren und Gestalten** (Designing and Form-Giving), Essen: Girardet, 1954

[427] Tsourikov, V., "Inventive Machine: Second Generation", *AI and Society*, **7** No. 1, 1993, p. 62--78

[428] Tuttle, S.B., **Mechanisms for Engineering Design**, New York: Wiley, 1967

[429] Ullman, D.G., Wood, S., and Craig, D., "The Importance of Drawing in the Mechanical Design Process", *Computers and Graphics* **14** No.2 1990

[430] Ullman, D.G., **The Mechanical Design Process**, New York: McGraw-Hill, 1992

[431] Vernon, P.E. (ed.), **Creativity**, Harmondsworth: Penguin, 1970

[432] Vidosic, J.P., **Machine Design Projects**, New York: Ronald Press, 1957

[433] Vidosic, J.P., **Elements of Design Engineering**, New York: Ronald Press, 1969

[434] von Fange, E.K., **The Creative Process**, 1954

[435] von Fange, E., **Professional Creativity**, Englewood Cliffs, NJ: Prentice-Hall, 1959

[436] Wakefield, L.P., **Dimensioning for Interchangeability**, Oxford: Pergamon, 1964

[437] Wales, C., Nardi, A., and Stager, R., **Professional Decision Making**, Morgantown, WV: Center for Guided Design, 1986

[438] Wales, C., Nardi, A, & Stager, R., **Thinking Skills: Making a Choice**, Morgantown, WV: Center for Guided Design, 1986

[439] Wallace, P.J., **The Technique of Design**, London: Pitman, 1952

[440] Wallas, G., **The Art of Thought**, London: Cape, 1926 (reprint 1931)

[441] Wankel, F., **Rotary Piston Machines**, London: Iliffe, 1965

[442] Warfield, J.N., **A Science of Generic Design**, Salinas, CA: Intersystems, 1989

[443] Warfield, J.N., **Societal Systems: Planning, Policy, and Complexity**, Salinas, CA: Intersystems, 1989 (reprint of 1976 ed.)

[444] Warman, E.A., "Object Oriented Programming and CAD", *Jnl. Engng. Design* **1** No. 1 1990, p. 37--46

[445] Watson, G.H., **The Benchmarking Workbook: Adapting Best Practices for Performance Improvement**, Cambridge, MA: Productivity Press, 1992

[446] Watson, S.R., & Buede, D.M., **Decision Synthesis: The Principles and Practice of Decision Analysis**, Cambridge: Univ. Press, 1987

[447] Wearne, S.H., **Principles of Engineering Organisation**, London: Edward Arnold, 1973

[448] Wearne, S.H., "A Review of Reports of Failures", *Proc.I.Mech.E.* **193**, 1979, p. 125--136

[449] Weisburg, R., **Creativity**, New York: W.H. Freeman, 1986

[450] Westerberg, A.W., "Synthesis in Engineering Design", Report EDRC 06--53--89, Pittsburgh, PA: Eng. Des. Res. Center, Carnegie-Mellon Univ., 1989

[451] Westerberg, A.W., "Process Engineering", Report EDRC 06--54--89, Pittsburgh, PA: Eng. Des. Res. Center, Carnegie-Mellon Univ., 1989

[452] Whiting, C.S., **Creative Thinking**, New York: (Van Nostrand) Reinhold, 1958

[453] Whyte, R.R., **Engineering Progress Through Trouble**, London: IMechE, 1975

[454] Whyte, R.R., **Engineering Progress Through Development**, London: MEP, 1978

[455] Wickelgren, W.A., **How to Solve Problems**, San Francisco, CA: Freeman, 1974

[456] Wiener, N., **The Human Use of Human Beings** (2. ed.), Garden City, NY: Doubleday Anchor, 1954

[457] Wilson, I.G., and Wilson, M.E., **From Idea to Working Model**, New York: Wiley, 1970

[458] Woods, D.R., **Problem-based Learning**: How to Gain the Most from PBL, Waterdown, ON: D.R. Woods, 1994

[459] Woodson, T.T., **Introduction to Engineering Design**, New York: McGraw-Hill, 1966

[460] Woodson, W.E., & Conover, D.W., **Human Engineering Guide for Equipment Designers** (2 ed), Univ. of California Press, 1970

[461] Wörgerbauer, H., **Die Technik des Konstruierens**, München: Oldenbourg, 1943

[462] Yoshikawa, H., "General Design Theory: Theory and Application", in **Conference on CAD/CAM Technology in Mechanical Engineering**, Cambridge, MA: MIT Press, 1981, p. 370--6

[463] Yoshikawa, H., "General Design Theory and CAD System", in T. Sata and E.A. Warman (ed.) **Man-Machine Communication in CAD/CAM**, Proc. of IFIP W.G. 5.2/5.3 Working Conference, Amsterdam: North Holland, 1981, p. 35

[464] Yoshikawa, H., and Warman, E.A., **Design Theory for CAD**, Proc. IFIP WG 5.2 Working Conference on Design Theory for CAD, Tokyo, Oct. 1985, Amsterdam: Elsevier, 1987

[465] Zwicky, F., **The Morphological Method of Analysis and Construction**, Courant Anniv. Vol., New York: Wiley-Interscience, 1948

# Names Index

# Subject Index